CONCEPTS IN MOLECULAR BIOLOGY

CONCEPTS IN MOLECULAR BIOLOGY

REVISED EDITION

S C Rastogi
Former Professor
Biological Sciences, Birla Institute of Technology & Science
Pilani, Rajasthan

V N Sharma
Biological Sciences Discipline
Birla Institute of Technology and Science
Pilani

Anuradha Tondon
Biological Sciences Discipline
Birla Institute of Technology and Science
Pilani

Publishing Globally

NEW AGE INTERNATIONAL (P) LIMITED, PUBLISHERS
LONDON • NEW DELHI • NAIROBI
Bangalore • Chennai • Cochin • Guwahati • Hyderabad • Kolkata • Lucknow • Mumbai
Visit us at www.newagepublishers.com

Published by New Age International (P) Ltd., Publishers
First Edition: 1993
Revised Edition: 1995
Reprint: 2017

GLOBAL OFFICES

- **New Delhi** **NEW AGE INTERNATIONAL (P) LIMITED, PUBLISHERS**
7/30 A, Daryaganj, New Delhi-110002, (INDIA)
Tel.: (011) 23253771, 23253472, **Telefax:** 23267437, 43551305
E-mail: contactus@newagepublishers.com • Visit us at www.newagepublishers.com
- **London** **NEW AGE INTERNATIONAL (UK) LTD.**
27 Old Gloucester Street, London, WC1N 3AX, UK
E-mail: info@newacademicscience.co.uk • Visit us at www.newacademicscience.co.uk
- **Nairobi** **NEW AGE GOLDEN (EAST AFRICA) LTD.**
Ground Floor, Westlands Arcade, Chiromo Road (Next to Naivas Supermarket)
Westlands, Nairobi, KENYA, **Tel.:** 00-254-713848772, 00-254-725700286
E-mail: kenya@newagepublishers.com

BRANCHES

- **Bangalore** 37/10, 8th Cross (Near Hanuman Temple), Azad Nagar, Chamarajpet, Bangalore- 560 018
Tel.: (080) 26756823, **Telefax:** 26756820, **E-mail: bangalore@newagepublishers.com**
- **Chennai** 26, Damodaran Street, T. Nagar, Chennai-600 017, **Tel.:** (044) 24353401
Telefax: 24351463, **E-mail: chennai@newagepublishers.com**
- **Cochin** CC-39/1016, Carrier Station Road, Ernakulam South, Cochin-682 016
Tel.: (0484) 2377303, **Telefax:** 4051304, **E-mail: cochin@newagepublishers.com**
- **Guwahati** Hemsen Complex, Mohd. Shah Road, Paltan Bazar, Near Starline Hotel
Guwahati-781 008, **Tel.:** (0361) 2513881, **Telefax:** 2543669
E-mail: guwahati@newagepublishers.com
- **Hyderabad** 105, 1st Floor, Madhiray Kaveri Tower, 3-2-19, Azam Jahi Road, Near Kumar Theater
Nimboliadda Kachiguda, Hyderabad-500 027, **Tel.:** (040) 24652456, **Telefax:** 24652457
E-mail: hyderabad@newagepublishers.com
- **Kolkata** RDB Chambers (Formerly Lotus Cinema) 106A, 1st Floor, S N Banerjee Road
Kolkata-700 014, **Tel.:** (033) 22273773, **Telefax:** 22275247
E-mail: kolkata@newagepublishers.com
- **Lucknow** 16-A, Jopling Road, Lucknow-226 001, **Tel.:** (0522) 2209578, 4045297, **Telefax:** 2204098
E-mail: lucknow@newagepublishers.com
- **Mumbai** 142C, Victor House, Ground Floor, N.M. Joshi Marg, Lower Parel, Mumbai-400 013
Tel.: (022) 24927869, **Telefax:** 24915415, **E-mail: mumbai@newagepublishers.com**
- **New Delhi** 22, Golden House, Daryaganj, New Delhi-110 002, **Tel.:** (011) 23262368, 23262370
Telefax: 43551305, **E-mail: sales@newagepublishers.com**

ISBN: 978-81-224-0494-4

C-16-12-9906

Printed in India at Kanishik Printing Press, Delhi.

NEW AGE INTERNATIONAL (P) LIMITED, PUBLISHERS
7/30 A, Daryaganj, New Delhi-110002
Visit us at **www.newagepublishers.com**
(CIN: U74899DL1966PTC004618)

PREFACE

The chemical processes in biological systems and the story of DNA have occupied centre-stage for more than two decades. Yet it is disconcerting to note that the teaching of biology courses continues to be on the conventional dull pattern utilizing qualitative approach. What should be done for improving first-year biology course in colleges and universities? It is imperative that an introductory biology course should now be designed to incorporate dynamic molecular approach to the cell, its ultrastructure and functions so that the impact of biology research can be appreciated as relevant to modern society.

Molecular biology is one of the most exciting disciplines of natural sciences and the thrill of its discoveries should reach the common man. Advances in molecular biology hold a lot of promise for improvement of the human life, hence we believe that the excitements of this field should be transmitted to the students. We have set our goals to prepare a chemical background to appreciate that biological sciences are making a hard progress that characterizes physics and chemistry, to provide a conceptual framework of quantitative biology that is more exciting, provocative and rational, and to emphasize relevance of biology for human welfare.

To achieve these goals, a selection of topics has been made so that molecular biology could be understood in its right perspective. The basic structural make-up of any organism is fundamental to its ability to cope with the environment, thus the structural design of organisms along with their functions has been given to emphasize the dynamic nature of life. The impact of biochemistry on biology has been exceedingly profound, hence an attempt has been made to present intimate relationship between biochemistry and biology, focussing on synthesis and functions of macromolecules and the unique mechanisms that regulate their activity. Maintenance of structure, growth and reproduction are fundamental properties of life and require some kind of energy input and its transformations. Continuity, growth and development depend mainly on how the hereditary material is transferred from parent to offspring. The concept of gene, chromosome behaviour, and variation have been presented in detail. Specifically, the text is intended to cover areas of molecular biology dealing with the synthesis of informational molecules (DNA and RNA) and proteins, gene expression and regulation, gene recombination and transfer mechanisms. Understanding of functional biology would be incomplete if the intimate relationship between organisms and their environment remains unexplained. How the basic principles involved in the field of genetics, are now being exploited to commercial advantages, have been described in the chapter on recombinant DNA technology. By emphasizing these principles, it is hoped that the reader will be able to interpret and assess any future developments in this rapidly advancing field.

The style of presentation is simple, lucid and easy-to-comprehend, backed by the vast experience of the authors in handling molecular biology course at the first-year of undergraduate level at BITS. It is hoped that this book will serve as a suitable text to provide the extra depth and stimulus which these students need.

We wish to express our sincere gratitude to Dr. S. Venkateswaran, Director of Birla Institute of Technology and Science, for much needed encouragement. We are also thankful to our colleagues for their many helpful suggestions on the subject matter.

S.C. RASTOGI
V.N. SHARMA
ANURADHA TANDON

CONTENTS

CONTENTS

INTRODUCTION: ATTRIBUTES OF LIFE

The ecosphere abounds in a variety of organisms which are composed of discrete units or packages. When the first living systems appeared on earth they were probably faced with the problem of exchanging materials with their environment. It is believed that the ancestral organisms obtained their energy from simple organic molecules (chemotrophs) and later specialized in synthesizing their own food materials by trapping sunlight (autotrophs). The two most important attributes of living organisms are *growth* and *reproduction*, the significance of which shall become clear as we proceed.

The absolute minimal structural requirements for the simplest organisms we know—the viruses—are two: a molecule containing the information necessary for reproduction, and a capsular coat around this molecule. Viruses, however, cannot reproduce without a larger host which has an efficient biological machinery. For this reason they do not conform to a true living system, and may be called quasi-living particles. Viruses have simple life histories. The capsular coat is made up of protein, while the information molecule is a nucleic acid, either DNA or RNA. When viral DNA enters the host cell, it induces the synthesis of a number of enzymes essential to the production of viral progeny and the coat protein is the last to be synthesized. It can be said that the orderly development of the virus is taped in the information molecule, which may be called the genetic tape.

The structural living unit other than the virus is the cell, which consists of smaller structures called organelles, each having a specialized function within the cell. A cell containing these organelles and a well defined nucleus is called a *eukaryotic cell*. The cells of bacteria and blue-green algae lack a definite nucleus and organelles like eukaryotes, and are called *prokaryotes*. Many plants and animals exist as single cells, hence the cell is the basic unit.

THE CELL DOCTRINE

The seeds of progress have frequently been sown in the past, often unconsciously, and history should be read not only for pleasure but profit also. Robert Hooke in 1665 introduced the term *cell* while describing the microscopic texture of the cork. This was followed by illustrations conforming to cellular nature of living organisms by a number of workers during the next two centuries. However, Schleiden and Schwann are generally accredited as the exponents of the *cell theory*, which states that both plants and animals are made up aggregations of basic units. The writings of Schledien and Schwann in the nineteenth century gave rise to the cell doctrine, which may be summarized as:

(a) Cells of organisms are differentiated into distinct cell types.

(b) A nucleus is a feature of all cells with the exception of viruses and red blood cells.

(c) The protoplasm is the living content of the cell and determines the activity of the cell, and thus, the whole organism.

(d) New cells originate from pre-existing cells through division. In any case, cells never arise *de novo*.

Before the mid-1940s, the approach to the cell tended to be largely descriptive and qualitative. In recent years, the approach to biology of cells has become more dynamic and quantitative as a result of developing techniques.

Life can be regarded as a unique and complex interaction of matter and energy that assumes a hierarchy of structural order. The biological continuum thus consists of different levels of biological organization. Atoms, the smallest units of matter, interact to form molecules that may be variously joined to form more elaborate macromolecules characteristic of living organisms. The interacting macromolecules often result into structures called organelles, and the next higher level of organization is the cell. This cell is the smallest unit of interacting matter and energy that exhibits properties of life.

The chemical components of the living cells are qualitatively quite different from those of the physical environment. Most of the molecules in living cells are organic compounds, which occur in great variety and complexity. In the bacterium *Escherichia coli,* for example, there are more than 5000 types of organic molecules, of which 3000 are proteins and 1000 are nucleic acids. A more complex organism such as a human being is known to have more than 100,000 types of proteins. This immense variety of organic compounds is reducible to a surprising simplicity: all proteins are composed of 20 types of linearly arranged amino acids. Similarly all nucleic acid molecules are composed of four types of nucleotides. These building blocks thus serve as the raw material for macromolecular synthesis and perform various functions. Amino acids for example, also serve as precursors of hormones, alkaloids, pigments etc. Strange as it may seem, the building blocks are identical in all living organisms. We can therefore formulate the following generalizations:

(a) There is an underlying simplicity in the molecular organization of all living cells.
(b) There is an underlying principle of molecular economy in living cells.
(c) All living organisms have a common ancestor.

CELLS OBEY THE LAWS OF THERMODYNAMICS

The molecular complexity and highly organized nature of cells, in contrast to the randomness of inanimate matter, do not constitute exceptions to the laws of thermodynamics. These most fundamental rules of nature deal with the energy relations of a system and its surroundings. The first law of thermodynamics, also known as the law of conservation of energy, states that energy can neither be created nor destroyed; it can however be transformed. *Living cells can only transform one form of energy into another*. They absorb the useful form of energy, i.e. *free energy,* and return equivalent amount of the less useful form of energy to the environment. The law, therefore, implies that energy can only be redistributed, or transformed.

Equally important is the second law of thermodynamics which states that all physical and chemical processes always proceed with an increase in the *entropy* or disorderliness of the universe. Since the cell absorbs the useful form of energy, it expels an equivalent amount of the non-useful form of energy to the environment resulting in increase of the entropy of the universe. Thus, cells follow the axiom: 'Living cells neither create nor destroy energy, but transform it into a useful form. They maintain their orderliness at the cost of their environment which they make more disordered

CELLS AS OPEN SYSTEMS

Cells function as chemical engines where reactions proceed under the influence of temperature, pressure and concentration of the reactants. External energy is needed to produce significant reaction rates. This external energy is called the *activation energy*. Chemical reactions in the physical world require a much greater amount of energy than can be provided by ordinary temperature. Most chemical reactions of biological interest require a fairly high amount of activation energy, which is obtainable through enzymes or *biocatalysts*. The special catalysts that occur in cells are enzymes or specialized proteins. There are thousands of enzymes in a living cell, each of which catalyzes a specific reaction at an amazing speed. The environment in which organisms live is essential to them as a source of energy and raw materials in the form of nutrients. Thermodynamically, a system can exist as an isolated *closed system* which does not exchange both matter and energy with its environment. It exchanges, either energy or matter i.e. one of the two. A living cell is an *open system* which exchanges both matter and energy with its environment, and in doing so, it may appear in equilibrium with its environment, representing a *steady-state* condition. We can, therefore, arrive at another axiom that cell is essentially an isothermal open system as all parts of the cell at any given time maintain the same temperature and pressure.

CELL AS SELF-REGULATING SYSTEM

The cells carry out in seconds a complex sequence of reactions that would require several days, weeks or even months in a laboratory. The remarkable character of enzyme-catalyzed reactions is that the yield is 100 per cent and no byproduct is formed as opposed to man-made reactions where one or more byproducts are produced. Each cell functions on a maximum economy basis, hence reactions must be self-regulatory and self-controlled. The product of one substrate becomes the substrate of a second reaction, because the enzyme-catalyzed reactions are linked through a reaction sequence consisting of anywhere between 2 to 20 or more reaction steps.

Basically cells are either *phototrophic* utilizing sunlight for synthesis of food or heterotrophic i.e. depending on organic molecules as nutrients. Although the two types of cells obtain energy in different forms, they transform it into chemical-bond energy in the form of adenosine triphosphate (ATP) molecules. ATP acts as a major energy source in all living organisms, which is regenerated through enzyme-catalyzed reactions. It is an interesting fact that the cells synthesize the required enzyme proteins by themselves.

CELLS AS SELF-REPLICATING SYSTEMS

The interacting molecules are structurally complementary thus providing a means of transfer of chemical energy from energy-yielding to energy-requiring processes. A fundamental function of life is its maintenance and extension in time and space. Extension of life is only possible through reproduction, which the organisms carry out with near perfect fidelity, not just once or twice, but for hundreds of generations. Reproduction is accomplished by a specific mechanism, but the unique resemblance between parents and offspring in terms of hereditary traits is rather surprising. The hereditary information for this purpose is encoded in a unique molecule, called deoxyribonucleic acid (DNA). This DNA is a double stranded macromolecule and its one strand acts as a template for the synthesis of another complementary DNA strand. The duplex DNA molecule is endowed with a precise mechanism for self-replication and information transfer through synthesis of ribonucleic acid (RNA). Sometimes errors or mutations are introduced during information transfer, which may give added advantage to the cell in adapting to its everchanging environment. The cells have remarkable ability to translate genetic information into specific proteins.

In the foregoing discussion we have examined certain attributes of the living, and at no stage could we find any departure from the basic laws of chemistry. In the following pages we shall concentrate on the basic theme that 'a cell is a self-assembling, self-regulating and self-replicating isothermal open system'.

CHAPTER 2

FUNCTIONAL ORGANIZATION: PARADIGM OF LIFE

In the first chapter we have stated that biological continuum consists of levels of biological organization. Thus life is organized in discrete packets whose survival depends upon interaction of matter and energy available in the inanimate environment. In order to utilize matter and energy, a complex organization is needed to funnel energy into channels useful for the maintenance of the organization. However, the complexity of the organization reflects the evolutionary status of an organism, for example, bacterial cells do not possess the elaborate organization found in the cells of higher organisms.

A useful way of studying the level of organization is to consider an organism's source of matter separately from its source of energy because all organisms do not obtain both from the same source. Energy utilized by organisms comes either from the sun or from exergonic oxidative chemical reactions. Similarly, the matter absorbed by organisms may be very simple, in the form of inorganic chemical substances or ions, or complex, as organic molecules. Thus the way organisms derive their nutrition and utilize energy may form an important basis for structural specialization. We may, therefore, propose a nutritional classification of organisms, which takes into consideration two factors: the source of energy and the source of matter.

There are two kinds of energy sources: light and chemical. *Autotrophs* synthesize their carbohydrates, lipids, proteins, nucleic acids and other constituents from inorganic molecules and ions; their carbon source is CO_2, and their sources of nitrogen, phosphorus and sulphur are nitrates, ammonia, phosphates and sulphates. *Photosynthetic autotrophs* include green plants, green protista, blue-green algae, and photosynthetic bacteria. Those depending entirely upon the energy liberated in oxidative chemical reactions are *chemosynthetic autotrophs*, which include all bacteria. All these autotrophs utilize CO_2 as their sole carbon source.

Heterotrophs are quite different and depend, to a greater or lesser extent, upon the molecules synthesized by autotrophs. They are also called *chemoorganotrophs*. There is a small group of *photosynthetic heterotrophs* (Photoorganotrophs) of purple sulphur bacteria which utilize organic molecules as a hydrogen source for photosynthetic reactions in which CO_2 is converted to carbohydrates. Thus, purple bacteria represent a typical kind of nutrition.

2.1 ORGANIZATION OF CELLS

Two groups of organisms have been recognized on the basis of their structural organization. These are prokaryotes (monera) and eukaryotes. The prokaryotes include all bacteria and blue-green algae while eukaryotes consist of protists, animals and plants. In an evolutionary sense, the prokaryotes appear to be more primitive, but they have remarkable adaptability in diverse environmental conditions. The primitiveness of bacteria is reflected in their structural organization. They have a relatively simple internal structure with distinct and physically separate regions of specialized metabolic activity, but are lacking in subcellular organelles derived from intracellular membranes.

The eukaryotic cell has an elaborate organisation including a number of membrane-bound organelles that carry out specialized functions. In bacteria, the principal subcellular units, the nuclear apparatus, respiratory or photosynthetic apparatus, are not bound by membranes, but the cells of higher organisms have a number of membrane-bound organelles that include lysosomes, vacuoles, food vacuoles, plastids etc. The membrane barrier is responsible for isolating the metabolic activities from the rest of the cell. Let us consider the important features of the two cell types.

The Prokaryotes

The prokaryotes represent a group of primitive cells which have a relatively simple organization. Some important features are the absence of a well-organized nucleus and membrane-bound organelles, the presence of well defined protein synthesizing ribosomal units of the 70 S type, and the presence of a well formed cell wall made up of peptidoglycans. The basic organization of phototrophic and chemotrophic prokaryotes is almost similar except that the phototrophs have a larger complement of enzyme catalysts and an elaborate photosynthesizing apparatus.

Prokaryotic chemotrophs. These include mycoplasmas and most bacteria. The structure can be best understood by examining a typical bacterial cell, i.e. *Escherichia coli*. The bacterial cell is surrounded by a distinct cell wall that is composed of peptidoglycans. On the basis of their cell walls, bacteria are often classified as gram-negative and gram-positive. The cell wall, besides giving shape to the cell, acts as a barrier to osmotic forces and some antibiotics. Below the cell wall lies the plasma membrane which is often thrown into small invaginations forming tubular complexes of different shapes and sizes, confined to a characteristic region of the cell (Fig. 2.1).

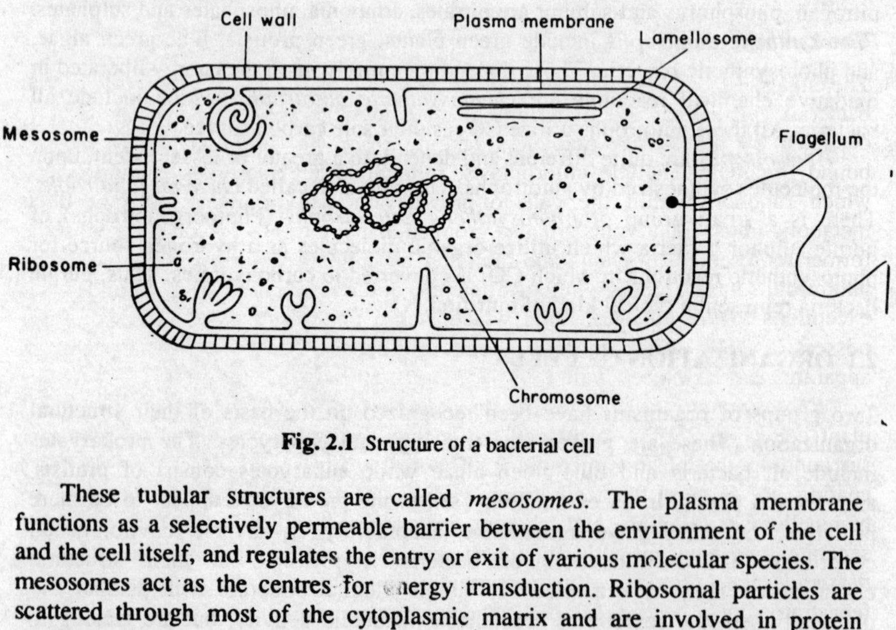

Fig. 2.1 Structure of a bacterial cell

These tubular structures are called *mesosomes*. The plasma membrane functions as a selectively permeable barrier between the environment of the cell and the cell itself, and regulates the entry or exit of various molecular species. The mesosomes act as the centres for energy transduction. Ribosomal particles are scattered through most of the cytoplasmic matrix and are involved in protein synthesis. The genetic apparatus consists of a double stranded DNA fibre of circular shape located near the central region of the cell and is not continuous with any structural element. The DNA is naked, hence called *nucleoid*. Some bacteria contain one or more flagellae for locomotion, and these originate from the basal granules located in the cytoplasm. Bacteria are often referred to as bags of enzymes, because a large number of enzymes involved in DNA replication, Krebs cycle and electron transport system are loosely associated with discrete structures.

Bacterial cells may exist freely (eg. *E. coli, Bacillus* species etc.), or occur as aggregates like *Streptococcus*. The chemotroph bacteria include chemolithotrophs which require organic energy sources and inorganic electron donors and chemoorganotrophs which need organic compounds both as energy sources and as electron donors.

Prokaryotic Phototrophs. The phototrophic prokaryotes include photosynthetic bacteria and cyanobacteria (blue-green algae). The blue-green algae are actually not blue-green, but are red, yellow, green, blue and black, and of various shades owing to several pigments. The pigments include *chlorophyll a*, phycocyanin (blue), and red phycoerythrin. The cell walls of cyanobacteria contain cellulose, and like bacterial cells, they lack a nuclear membrane.

Cyanobacteria carry out photosynthesis, hence they were earlier grouped as algae, but on the basis of their structure they are now included under bacteria. Among the prokaryotes, cyanobacteria were the first to exhibit definite membranous vesicles which fill the cell interior. Sometimes a stack of lamellae similar to those in chloroplasts are also found which form physically separate photosynthetic systems. Besides other pigments, a group of unique pigments called phycobilins are found, which include phycocyanins and allophycocyanins. Phycobilins are located in phycobilisomes associated with photosynthetic algae containing chlorophyll *a*. The blue-greens grow as single cells or in colonies and in the latter case many complex shapes are formed which remain embedded in a gelatinous mass secreted by the cells.

The Eukaryotes

The eukaryotic cells are more advanced and show striking characters in contrast to bacterial cells. There is a definite nucleus with a nuclear membrane that is occupied by pores with diameters ranging from 400-800 Å. Other membrane bound organelles include chloroplasts, mitochondria, lysosomes, vacuoles etc., which function within the cell to provide microenvironments to subcellular metabolic units. Eukaryotes are also grouped as phototrophs and chemotrophs, the former represented by plants and the latter by animals and most protozoans. The phototrophs have a photosynthetic apparatus and a rigid cell wall of cellulose, but exceptions to this rule are fungi which lack a photosynthesizing apparatus and possess a cell wall. Similarly, some protozoans have a photosynthesizing apparatus, but lack a cell wall.

Photosynthetic Eukaryotes. As an example of this, we can consider a typical plant cell (Fig. 2.2). A plant cell has a thick cell wall composed of cellulose, surrounding the plasma membrane. The cell wall is actually an extracellular coating to protect the cell from various osmotic stresses. It consists of hemicelluloses, lignins, and pectins, in addition to cellulose. The cell wall separates two adjacent cells, though at several locations intercellular channels or *plasmodesmata* provide cellular communication. The cytoplasm of a plant cell is fenestrated by a well-developed vacuolar system, which is associated with an endoplasmic reticulum or network. The vacuoles are bound by a membrane called *tonoplast*. The endoplasmic reticulum (ER) forms a network of anastomosing tubules whose lumen is connected with the vacuolar system. Some of the ER lamellae show cisternal orientation with ribosomes attached on their surface giving an appearance of *granular endoplasmic reticulum* (GER), while others which do not have ribosomes are known as smooth endoplasmic reticulum (SER).

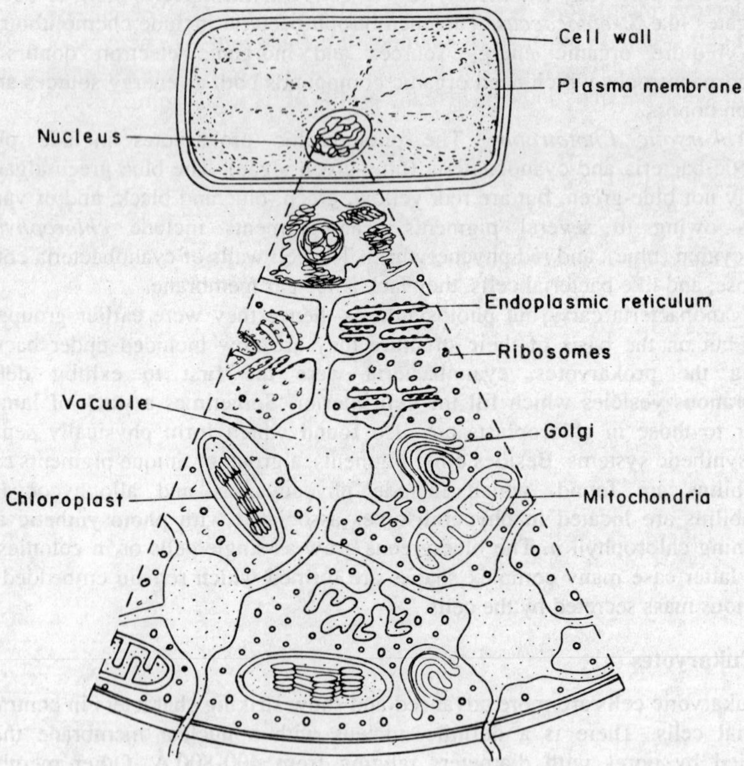

Fig. 2.2 A typical plant cell

In addition to ER, another system of membranous lamellae, called Golgi complex is present in the vicinity of the nucleus. The Golgi complex is highly secretory in function and maintains communication with the endoplasmic reticulum also. A number of other membrane-bound organelles are also found in the cell matrix which include: (a) *lysosomes* containing hydrolytic enzymes for digestion, (b) *peroxysomes* which neutralize metabolic peroxides in the cell, and (c) *glyoxysomes* which are involved in glyoxylate metabolism.

The most important energy transducing systems in a plant cell are mitochondria and chloroplasts, both of which are autonomous and bound by twin membranes. The chloroplasts are able to synthesize organic molecules and carry out oxidative phosphorylation in the presence of sunlight, whereas mitochondria carry out oxidative phosphorylation in the absence of light by breaking down organic molecules.

Chemosynthetic Eukaryotes. Animal cells represent such chemotrophs which are unable to synthesize their food in the presence of sunlight. They are devoid of a cell wall and a vacuolar system, but possess all membrane-bound organelles similar to those of plant cells, with the exception of chloroplasts.

We shall examine here the functional organization of a typical cell as revealed by the electron microscope.

2.2 ANATOMY OF THE EUKARYOTIC CELL

The electron microscope reveals a great deal of structural organization in an animal cell, which is taken as an example of a eukaryote. The activity of a cell is considered to be the sum total of the activities of different parts or organelles of the cell, hence we shall describe here the profile of these structures (Fig. 2.3).

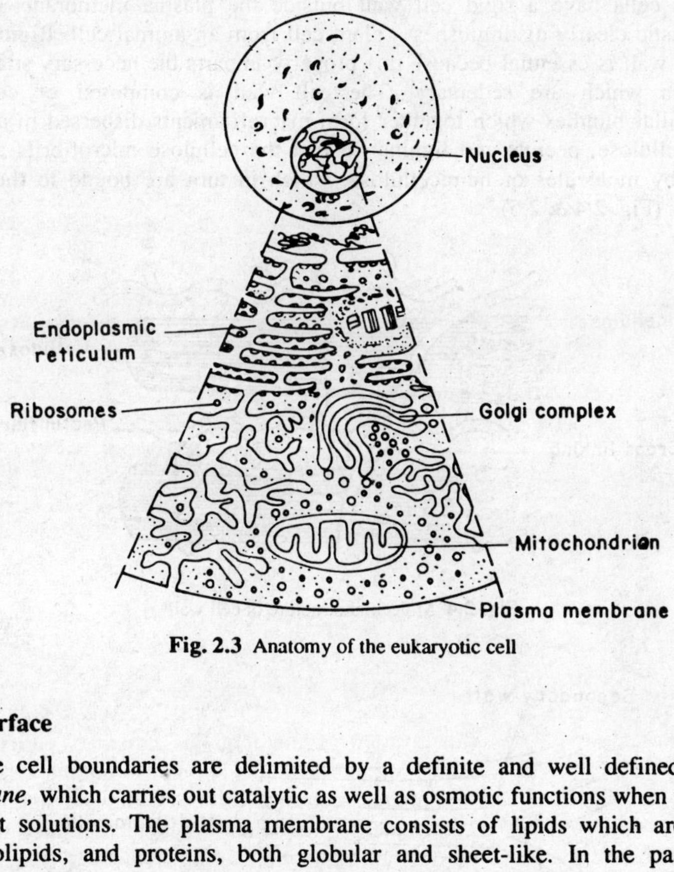

Fig. 2.3 Anatomy of the eukaryotic cell

Cell Surface

The cell boundaries are delimited by a definite and well defined *plasma membrane,* which carries out catalytic as well as osmotic functions when placed in different solutions. The plasma membrane consists of lipids which are mainly phospholipids, and proteins, both globular and sheet-like. In the past, many molecular models of the plasma membrane have been proposed, but the current and most favoured model is the *'fluid-mosaic model'* suggested by Singer and Nicolson in 1972. The model proposed by them is based on the freeze-fracture technique and scanning electron microscope, and accounts for all biological properties of the membranes.

In the fluid mosaic model, there is a lipid bilayer which exhibits fluid-like consistency owing to the constituent phospholipids, in which a variety of proteins are distributed. The protein molecules show movements within the bilipid layer. Some proteins remain fully embedded and form *integral proteins,* while others which are partially embedded are called *extrinsic proteins*. The outer surface of the plasma membrane carries carbohydrate molecules, generally oligosaccharides, which either bind with lipid molecules to form glycolipids or with proteins to form glycoproteins. These structures have a branched appearance and project out from the surface of the membrane in the form of *glycocalyx*.

Increase in cell complexity stems from modifications of its component structural units, and the plasma membrane is no exception. The membrane gives shape and size to the cell and regulates the molecular traffic of substances by selective permeability. The cell membranes owe their capacity to act as agents of nervous transmission and to transduce signals and stimuli to and from the environment, to their selective permeability. The membrane also functions in cellular association to produce a multicellular system.

Plant cells have a rigid cell wall outside the plasma membrane and this characteristic clearly distinguishes a plant cell from an animal cell. Rigidity in a plant cell wall is essential because this property imparts the necessary strength to the plants which are sedentary. The cell wall is composed of cellulosic microfibrillar bundles which together form macrofilaments dispersed in a matrix of hemicellulose, pectins and lignins. In fact the cellulose microfibrils are held together by molecules of hemicellulose, which in turn are bound to the pectin molecules (Fig. 2.4 & 2.5).

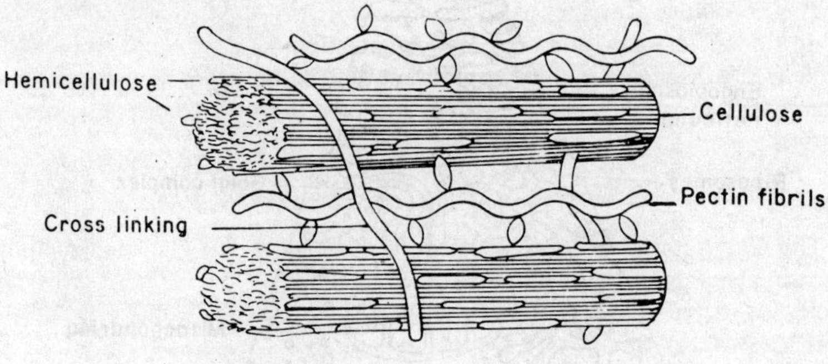

Fig. 2.4 Molecular structure of cell wall

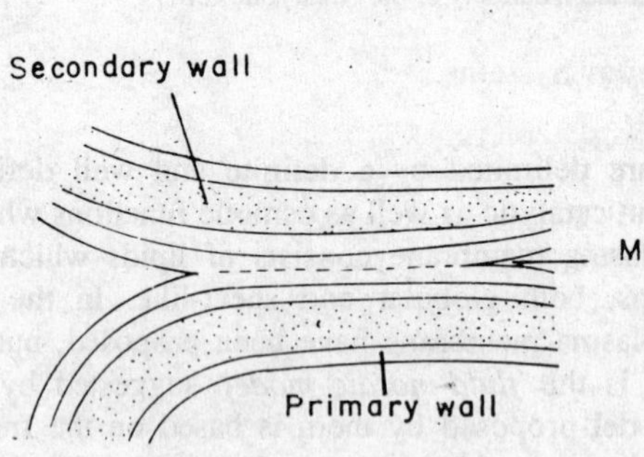

Fig. 2.5.

The cell wall has a limited number of pores through which the plasma membranes of two adjacent cells fuse to form channels called plasmodesmata. These tubular channels serve as intercellular routes for transport of cellular materials. Water and minerals from the root tip to shoot tip, and carbohydrates from the site of photosynthesis to the root tip cells, are transported through such channels. Unlike the plasma membrane which is selectively permeable the cell wall itself is permeable to a wide variety of substances. The cell wall restricts osmotic water uptake and protects the cell from bursting open as a result of the build up of internal turgor pressure.

Endoplasmic Reticulum

The cytoplasmic matrix of the eukaryotic cells has an extensive network of membrane bound channels called *endoplasmic reticulum* (ER) which link the external and internal regions. These channels are highly branched and divide the matrix into cytosol and intrachannel spaces containing materials synthesized at ER sites. The ER system is divided into two types based on morphological characters (Fig. 2.6), those having ribosomes on the lamellar surfaces are called *granular endoplasmic reticulum* (GER) and those devoid of them are termed *smooth endoplasmic reticulum* (SER). The GER represent the sites of protein synthesis whereas SER act as the sites for many non-protein metabolic activities.

Ribosomes

The ribosomes are small particles of definite geometrical configurations. They contain ribonucleic acid and proteins forming ribonucleoprotein complexes. They may be freely dispersed in the cytoplasm or may be bound to the ER membrane. In prokaryotes they are always free. Ribosomes are the sites where protein synthesis occurs.

Golgi Complex

The Golgi complex consists of stacks of smooth and flattened membranous cisternae, generally arranged in a parallel fashion. They are usually surrounded by vesicles of different sizes, some of which are discharged from the tips of the cisternae. Several functions have been assigned to the Golgi complex which include cellular secretion, sorting and glycosylation of proteins synthesized in GER, proliferation of additional membranes for the formulation of intercellular organelles and the plasma membrane.

Membrane-Bound Vesicular Systems

The cytoplasm contains several types of vesicles bound by single membranes, the major types include *lysosomes, peroxysomes* and *glyoxysomes*. Lysosomes contain a number of hydrolytic enzymes capable of hydrolyzing most biomolecules and non-functional cell organelles. These bodies are responsible for intracellular digestion, scavenging of wornout organelles and lysis of damaged cells. At cell death, these lysosomes break down and release the digestive enzymes to lyse the cell.

The peroxysomes and glyoxysomes are spherical microbodies that carry out specialized functions in the cell. The peroxysomes contain enzymes for peroxide detoxification, and glyoxysomes carry out glyoxylate metabolism. Glyoxysomes are found in abundance in germinating seeds and early embryo.

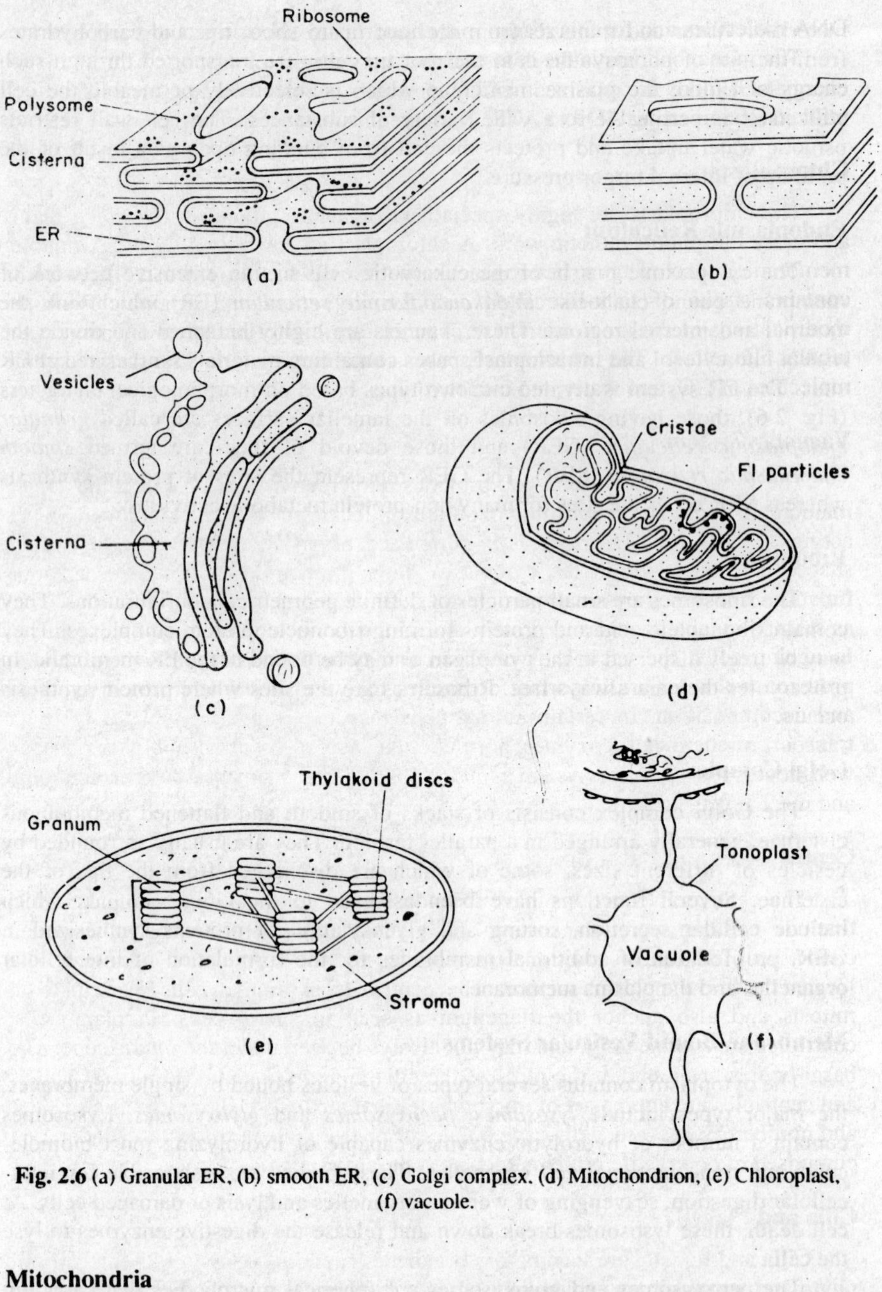

Fig. 2.6 (a) Granular ER, (b) smooth ER, (c) Golgi complex, (d) Mitochondrion, (e) Chloroplast, (f) vacuole.

Mitochondria

The mitrochondrion is a double-membraned structure with the outer smooth membrane stretched tightly around it and the inner membrane thrown into folds projecting into the matrix. These folds called *cristae*, serve to increase the surface area of the inner membrane. The location and number of mitochondria is related to the cell function. The inner membrane facing the matrix is studded with minute particles called F_1-particles. The mitochondrial matrix also contains several circular

DNA molecules and for this reason mitochondria are autonomous organelles.

The mitochondrion is the centre of the cell's respiratory metabolism where the energy-rich foods are oxidized to CO_2 and water, and the bond–energy released is utilized in converting ADP to ATP. They are thus, the power houses of the cell.

Chloroplasts

The chloroplasts are highly specialized organelles found in the green plants and carry out photosynthetic work. A chloroplast is generally ellipsoidal, double-membraned, enclosing a heterogenous matrix called the stroma. The stroma contains stacks of sack-like *thylakoids* forming grana and originate from the modifications of the inner membrane. The grana are interconnected by intergranal tubular lamellae. The chloroplasts use solar energy and transform simple molecules of CO_2 and water into carbohydrates.

Vacuolar System

The vacuolar system of a cell provides a microenvironment within the cell to maintain differential composition of solutes in localized area bound by membranes acting as diffusional barriers. It is comprised of vacuoles that exist in various shapes and sizes within the cytoplasm. It is impossible to designate a single function to them. They are more common and well-developed in plants, and serve as storage organs for water and salts, and as a sink for unused metabolic products. In ageing cells, the vacuole is enlarged and occupies almost the entire cell. In protozoa, vacuoles are important organelles for water elimination, food retention and decomposition. In certain animal cells they may serve as vehicles for bulk transport of chemicals, enzymes, hormones etc. across the cells. In higher plants, a well developed vacuolar system is indispensable for maintaining water economy and mechanical strength.

Centrosomes and Basal Bodies

The centrosomes are complex structures with a dense outer granular mantle that encloses a less granular material around two pairs of cylindrical bodies called *centrioles*. The centrosomes are common in animal cells, usually one per cell, located near the nuclear surface. The centrioles in animal cells participate in mitosis and also anchor the flagellum as seen in sperm cells. In plant cells, centrioles are rudimentary, and may not always be seen under the microscope. The basal bodies are found at the roots of the cilia and the flagella. Both basal bodies and centrioles are composed of microtubules that are arranged in a specific manner and number. Eukaryotic cells have 9 pairs of microtubules forming a cylinder, and composed of contractile proteins such as actin, myosin and dynein.

Cilia and Flagella

Cilia and flagella are locomotory organelles found in many free-living cells, in phyto and zooplanktons, epithelial cells and intestinal lining of animals. These are motile organs with identical structure, except that the cilia are much shorter than the flagella. A typical flagellum is covered by a plasma membrane, enclosing an array of microtubules (Fig. 2.7). These microtubular structures fuse or anchor to the basal bodies (Kinetochore) or to the centriole.

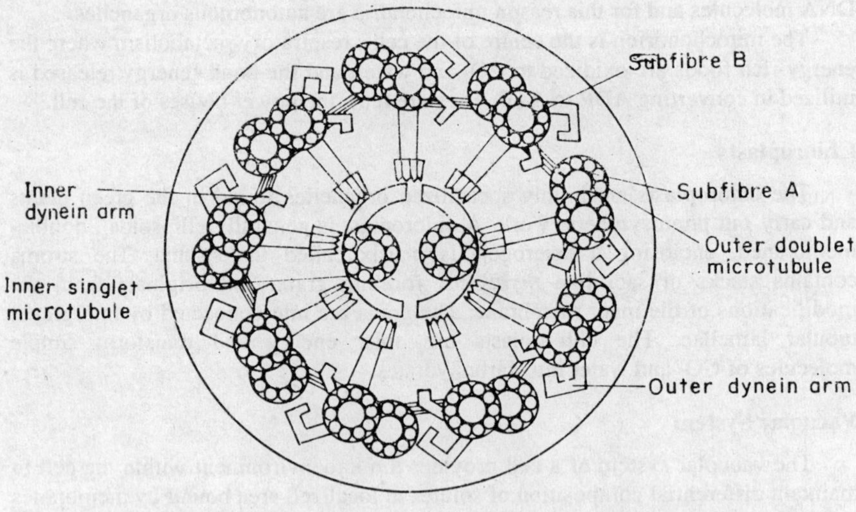

Fig. 2.7 Cross section of a flagellum

Cytoskeleton

The cytoplasmic matrix consists of a complex network of microtubules and microfilaments, forming the cytoskeleton. In addition, there are intermediate filaments also. The microtubules are cylindrical structures composed of heterodimers of α and β tubulin proteins. These are constituents of centromeres, spindle fibres, flagella and cilia.

The microfilaments are rods of indefinite length and are composed of actin and myosin. These structures give rise to microvilli and help in the movement of cell organelles, change in celi shape, amoeboid locomotion, and cytokinesis etc.

Nucleus

The nucleus is a crucial organelle of the cell, which is highly organized and carries the genetic material that exerts control over the vital activities of the cell. If a nucleus is removed from the cell, all its metabolic functions come to a standstill leading to cell death. The nucleus is enclosed in a nuclear envelope which is a double membranous structure, the outer membrane is continuous with the GER while the inner one is smooth and shows attachment with the chromatin at various places. The nuclear envelope is punctured by a large number of *nuclear pores* which serve as channels of communication and transport of materials between the cytoplasm and the nucleus (Fig. 2.8). The nuclear envelope encloses a fluid-filled matrix called the *nucleoplasm* which contains delicate thread-like chromatin fibres. At the time of cell division these chromatin fibres condense into chromosomes. The chromatin material is composed of deoxyribonucleic acid (DNA), histones and some non-histone proteins. DNA is the hereditory material which expresses itself through protein synthesis.

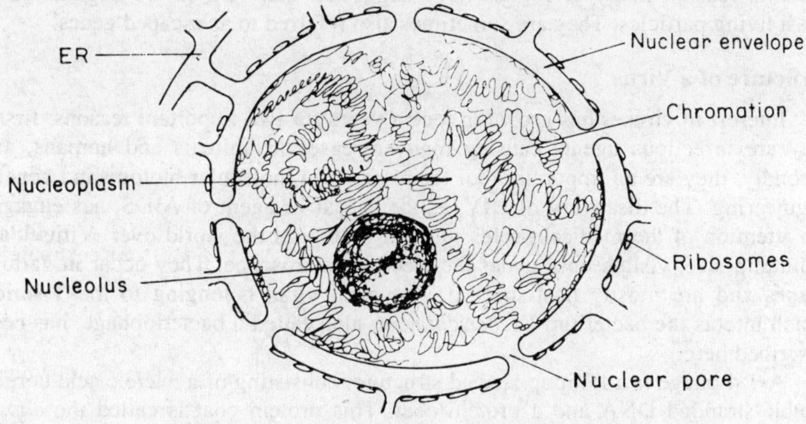

Fig. 2.8 Ultrastructure of nucleus in non-dividing phase.

The nucleus usually contains one or more spherical bodies called nucleoli. The nucleolus is found attached to a specific region on the chromosome, called the nucleolar organizing centre the primary function of which is the synthesis of ribosomes.

It has been our attempt in this chapter to discuss the structure and behaviour of the cell and to understand the level at which structure and functions are related. The eukaryotic cell which has been described here with respect to subcellular architecture does not exist, because certain cells develop specialized and accessory structures to perform a particular function. Nevertheless, the concept of a typical cell can be formulated which is that a cell is a self-assembling, self-regulating and self-replicating isothermal open system, and in order to achieve these capabilities a highly organized molecular order is necessary. We have also to consider the organization of prokaryotes which represent a low molecular order compared to eukaryotes.

There are, however, certain parasitic structures which are neither organisms nor cells, but have assumed tremendous significance in the study of molecular biology and evolution. These structures are the *viruses*.

2.3 VIRUSES

Viruses are neither organisms nor cells, but consist of a core of nucleic acids enclosed in an external protein mantle. They are exclusively parasitic and in order to reproduce, they require a host cell which supplies not only nutrition but also the metabolic machinery. Thus viruses have three fundamental properties; they have a limiting boundary, a genetic control system, and the capacity to reproduce within a host cell. Though viruses carry nucleic acid which is the genetic information system, they are not considered as living organisms since they lack a matrix of cytoplasm, which is essential for synthesis of proteins and release of energy.

Viruses exhibit symptoms of life when placed in a suitable environment. Reproduction is a fundamental characteristic of life, and when a virus infects a living host cell, the viral nucleic acid takes over the metabolism of the infected cell and parasitizes the resources of the cell for formation of more viruses; the nucleic acid core reproduces and new protein coats are formed surrounding the nucleic acid giving rise to new viruses which escape the host cell upon its disintegration.

For this reason, many biologists have suggested that viruses are degenerate or quasi living particles. They are sometimes also referred to as escaped genes.

Structure of a Virus

Interest in viruses has grown in recent years for two important reasons; firstly they are infectious agents causing many diseases in animals and humans, and secondly, they are an important tool for research in molecular biology and genetic engineering. The discovery of HIV, as the causative agent of AIDS, has engaged the attention of the medical profession and biologists the world over. Viruses are minute in size, visible only under the electron microscope. They occur in various shapes and are mostly host-specific. A typical virus belonging to the T-series, which infects the bacterium *Escherichia coli* also called a bacteriophage, has been described here.

A T 4 phage is a lollipop-shaped structure, consisting of a nucleic acid core of double-stranded DNA and a protein coat. This protein coat is called the *capsid* (Fig. 2.9). In simple viruses, instead of a capsid there are subunits of a single protein which wrap around the nucleic acid core; these subunits are called *capsomeres,* as in the case of *Adenovirus*. The capsid is composed of proteins that are arranged in two patterns: (1) helical arrangement as found in rod-shaped viruses (eg. TMV), and (2) isometric pattern as in globular or lolli-pop shaped viruses (eg. T 4). Many viruses have capsids composed of lipoproteins, as in the influenza virus, and the herpes and small-pox viruses.

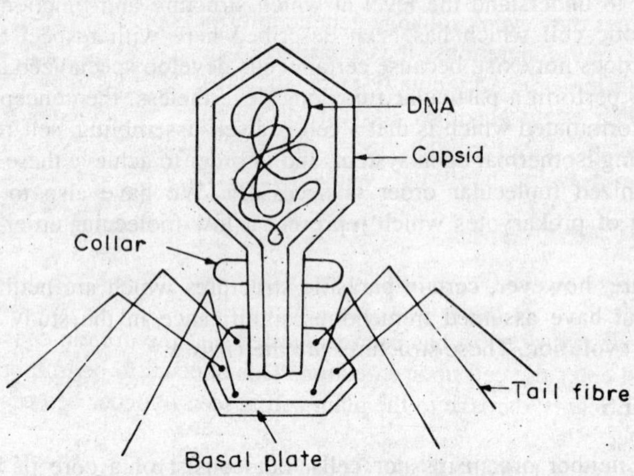

Fig. 2.9 Structure of bacteriophage T4.

The nucleic acid core forms the basis of modern classification of viruses. Viruses which infect bacterial cells *(bacteriophages)* mostly contain DNA; plant viruses generally contain RNA, and animal viruses may contain either RNA or DNA. The chemical nature of nucleic acid has important implications, because it explains the transmission of genetic information and also shows that both DNA and RNA could be used as molecules for storing genetic information.

Mode of Viral Infection

The mode of viral infection has been studied with the help of bacteriophages. Bacteriophages are introduced into a bacterial culture where they come in contact with the bacteria. The attachment of the phage takes place through the tail region, the phage adheres to the cell wall. The sequence of events during infection are: (1) the tail fibres attach to the cell surface, (2) the tail fibres contract and pull the tail plate towards the cell surface, (3) the cell wall lyses at the point of contact, and (4) finally the tail sheath contracts to inject viral DNA into the bacterial cell, while the protein coat remains outside. Plant viruses do not have any specific attachment mechanism and are sometimes injected into the plant cells by insects which act as vectors.

Once the phage DNA is injected into the bacterial cell, it takes over control of the bacterial cell and destroys its structural integrity. The cellular activities of the bacteria are directed to making copies of viral DNA. The nucleic acids of the host cell are depolymerized and become a part of the nucleotide pool to be used for phage DNA replication. The cellular proteins are also depolymerized and the amino acids are incorporated into the viral coats. After a new generation of viruses is formed, the bacterial cell lyses and the new viruses are released which may then infect the population of uninfected bacteria. This kind of virus which destroys the host cell is called a *lytic phage*.

Lysogenic Bacteriophages. An interesting mode of infection has been found in bacterial systems in which, upon injection into the host, the viral nucleic acid becomes integrated with the host chromosome. The phage DNA is replicated along with the bacterial DNA and is transmitted into the daughter cells through many generations. Such a virus is called a *provirus* and is said to undergo a lysogenic phage. The lysogenic phage attaches itself to a specific site on the bacterial DNA and sometimes may even exchange genetic information.

After a few generations, it is possible to detach the phage DNA from the bacterial chromosome accidentally or by special treatment. Once the phage DNA is detached, it again becomes active and continues with the lytic cycle. In the lysogenic phage, however, the virus DNA can go on replicating along with the bacterial chromosome indefinitely without any symptoms showing in the host cell.

Infection by Plant Viruses

The mode of infection exhibited by plant viruses is not the same as in case of bacteriophages. Since there is no tail structure in the plant viruses to inject viral DNA into the plant cell, it has been suggested that the protein coat as well as the nucleic acid enter the cell upon infection. This appears to be true because in virus infections in plants, damage to the plant cell is seen to occur before infection takes place.

Plant viruses contain RNA. Tobacco mosaic virus (TMV), which infects tobacco leaves causing mottling, is one of the best studied viruses. W. Stanley was the first to crystallize TMV virus and to study the mode of infection. The earlier chemical studies on TMV virus proved to be confusing because in this case the informational material was RNA and not DNA as had been universally accepted.

Infection by Animal Viruses

Animal viruses occur in various geometrical shapes, but none has a tail so characteristic of bacteriophages. The mode of infection in animal viruses has been studied extensively because of their importance in causing diseases. It is only the nucleic acid component that enters the host cell, but the viral multiplication is very fast. Animal virus enters the host cell through direct contact or indirectly through food, drink or faeces. Certain animal viruses are transmitted through the placenta, and in such cases the effect on the foetus is very damaging while the mothers remain relatively unaffected. One such example is the small pox virus which is often fatal to the developing foetus.

Animal viruses affect the host cells in various ways. The growth of the virus may take place in the nucleus or cytoplasm, leading to the disruption of normal metabolism of the cell. This is called *cytopathic effect* resulting is shrinkage of cells, or giant multinucleate cells or excessive proliferation of cells.

2.4 VIRIOIDS

Virioids are naked circular forms having RNA molecules consisting of 300-400 nucleotides. They are found in many plant cells. Since they are naked RNA molecules, the mode of transmission is through damaged cells. The virioid replicates by using the plant cell enzymes. They are known to cause a number of plant diseases, but there is a growing suspicion that they infect animals also.

2.5 PRIONS

Prions have been recently described as infectious agents causing many diseases in animals and humans. A prion consists of a naked protein rod without any nucleic acids associated with it. They were first reported from the brain of goat and sheep suffering from scrapie, disease showing neuronal degeneracy. They have been reported from the brain tissue of human suffering from neural disorders like *kuru* and creutzfeldt Jakob disease.

CHAPTER 3

MOLECULES OF LIFE

An examination of the chemical composition of a cell reveals that life is composed of large size complex molecules, small molecules, inorganic ions and water. Water, of course, is most abundant in living organisms because the chemical reactions always proceed in aqueous medium. Quite a large amount of water in cells exists in the form of individual molecules, and only about 5% is found in bound form in association with other molecules. This relatively small amount of water contains chemical substances of different types in soluble or insoluble form while some chemical substances may be present in colloidal suspension. Besides water, the living cells contain molecules of large size which can be classified according to their functions. For example, there are certain molecular species of compounds that are involved in various metabolic functions, such as production, utilization and storage of energy. There are other classes of compounds which participate in structure building, repairs, catalytic functions, storage and transmission of hereditary information, and a host of other functions. Therefore, it is necessary that we learn about the structure and functions of these chemical molecules.

3.1 PROPERTIES OF MOLECULES

In 1920s Schrodinger coined the term 'macromolecules' to describe chemical compounds of relatively large size and large molecular weights. Chemists tried to elucidate the chemical nature of macromolecules by synthesizing them from smaller molecules. The chemical 1,4-isoprene, for example, can be condensed to produce a linear structure, polyisoprene which has a repeating sequence of isoprene units. The repeating isoprene unit or building block is termed a *monomer*, whereas the condensation product or the polymer is called a *macromolecule*. Biologically significant macromolecules are mostly linear polymers, but branched or cross-linked polymers also exist which are endowed with a variety of functions. Biological polymers may be composed of identical monomers (homopolymers) or different monomers (heteropolymers), but in the majority of cases they are irregular polymers. The properties of the polymers are entirely different from their constituent monomer units. Some of these are considered here.

1. Many types of macromolecules show elasticity and can undergo deformation. Cell membranes and muscle tissues are excellent examples which demonstrate this property.

2. A most important property of macromolecules is their ability to provide a definite structure and form even when they remain dissolved in a solvent.

3. Being large in size, macromolecules cannot cross the membrane barrier, hence do not get lost to the environment.

3.2 CHEMICAL NATURE OF MACROMOLECULES

The class of macromolecules includes three types of compounds. These include carbohydrates, nucleic acids and proteins. Besides these compounds there are molecules which come under the category of lipids. These are structural components of membranes and also form precursor molecules of the steroid hormones. We shall consider the chemical nature and functions of these macromolecules.

Carbohydrates

A number of monosaccharides and disaccharides are found in living organisms, some of which produce energy needed to drive metabolic reactions. The amount of such simple sugars present in the cell is quite small, hence most of the carbohydrates are present in the form of polysaccharides. These polysaccharides are composed of different types of sugar moieties bound together by means of chemical bonds. In times of need the cell obtains sugar moieties from the polymer by a process of depolymerization and oxidizes them to produce energy. Thus, the carbohydrate polymers store excess amount of these simple sugars.

Though there are many types of simple sugars or monosaccharides present in a cell in solution, the commonest ones are those which have 5- and 6-carbons having a general formula $(CH_2O)_n$. The simplest monosaccharides are three carbon compounds such as glyceraldehyde and dihydroxyacetone.

glyceraldehyde dihydroxyacetone

Besides simple monosaccharides, there are a number of naturally occurring sugar derivatives containing less oxygen and having groups such as amino or carboxylic groups.

These 3-carbon sugars (trioses) are important intermediates in the oxidation of hexoses in organisms and occur as linear molecules. Sugars containing four and more than 4 carbon atoms generally occur in solution in cyclic forms that can be derived from the linear molecules (Fig. 3.1). Monosaccharides may be classified into reducing and non-reducing sugars, the former having a potential aldehyde group (–CHO), while the latter have a corresponding ketone group ($>C{=}O$). Glucose and fructose are the common examples which can be shown in the pyranose and furanose ring forms (Fig. 3.2). Another group of monosaccharides includes 5-carbon pentoses, ribose and deoxyribose which are constituents of nucleic acids (Fig. 3.3).

Disaccharides are formed by condensation of two monosaccharide molecules which may be identical or non-identical. The linkage between two monosaccharide units is very specific and may be cleaved by hydrolysis, Most important and commonly occurring disaccharides include maltose, sucrose (cane sugar), and lactose (milk sugar), each having a specific linkage (Fig. 3.4). Maltose is a reducing sugar and on hydrolysis yields two molecules of glucose. Sucrose is a non-reducing sugar consisting of glucose and fructose units linked through glycosidic bond between C-2 of fructose and C-1 of the glucose unit. Lactose is also a non-reducing sugar and consists of galactose and glucose moieties linked through a β-1, 4-glycosidic bond.

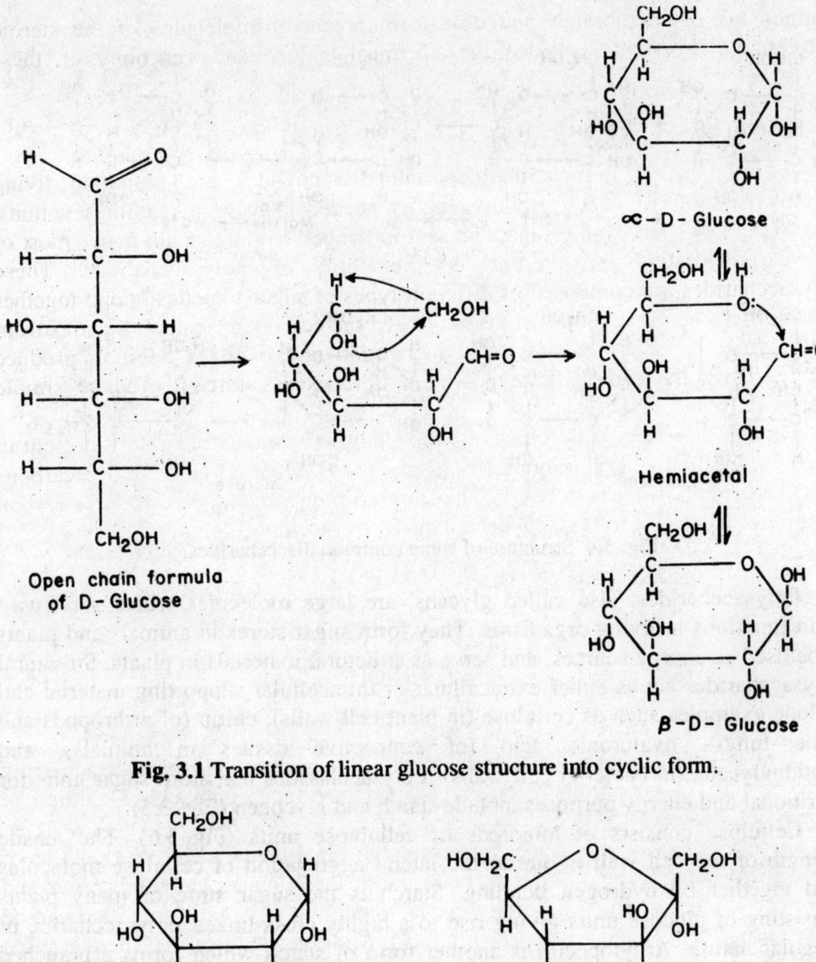

Fig. 3.1 Transition of linear glucose structure into cyclic form.

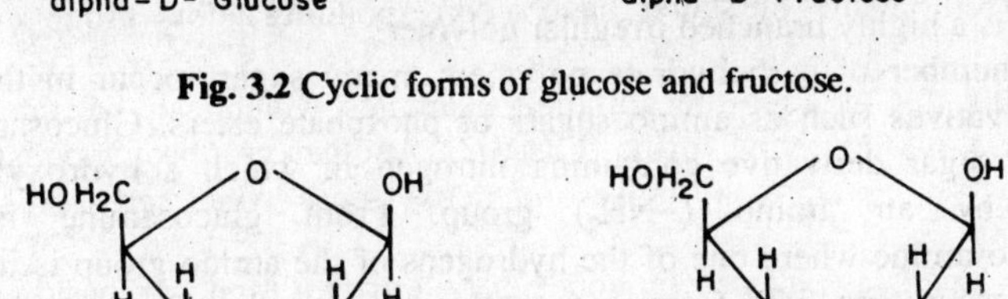

Fig. 3.2 Cyclic forms of glucose and fructose.

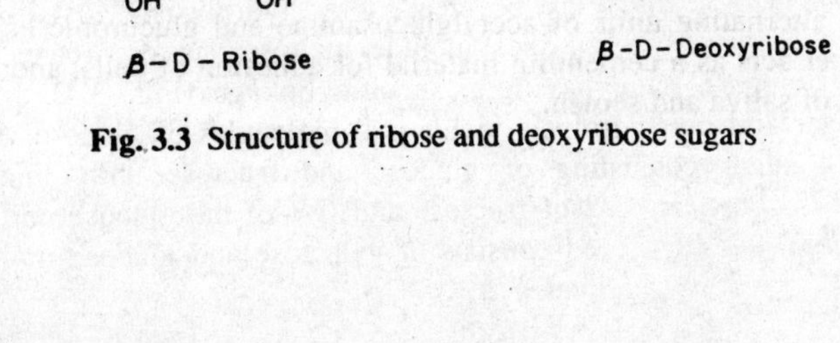

Fig. 3.3 Structure of ribose and deoxyribose sugars

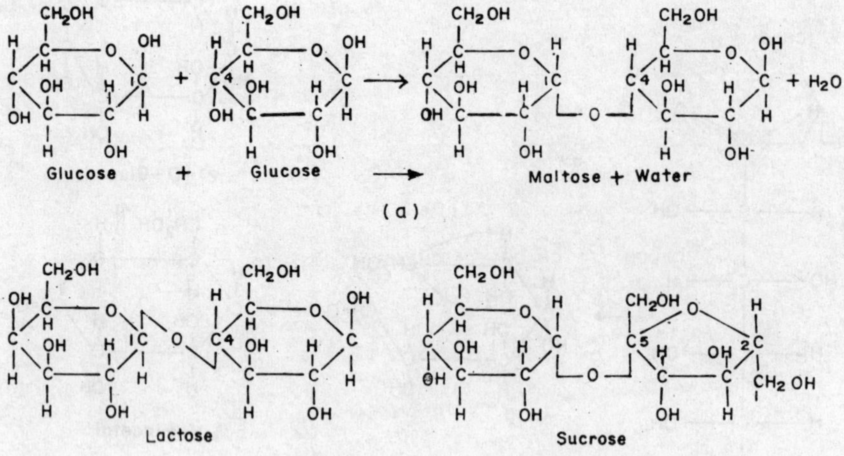

Fig. 3.4 Structure of some common disaccharides.

Polysaccharides, also called glycans, are large molecules which serve two main functions in living organisms. They form sugar stores in animals and plants to be used as energy sources, and serve as structural materials in plants. Structural polysaccharides act as either extracellular or intracellular supporting material and include examples such as cellulose (in plant cell walls), chitin (of arthropods and some fungi), hyaluronic acid (of connective tissues in animals), and peptidoglycans (in bacterial cell walls). Polysaccharides that store sugar units for nutritional and energy purposes include starch and glycogen (Fig. 3.5).

Cellulose consists of hundreds of cellobiose units (Fig. 3.6). The tensile strength of the cell wall is due to the lateral aggregation of cellulose molecules held together by hydrogen bonding. Starch is the sugar store of many plants consisting of glucose units giving rise to a highly cross-linked polysaccharide of irregular nature. Amylopectin is another form of starch which forms a branched chain. In animals the storage form of polysaccharide is called glycogen where glucose units are joined together as in case of amylopectin, but with side chains. Glycogen is a highly branched irregular polymer.

In a number of carbohydrate polymers, many sugars occur in the form of sugar derivatives such as amino sugars or phosphate esters. Glucosamine is an important sugar derivative containing nitrogen in which a hydroxyl group is replaced by an amino ($-NH_2$) group. From glucosamine is derived acetylglucosamine where one of the hydrogens of the amino group is replaced by an acetyl group (Fig. 3.7), forming the polymer chitin that forms the exoskeleton of arthropods and cell walls of fungi, Hyaluronic acid is another polymer which consists of alternating units of acetylglucosamine and glucuronic acid (Fig. 3.8). This polymer acts as a cementing material for adhesion of cells. and also forms a component of saliva and semen.

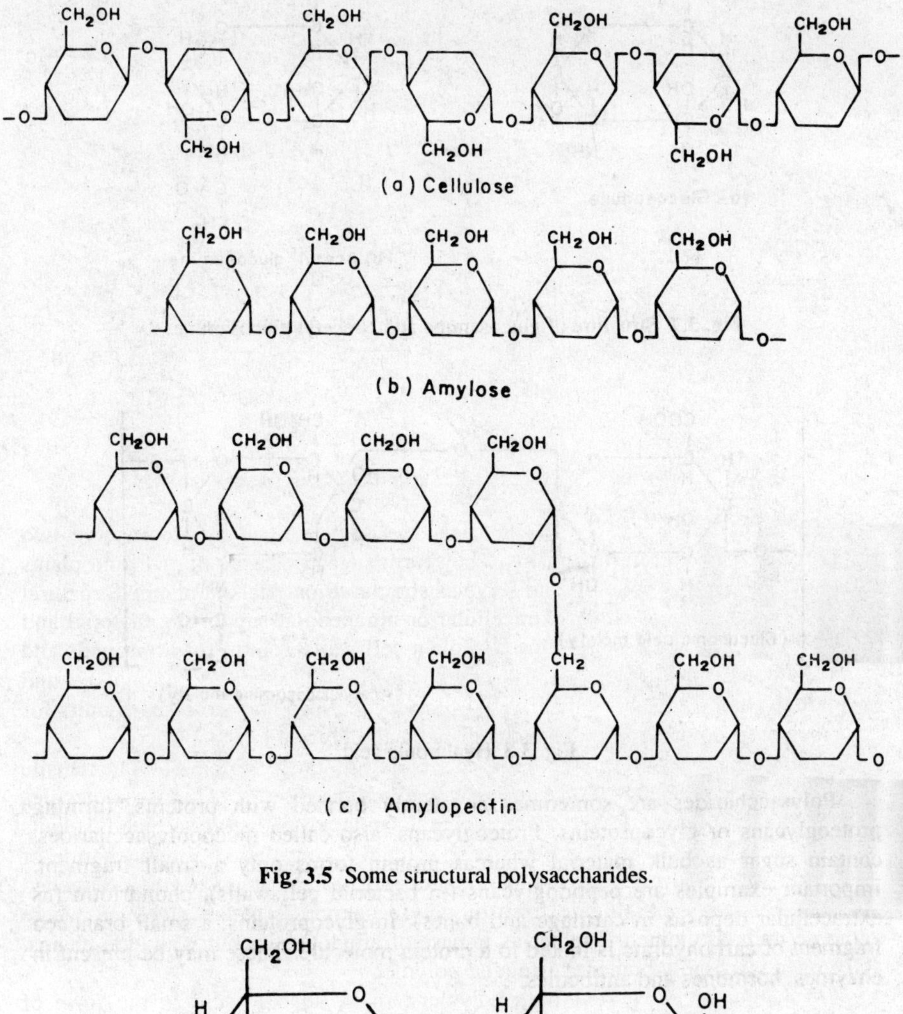

Fig. 3.5 Some structural polysaccharides.

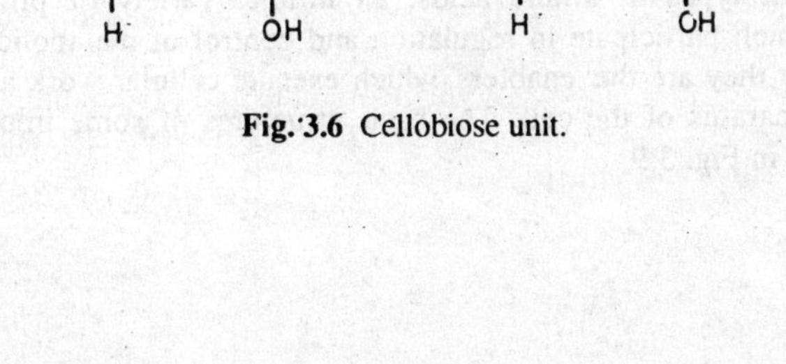

Fig. 3.6 Cellobiose unit.

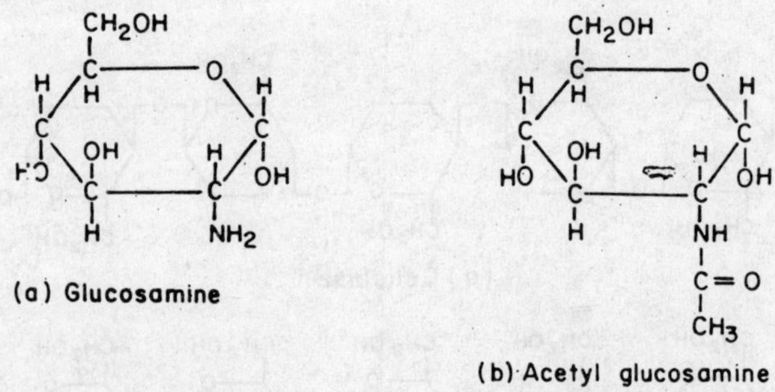

Fig. 3.7 Structure of glucosamine and N-acetyl glucosamine.

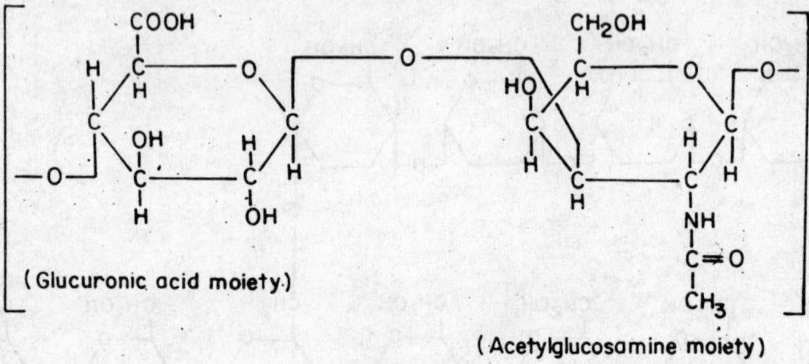

Fig. 3.8 Hyaluronic acid.

Polysaccharides are sometimes covalently bonded with proteins, forming proteoglycans or glycoproteins. Proteoglycans, also called mucopolysaccharides, contain sugar as bulk material whereas protein forms only a small fragment. Important examples are peptidoglycans (in bacterial cell walls), chondriotin (as extracellular deposits in cartilage and bones). In glycoproteins, a small branched fragment of carbohydrate is linked to a protein molecule. These may be present in enzymes, hormones and antibodies.

Proteins

In general, proteins are irregular polymers which are made of some twenty different kinds of monomer units, called L-amino acids. With the help of these twenty different types of amino acids, an infinite variety of proteins can be synthesized which participate in regulation and control of metabolic activities of the cell. In fact they are the 'enablers' which execute cellular work as directed by the genetic apparatus of the cell. The basic structures of some important amino acids are given in Fig. 3.9.

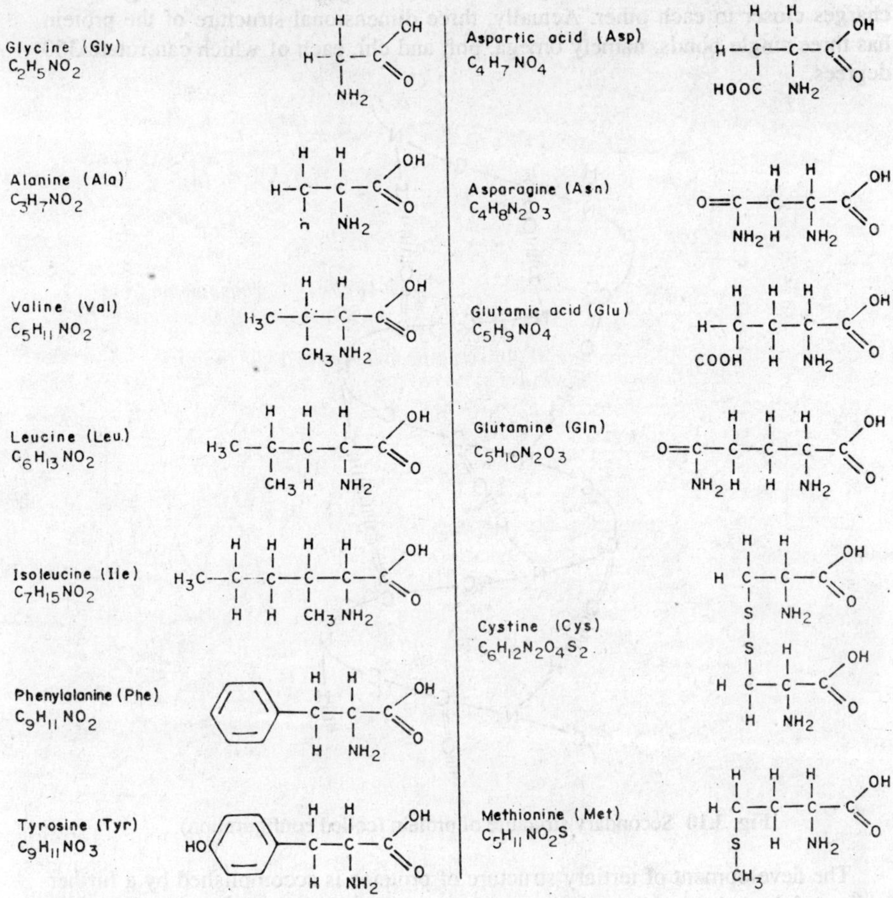

Fig. 3.9 Structure of some important amino acids.

Proteins may participate in providing the structural parts of a cell, or they may serve as functional components. Structural proteins are generally fibrous in nature, such as actin and myosin of the muscle cells, cilia and flagella etc.; collagens in connective tissues; keratins in nails, hair, hooves, skin etc. Functional proteins, are globular structures that include all enzymes, antibodies, surface receptors, and some hormones. Four basic protein structures, the primary, secondary, tertiary and quaternary have been described.

The primary structure of a protein is a linear chain of amino acids that occur in a definite order and number. Individual amino acids are linked together by peptide bonds to form a polypeptide chain. Each polypeptide has a unique sequence of amino acids. Some proteins may have cross linkages in the form of disulphide bonds.

The secondary structure of proteins has a specialized three-dimensional configuration in which the polypeptide chain undergoes a tight coiling. The coiled conformation is achieved as a result of the hydrogen bond formation between carbonyl and imido groups of the individual amino acids (Fig. 3.10). A folded three-dimensional model of secondary protein may be in the form of an α-helix in which the charges on the side groups would bring the groups bearing similar charges closer to each other. Actually, three dimensional structure of the protein has three single bonds, namely omega, phi, and chi, each of which can rotate 360 degrees.

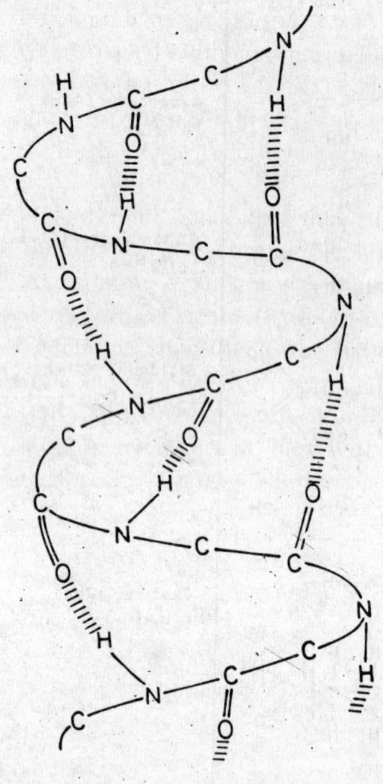

Fig. 3.10 Secondary structure of protein (coiled configuration).

The development of tertiary structure of proteins is accomplished by a further folding of the polypeptide chain in a specific manner so that the configuration and orientation of the chain confers a specialized function. The tertiary structure is stabilized by various secondary bonds that exist between specific R groups of the constituent amino acid residues. Three types of iinkages are generally found, which are, disulphide linkages holding two polypeptide chains together, hydrogen bonding and the electrostatic bonds between positively and negatively charged groups. Some common examples of tertiary proteins are myoglobin and insulin molecules. Myoglobin has a definite shape which is determined by the nature of the side groups on the amino acid residues. Insulin has two polypeptide chains held in position through disulphide bonds. The final configuration of tertiary proteins can be altered by variations in pH and temperature.

We have earlier said that the proteins possess a highly complicated three-dimensional configuration which is determined by the sequence of amino acids and the secondary bonds that stabilize the shape. There is yet another configuration of proteins which refers to aggregation of a number of polypeptide chains to produce a complex molecule. The chains may be identical in their amino acid composition. For example, haemoglobin is a quaternary protein which attains a greater complexity of structure since the molecule consists of four subunits, two alpha and two beta chains, held together in a spatial arrangement (Fig. 3.11).

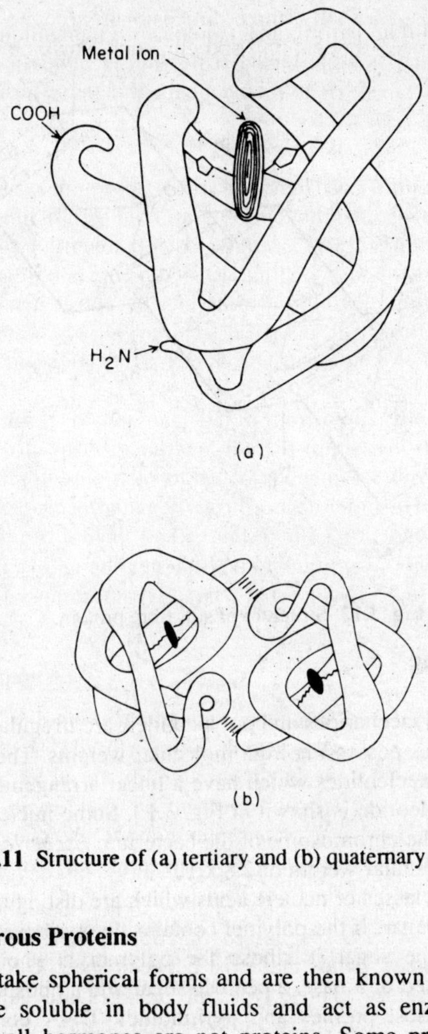

Fig. 3.11 Structure of (a) tertiary and (b) quaternary proteins.

Globular and Fibrous Proteins

Proteins may take spherical forms and are then known as globular proteins. These proteins are soluble in body fluids and act as enzymes, antibodies or hormones, though all hormones are not proteins. Some proteins are fibrous in nature, hence insoluble in solvents or body fluids. Most of the proteins that occur in muscles, skin and teeth form bulk of the body proteins and are fibrous in nature (Fig. 3.12). Fibrous proteins may be contractile or nonelastic. Muscles are composed of contractile proteins such as actin and myosin, while collagen and silk fibroin are nonelastic proteins. Silk fibroins predominantly contain glycine, alanine and serine, besides other amino acids, and occur as pleated sheets in which the side chains project upwards and downwards. There is another protein present in the blood called fibrinogen which remains dissolved in the plasma. The peculiarity about this protein is that during clotting of blood, fibrinogen is converted to fibrin forming a network of fibres, which are insoluble in plasma.

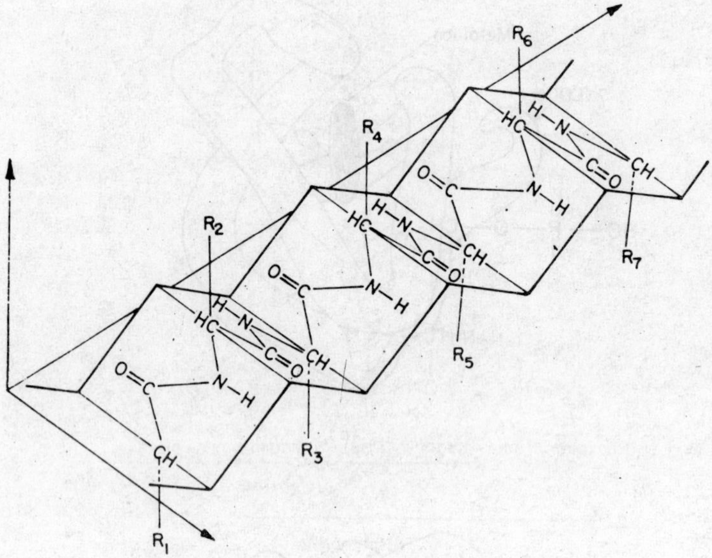

Fig. 3.12 Structure of silk fibre protein.

3.3 NUCLEIC ACIDS

Nucleic acids, like polysaccharides and polypeptides, are irregular polymers which are very large molecules possessing high molecular weights. They are made up of monomer units called nucleotides which have a linear arrangement. The chemical representation of a nucleotide is shown in Fig. 3.13. Some nucleic acid molecules are very long, such as the chromosome of the bacterium *Escherichia coli* is about 1 mm long and has a molecular weight of 2.8×10^9.

There are two subclasses of nucleic acids which are distinguishable mainly on the basis of a pentose sugar. If the polymer contains deoxyribose-nucleic acid, it is called DNA, and if the sugar is ribose the polymer is known as RNA. All nucleotides are composed of a base, a pentose sugar and a phosphate group. There are two classes of bases, purines and pyrimidines. DNA contains four bases namely adenine, guanine, cytosine and thymine, and in RNA thymine is replaced by uracil (Fig. 3.14). When a base is linked to a pentose sugar the resulting compound is called a nucleoside. The nucleoside is phosphorylated by a phosphoric acid molecule in the fifth carbon of pentose sugar through an ester bond, resulting in the formation of a nucleotide. The nucleotides of both DNA and RNA also occur as diphosphates and triphosphates, some of which act as coenzymes and as high energy compounds.

Nucleic acids are linear and unbranched molecules and their monomer units are linked together through phosphodiester bonds occurring between 3' position of one sugar and 5' position of the next. Thus a large number of nucleotides can be polymerized to produce a long polymer (Fig. 3.15). The three dimensional structure of DNA is represented by a double helix in which two chains are intertwined around each other so that adenine is bonded to thymine and cytosine is bonded to guanine by noncovalent, weak hydrogen bonds. These week interactions are present between chains as well as within chains (Fig. 3.16).

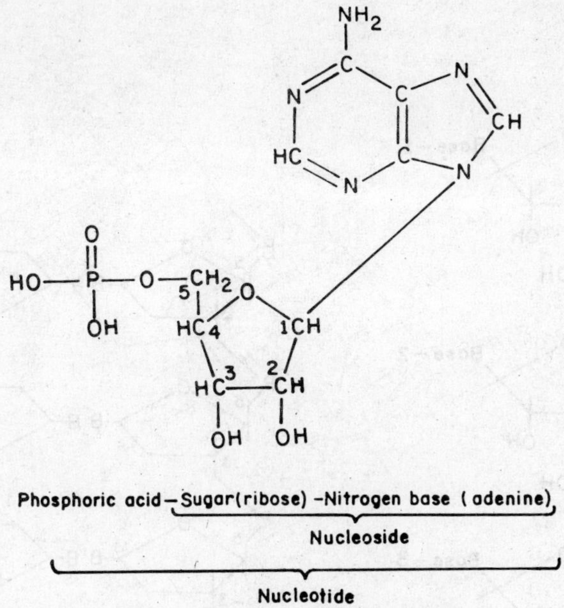

Fig. 3.13 Structure of a nucleotide.

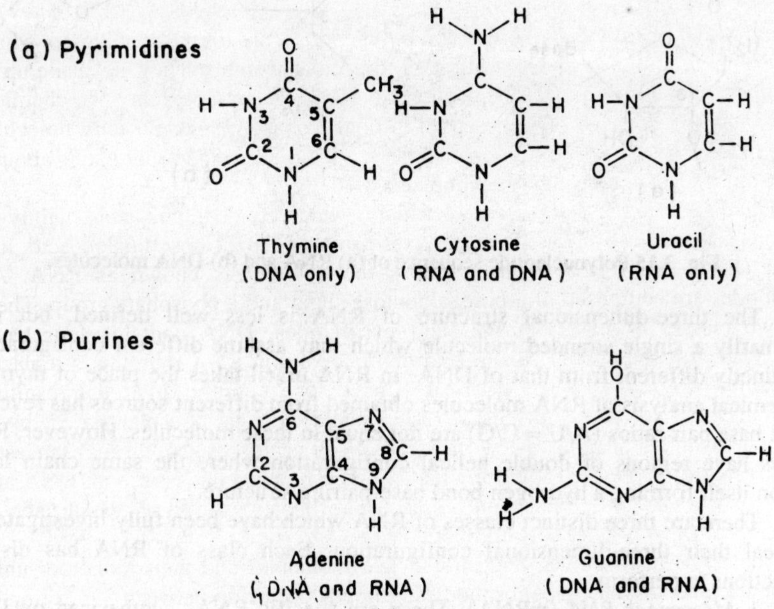

Fig. 3.14 Structure of pyrimidine and purine bases.

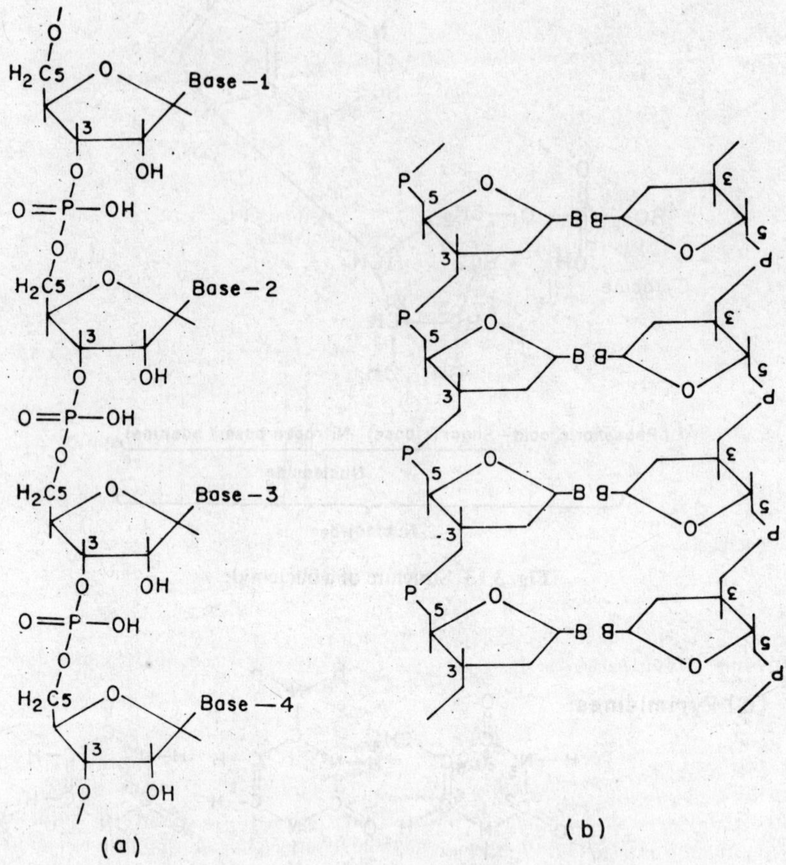

Fig. 3.15 Polynucleotide sequence of (a) RNA and (b) DNA molecules.

The three-dimensional structure of RNA is less well defined, but it is primarily a single stranded molecule which may assume different configurations distinctly different from that of DNA. In RNA uracil takes the place of thymine. Chemical analysis of RNA molecules obtained from different sources has revealed that base pair ratios (A/U = C/G) are not equal in these molecules. However, RNA does have regions of double helical configuration where the same chain loops upon itself forming a hydrogen bond base-pairing structure.

There are three distinct classes of RNA which have been fully investigated to reveal their three-dimensional configuration. Each class of RNA has distinct functions to perform.

1. *Messenger RNA* (*m*RNA): These are specific RNAs synthesized on DNA templates and truely representing a blue print of the DNA. Thus *m*RNA is an exact complementary sequence of DNA which is decoded and translated into a protein molecule. These molecules are unstable and have a short half-life.

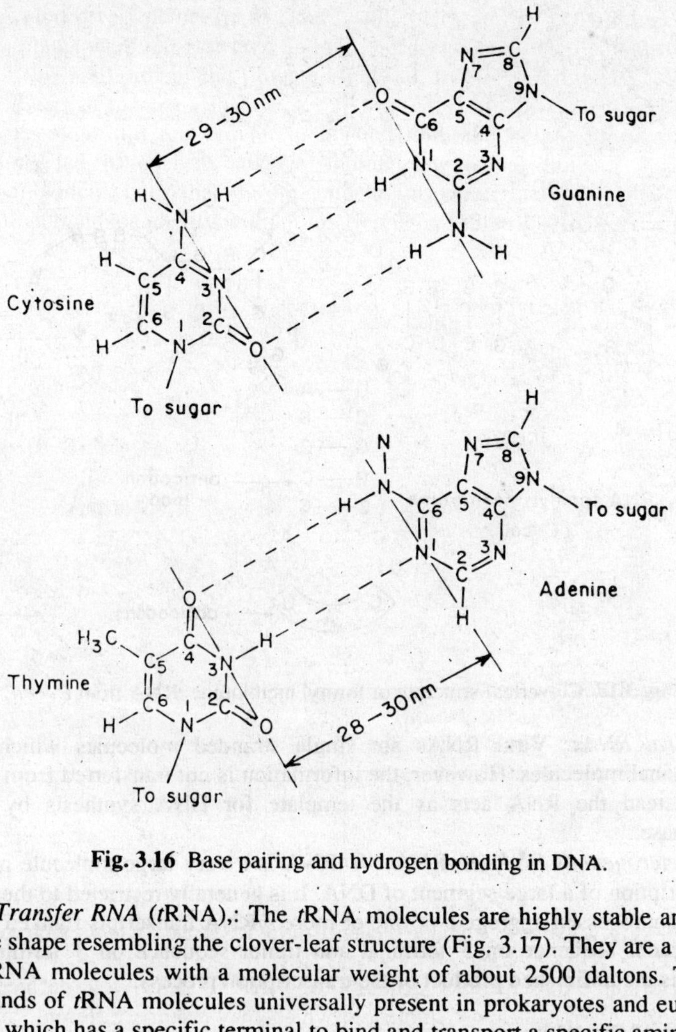

Fig. 3.16 Base pairing and hydrogen bonding in DNA.

2. *Transfer RNA* (*t*RNA),: The *t*RNA molecules are highly stable and have a definite shape resembling the clover-leaf structure (Fig. 3.17). They are a family of small RNA molecules with a molecular weight of about 2500 daltons. There are sixty kinds of *t*RNA molecules universally present in prokaryotes and eukaryotes, each of which has a specific terminal to bind and transport a specific amino acid. It should be noted that three necleotides on the mRNA form hydrogen bonds with the three corresponding bases on the anticodon end of the *t*RNA.

3. *Ribosomal RNA* (*r*RNA): Ribosomal RNA forms about 80% of the total cellular RNA and is quite stable. Ribosomal RNA is associated with ribosomes which are factories of protein synthesis. Four kinds of *r*RNAs are found in animals, plants and prokaryotes which are distinguished on the basis of their sedimentation coefficients (*S*)*. In prokaryotes the molecular species of *r*RNAs include 16*S*, 5*S* and 23*S* types, whereas eukaryotes have 18*S*, 5*S*, 5.8*S* and 28*S* species. In ribosomes *r*RNAs are found to be complexed with 50 different protein molecules.

*** *S* stands for Swedberg units which give quantitative data on sedimentation rates of molecules on centrifugation.**

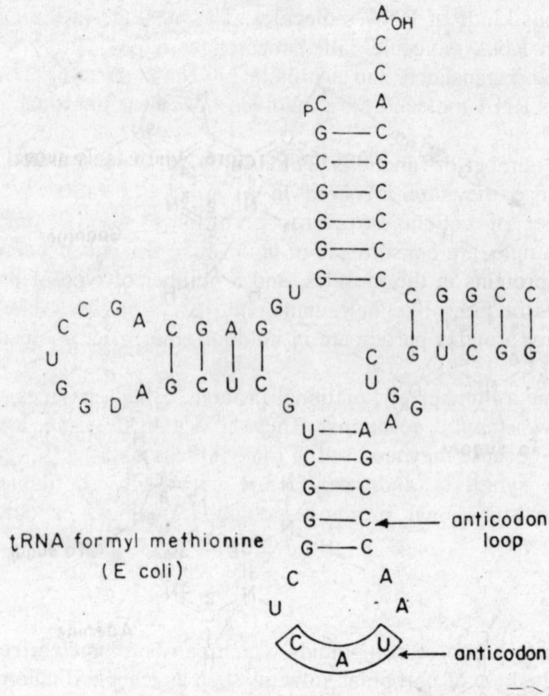

Fig. 3.17 Cloverleaf structure of formyl methionine *t*RNA from *E. coli*.

4. *Viral RNAs:* Viral RNAs are single stranded molecules which act as informational molecules. However, the information is not transferred from RNA to RNA, instead the RNA acts as the template for DNA synthesis by reverse transcriptase.

5. *Heterogenous RNA* (*Hn*RNA): *Hn*RNA is a very large molecule produced by transcription of a large segment of DNA. It is generally restricted to the nucleus forming unprocessed aggregate of one or more *m*RNA transcripts flanked by non-coding leader sequence on 5' terminal and trailor sequence on 3' terminal. The *Hn*RNA is the unfinished product of the transcription process.

Functions of Biopolymers

The biopolymers discussed above serve many specialized functions. In general, polysaccharides have two important roles in living organisms; one is structural and the other is nutrition to provide energy. Starch in plants and glycogen in animals are examples of storage polysaccharides which provide metabolizable sugars. Cellulose is an important structural polysaccharide found in plant cell walls. The prokaryotes have a heterogenous polysaccharide in the form of neuraminic acid.

Biological functions of nucleic acids can be considered separately for DNA and RNA. The structure of the DNA explains the self-duplication mechanism of the molecule, which points to a single responsibility of transmitting the genetic message of a cell from one generation to the next. The DNA molecules are fairly stable, hence the genetic information encoded in the nucleotide sequence remains stable unless mutations alter it. This genetic message is transmitted in a specific manner involving various kinds of RNA molecules. The message in the DNA is first transcribed into an RNA molecule called messenger RNA. The messenger RNA is then decoded and translated into a protein molecule through ribosomes and transfer RNA. Thus RNA molecules are involved in information transfer and protein synthesis.

Proteins perform a number of functions. The infinite variety of proteins has a special significance since they are involved in a variety of functions. They participate in a number of cellular structures performing specific functions. Collagen fibres, for example, are constituents of connective tissue cells, actin and myosin are contractile proteins in the muscles, and a number of type of proteins may form specialized structures like hair, nails, hooves, horns etc. These are structural proteins. Many proteins participate in holding other macromolecules in position.

Living systems have a number of functional proteins called enzymes, which catalyze and regulate metabolic reactions. These biocatalysts have assumed tremendous importance because they are used in many industrial processes such as fermentation, antibiotic synthesis, and manufacture of proteins, hormones etc. Besides enzymes, other functional proteins include antibodies, or defensive proteins and some hormones.

3.4 LIPIDS

Lipids are a heterogenous group of compounds which can be extracted from the living organisms with the help of non-polar solvents such as ether and chloroform. Though they do not form biopolymers, they are present in every cell as structural and functional components. They are insoluble in water, hence hydrophobic in character. This property of lipids can be attributed to the predominance of long linear hydrocarbon chains which are non-polar. Lipids include three principal types of compounds, namely glycerides, phospholipids and steroids.

In order to gain some insight into the nature of lipids, let us consider briefly one of the simpler types of lipids called glyceride, which is a fatty acid ester with glycerol as alcoholic component. The general formula is given here:

$$\begin{array}{l} CH_2O-CO-R' \\ \;| \\ CH-O-CO-R'' \\ \;| \\ CH_2O-CO-R''' \end{array}$$

The above mentioned molecule is a triglyceride containing three fatty acid molecules which may or may not be identical. Fatty acids have long hydrocarbon chains with the general formula

$$\begin{array}{c} COOH \\ | \\ (CH_2)_n \\ | \\ CH_3 \end{array}$$

Most of the fatty acids that occur in living organisms have even number of carbon atoms since they are synthesized by joining acetyl fragments. These are saturated fatty acids such as 16-carbon palmitic acid, and 18-carbon stearic acid.

Sometimes the hydrocarbon chains have double bonds e.g. 18-carbon oleic acid, and these are called unsaturated fatty acids. The double bond is biologically important as it increases the flexibility of the hydrocarbon chain as well as the fluidity of the biological structure.

Triglycerides are either solids or liquids. Unsaturated fatty acids are liquids at room temperature, whereas saturated fatty acids are solids. Triglycerides are also called neutral fats which are stored in the fat cells or adipose tissue of the animal body as a convenient source of energy. Neutral fats are a bad conductor of heat and thus act as a source of insulation. Waxes are also fats which are esters of fatty acids and higher monohydric alcohols, usually found outside the organisms or their cells in the form of a waterproof layer.

Not all lipids act as energy reserves, though some are important structural constituents of cell membranes. This category of lipids is known as phospholipids:

$$\begin{array}{l} CH_2-O-CO-(CH_2)_n-CH_3 \\ | \\ CH-O-CO-(CH_2)_n-CH_3 \\ | \\ CH_2-O-CO-(CH_2)_n-CH_3 \end{array}$$

A triglyceride

$$\begin{array}{l} CH_2-O-CO-(CH_2)_n-CH_3 \\ | \\ CH-O-CO-(CH_2)_n-CH_3 \\ \quad\quad\quad\quad\;\; OH \\ | \quad\quad\quad\quad\;\;\; | \\ CH_2-O-P-OH \\ \quad\quad\quad\quad\;\; \| \\ \quad\quad\quad\quad\;\; O \end{array}$$

Phosphatidic acid

Phospholipids are triglycerides in which one of the fatty acids is replaced by a phosphoric acid molecule and either choline, ethanolamine or serine. These are derivatives of phosphatidic acid (Fig. 3.18).

A phospholipid molecule consists of two hydrophobic fatty acid hydrocarbon chains forming the tail (non-polar) and a hydrophilic (polar) phosphate head. Thus lipids are amphipathic, that is they have a hydrophobic as well as a hydrophilic region, and this accounts for many properties of biological membranes.

The cell membranes are actually bilayers of phospholipids with hydrophobic ends directed upwards (away from water) and hydrophilic ends directed (polar groups) towards the water-interface. When phospholipids are mixed with water, the bilayer arrangement is adopted spontaneously. This principle of self-assembly in which arrangement of a complex structure arises exclusively from the physico-chemical properties of chemical molecules is a characteristic of living systems.

Another type of lipid called cholesterol is also found to be associated with the cell membranes -which alternate with the phospholipid molecules. Besides participating in the membrane structure, cholesterol also acts as a precursor of steroid hormones and vitamin D. The steroid hormones include testosterone in the male and estrogen and progesterone in the female. The sex hormones stimulate the development of sex characters and sexual cycles in animals including man.

Lipids are vital for initiating and stimulating many physico-chemical processes in cells. Sphingolipids are such compounds which are associated with the myelin sheath of nerve cells. Sphingolipids are absent in bacteria, but present in yeast, animal cells and plant tissues. The structural units of sphingolipids is either a ceramide or a sphingosine (4-sphingenine) unit.

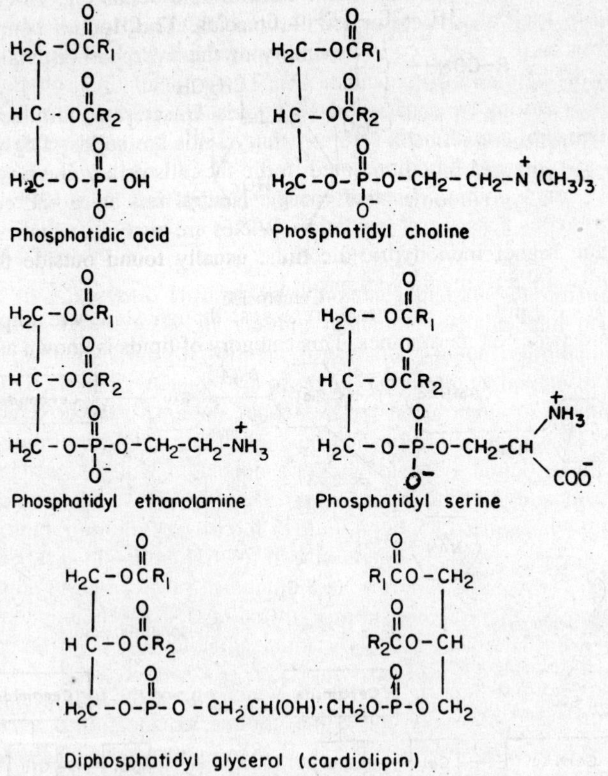

Fig. 3.18 Structure of some phospholipids.

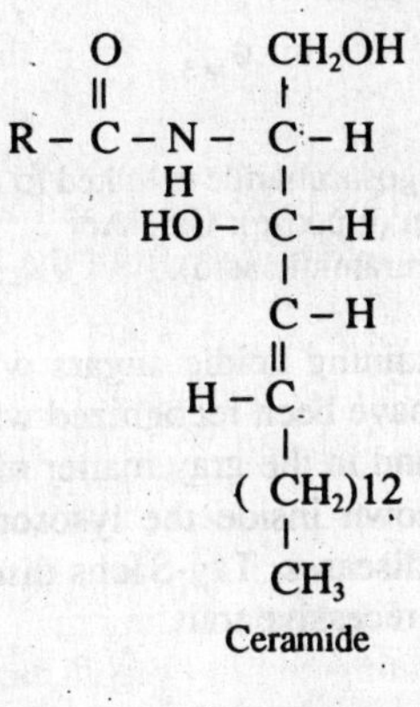

Ceramide

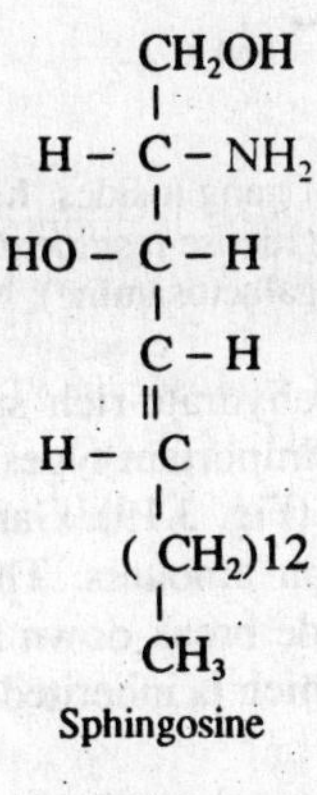

Sphingosine

Some important derivatives of sphingosine are cerebrosides and gangliosides. In cerebroside glucose or galactose is linked to the terminal hydroxyl group of ceramide:

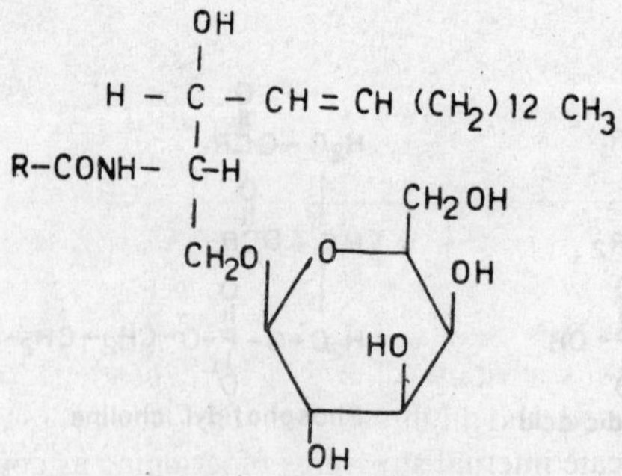

Cerebroside

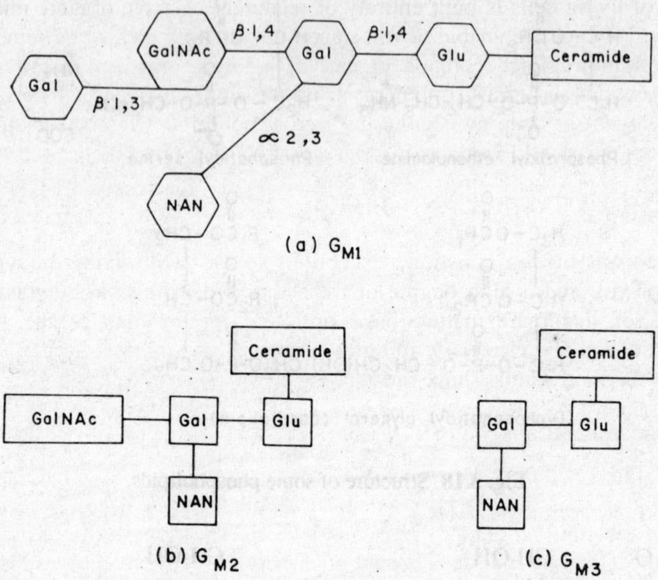

Fig. 3.19 Structure of gangliosides. In ganglioside an oligosaccharide is linked to a ceramide by a glucose residue. Gal (galactose); Glu (glucose); GalNAc (N-acetylgalactosamine); NANA (N-acetylneuraminic acid).

Gangliosides are carbohydrate-rich sphingolipids containing acidic sugars which are sialic acids. Three important types of gangliosides have been recognized which are G_{M1}, G_{M2}, and G_{M3} (Fig. 3.19). Gangliosides are found in the gray matter of the nervous system in high amounts. They are broken down inside the lysosomes. Disorders in ganglioside break down result in serious diseases. Tay-Sachs disease is one such disorder which is inherited as an autosomal recessive trait.

CHAPTER 4

ENERGY METABOLISM

The most conspicuous attribute of life is its dependence on energy. Living organisms are complicated and highly organized. The cells, of which they are composed, possess intricate internal structures functioning as complex biochemical machinery inside which most of the complex molecules (lipids, proteins, carbohydrates and nucleic acids) are processed. The energy transforming machinery of living cells is built entirely of relatively unstable organic molecules (see chapter 3) which are unable to withstand high temperatures, or extremely acid or basic conditions. Each component part of a living organism has a specific function to extract and transform energy from the environment in which it exists. Most of an organism's energy is either utilized to build its own intricate structures or to carry out purposeful work, or it is dissipated to the environment.

4.1 THERMODYNAMICS AND LIFE

Living organisms do not constitute exceptions to the laws of thermodynamics. Their complexity and a high degree of molecular orderliness, in contrast to the randomness of inanimate matter, have profound implications of the laws of physics and chemistry, especially in terms of energy and its transformations. The two laws of thermodynamics have outstanding significance in this context.

Energy Transformations

The first law of thermodynamics states that in ordinary physical and chemical processes energy can neither be created nor destroyed, but it can be converted from one form into another. This law is also known as the 'principle of conservation of energy'. Various forms of energy exist, such as chemical, heat, light, electrical energy, kinetic energy or potential energy, which only undergo transformation in physical processes, without affecting a change in the total quantity of energy. In autotrophs, for example, light energy is converted into chemical-bond energy to be eventually used by the heterotrophs.

The second law of thermodynamics deals with energy and its transformations and states that all physical and chemical processes involving transformation of energy, i.e. work will not occur spontaneously unless the total products of transformation are more random than their precursors. The law explains that spontaneous processes go in the same direction—towards an equilibrium of greatest disorder, or randomness. Energy in this form is called *entropy* and is unusable.

On the basis of the first law, living organisms cannot consume or use up energy. They can only transform it. From their environment they absorb a form of energy that is useful to them under special conditions of temperature and pressure in which they live and they then return to the environment an equivalent amount of energy in a less useful form. The useful form of energy, which is taken in by the cells is called *free energy*. The less useful form of energy that is returned to the environment consists of heat and other forms of energy that become quickly randomized in the environment and thus increase its disorder or entropy.

In an organism, the energy needed to enable all chemical reactions to proceed can come in various forms. Chemical energy is especially important as it is an intrinsic form, i.e. an inherent property of matter. Organisms have developed mechanisms for storing chemical energy and using it to drive energy-consuming reactions.

Energetics of Open Systems

Energy changes in isolated chemical reactions dealing with equilibrium thermodynamics are closed systems which do not exchange matter with their surroundings. On the other hand, the environment is absolutely essential to living organisms as it is not only a source of raw materials but also a source of free energy for them. Normally, organisms exchange both energy and matter with the environment in which they live. Therefore, thermodynamically these organisms are open systems. During this transaction both energy and matter are transformed. It is characteristic of open systems that they appear to be in equilibrium with the environment, but actually they are not and usually exist in a steady state. A steady state can be defined as that condition in which the rate of transfer of matter and energy from the environment into the system is exactly balanced by the rate of transfer of matter and energy out of the system. Therefore, in a true sense, the living cell is a non-equilibrium open system that extracts free energy from the environment, which it causes to increase in randomness. Analysis of the magnitude and efficiency of energy changes in steady-state systems is much more complex than closed systems.

Exergonic Versus Endergonic Reactions

Chemical reactions in an organism represent a form of work and according to the laws of thermodynamics, usable energy must be available for all reactions to occur. Such reactions can be divided into exergonic and endergonic types. Exergonic reactions are those which occur spontaneously without energy being available from an external source. They do so with a decrease in free energy and have a negative ΔG. On the other hand, endergonic reactions proceed when external energy is available as a driving force. Such reactions have a positive ΔG.

The free energy content (G) of a substance cannot be measured experimentally, but it is possible to speak of the change in free energy (ΔG) during the conversion of a substrate to a product in a chemical reaction. If the free energy of the product is less than the free energy content of the reactant, ΔG will be a negative quantity, i.e. the reaction occurs with a decrease in free energy. However, if the product is converted back to reactant, the reaction involves increase in free energy, i.e. ΔG will be positive.

In a living organism reactions which involve breaking down of complex organic molecules into simple ones are called exergonic. For example, glucose is oxidized by molecular oxygen to carbondioxide and water:

$$C_6H_{12}O_6 + 6\,O_2 \longrightarrow 6\,CO_2 + 6\,H_2O$$

ΔG for this reaction is a very large negative quantity, approximately −686,000 calories/mole of glucose oxidized. The large negative ΔG has no relationship to the rate of reaction. In most organisms the reaction goes on at varying rates. However, the same oxidation may occur in a matter of a few seconds in the presence of a catalyst. The factor which determines the rate at which a reaction proceeds is the activation energy of that process.

Reactions involving the synthesis of complex molecules from simple precursors are typically endergonic. Photosynthesis is a classic example, involving conversion of CO_2 and water to glucose:

$$6\ CO_2 + 6\ H_2O \longrightarrow C_6H_{12}O_6 + 6\ O_2$$

The reaction requires an input of some amount of energy that is released when glucose is oxidized to carbondioxide and water. ΔG for this reaction is a positive quantity, approximately +686,000 calories/mole of glucose synthesized and as the term photosynthesis implies, the energy for this process is provided by the radiant energy of the sun, which is the ultimate source of energy for the vast majority of endergonic reactions in organisms. Photosynthetic organisms are rather like huge, complex, endergonic reaction centres, in which ΔG can only be made favourable by coupling them to an external energy supply i.e. light. The heterotrophic organisms, on the other hand, utilize energy inherent in organic nutrients obtained from their surroundings. These external forms of energy are used directly or indirectly to generate high-energy intermediates, which are the immediate power-donors for endergonic reactions.

4.2 HIGH ENERGY INTERMEDIATES

Organisms are remarkably conservative in their choice of high-energy intermediates, relying principally on just two main categories: phosphorylated nucleotides and reduced dinucleotides.

Phosphorylated Nucleotides

In all living forms one compound repeatedly functions as a common reactant linking endergonic processes to exergonic ones. This compound, a phosphorylated nucleotide, *adenosine triphosphate* (ATP) functions as a carrier of energy to those processes in cells that require energy input (Fig. 4.1).

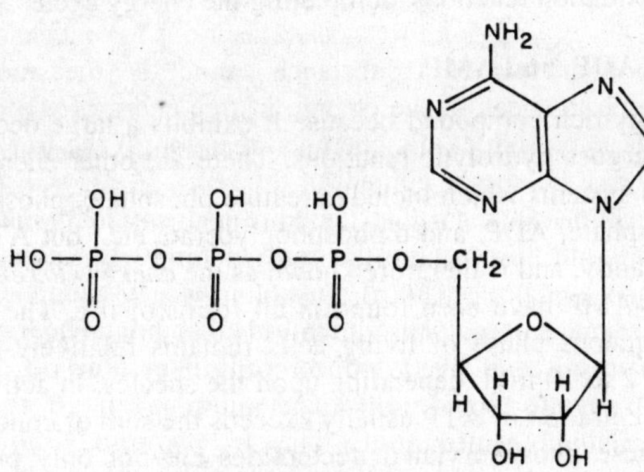

Fig. 4.1 Adenosine triphosphate, made from an organic base adenine, a C-5 ribose sugar, and three phosphates.

The ATP Cycle

In 1941 Fritz Lipmann gathered evidence to explain energy transfer in living cells. Lipmann postulated that ATP functions in a cyclic manner as a carrier of chemical energy from catabolic reactions to various cellular processes that require energy input (Fig. 4.2).

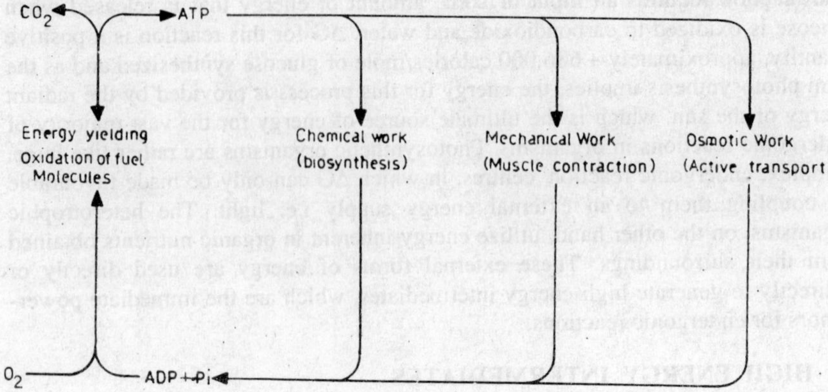

Fig. 4.2 Major uses of ATP energy.

ATP is generated from ADP by coupled or linked phosphorylation reactions at the expense of energy yielded by degradation of fuel molecules. The ATP so generated was postulated to donate its terminal phosphate to specific acceptor molecules, to energize them for carrying out energy-requiring functions in the cell. For example, biosynthesis of macromolecules, active transport of inorganic ions, and cell nutrients across the membranes against concentration gradients, and muscle contraction, are functions which are ATP-dependent. As ATP delivers energy to these energy-requiring processes, the ATP molecule undergoes cleavage to ADP and inorganic phosphate (Pi). ADP is then rephosphorylated at the expense of energy-yielding oxidation reactions, completing the energy cycle.

Properties of ATP, ADP, and AMP

ATP is an energy-rich compound because it exhibits a large decrease in free energy when it undergoes hydrolytic reactions. There are other energy-rich compounds in biological systems which include creatine phosphate, phosphoenolpyruvate, glucose-6-phosphate, ADP, and diphosphoglycerate etc., but ATP is the one which is used abundantly, and is therefore known as the *energy currency* of cells.

ATP, ADP and AMP have been found in all forms of life. The sum of their concentrations in aqueous phase of living cells remains relatively constant and amounts to between 2 to 10 mM, depending upon the species. In actively metabolizing cells, the concentration of ATP usually exceeds the sum of concentrations of ADP and AMP. These phosphorylated nucleotides are not only present in the soluble cytoplasm but also in organelles like mitochondria and nuclei.

At pH 7.0 in aqueous solution, ATP and ADP are anions. ATP has four ionizable protons in its condensed phosphate groups and ADP has three. In intact cells where a relatively higher concentration of Mg^{2+} exists in the intracellular fluid, both ATP and ADP are complexed with this cation in a one-to-one ratio to form divalent and monovalent complexes, respectively (Fig. 4.3). The affinity of ATP for Mg^{2+} is about 10 times that of ADP. In most enzymatic reactions in which ATP participates as phosphate donor, its active form is $[Mg\text{-}ATP]^{--}$ complex. Hydrolysis of the terminal phosphate of ATP, which is called orthophosphate cleavage, yields a good amount of energy which is utilized for purposeful work. ADP and Pi are the byproducts of this reaction. This reaction is very common and is represented as: ATP ——→ ADP + Pi. ΔG' or energy of hydrolysis at pH 7.0 has been estimated to be –7300 cal/mole.

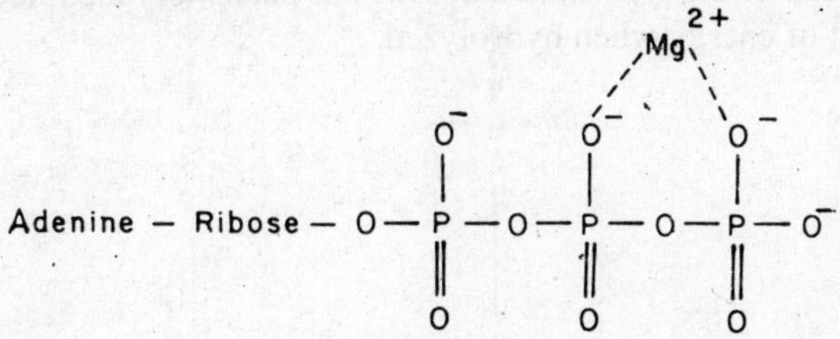

$[ATP - Mg]^{2-}$ complex

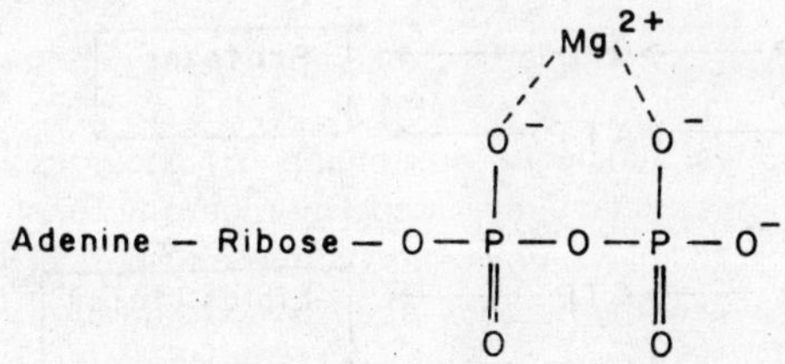

$[ADP - Mg]^{-}$ complex

Fig. 4.3 Metal complexes of ATP and ADP.

Similarly, ADP may be hydrolyzed to AMP and Pi, where the observed ΔG' at pH 7.0 is –6500 cal/mole. On the other hand, ΔG' of hydrolysis of AMP to adenosine and H_3PO_4 is much less, –2200 cal/mole at pH 7.0.

$$\text{ADP} \longrightarrow \text{AMP} + \text{Pi}$$
$$\text{AMP} \longrightarrow \text{Adenosine} + H_3PO_4$$

Although ATP is converted to ADP in many reactions of intermediary metabolism, there are a number of important reactions in which the interior phosphate bond of ATP is cleaved to yield AMP and inorganic pyrophosphate (PPi), ΔG' being –8600 cal/mole at pH 7.0. This type of cleavage is known as the pyrophosphate cleavage:

$$\text{ATP} \longrightarrow \text{AMP} + \text{PPi}$$

During respiration and phostosynthesis, ATP is resynthesized from ADP and Pi, thus the same molecules can be used over and over again. They can be made in one part of the cell and consumed in another.

Channeling of Phosphate Groups

Although an ATP/ADP system is the obligate carrier in the main stream of energy transfers in the cell, 5′-diphosphates and 5′-triphosphates of other ribonucleotides also participate in energy transfers by channeling high-energy phosphate groups into other specific biosynthetic reactions (Fig. 4.4). All these channels connect with ATP through the action of enzyme nucleoside diphosphate kinase, which is found in mitochondria and the soluble cytoplasm. It catalyzes reversible reactions of the type shown below. All phosphorylated nucleotides yield a similar amount of energy when hydrolyzed.

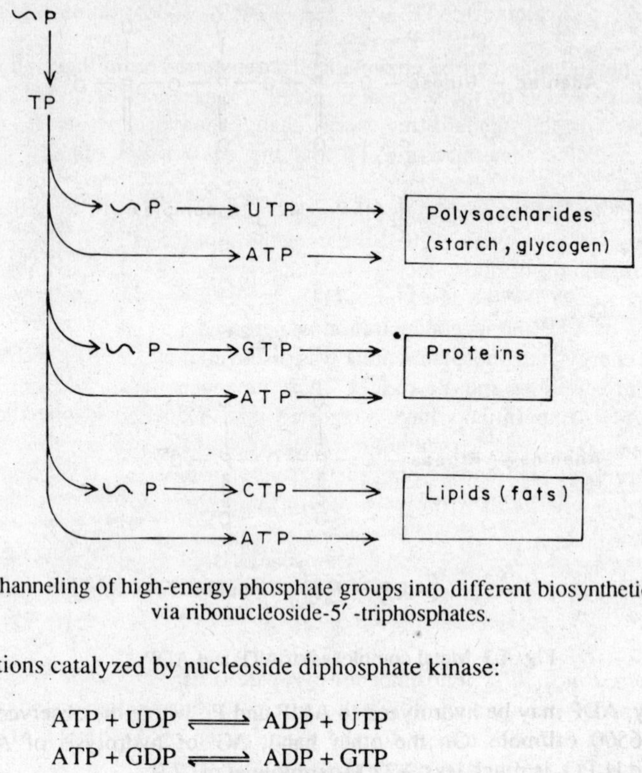

Fig. 4.4 Channeling of high-energy phosphate groups into different biosynthetic pat via ribonucleoside-5′ -triphosphates.

Reactions catalyzed by nucleoside diphosphate kinase:

$$ATP + UDP \rightleftharpoons ADP + UTP$$
$$ATP + GDP \rightleftharpoons ADP + GTP$$
$$ATP + CDP \rightleftharpoons ADP + CTP$$
$$GTP + UDP \rightleftharpoons GDP + UTP$$

Other High-energy Intermediates

Guanidium Phosphates. Another group of energy-rich compounds that play an important role in energy transfer and storage are guanidium phosphates. For example, phosphocreatine is found in muscles and nerve tissues of vertebrates, and phosphoarginine in invertebrates. These compounds are also known as *phosphagens.* In both phosphagens the phosphorus atom is bonded directly to a nitrogen atom of a guanido group. Phosphagens are formed by phosphorylation of creatine or arginine with ATP in the presence of an appropriate enzyme.

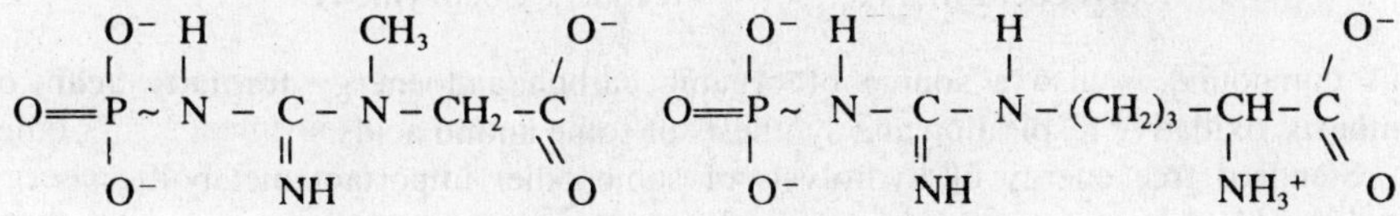

Phosphocreatine Phosphoarginine

$$\text{Creatine} + \text{ATP} \xrightleftharpoons{\text{Creatine kinase}} \text{Phosphocreatine} + \text{ADP}$$

$$\text{Arginine} + \text{ATP} \xrightleftharpoons{\text{Arginine kinase}} \text{Phosphoarginine} + \text{ADP}$$

The phosphate group can be enzymatically transferred from these phosphorylated compounds to ADP by the reversal of creatine kinase and arginine kinase to yield ATP. However, the standard free energy change on hydrolysis of these compounds is more negative than that of ATP and the equilibrium actually favours ATP production.

The phosphagens perform a physiological role by furnishing a site for storage of energy-rich phosphate. When the concentration of ATP is high, the reaction favours formation of phosphocreatine which is then stored as an energy-rich form of phosphate. When the level of ATP is depleted the reaction proceeds for the formation of ATP and its concentration is increased.

The energy-rich phosphocreatine was discovered by C. Fiske and Y. Subbarow in 1927 in vertebrate muscle tissues where its concentration is 5 times more than that of ATP. Experiments have suggested that ATP may not be the immediate energy source for muscle contraction. However, it is phosphocreatine which serves as a reservoir of high-energy phosphate groups for rephosphorylation of ADP. The standard free energy of hydrolysis, $\Delta G'$ of phosphocreatine is –10,300 cal/mole at pH 7.0.

Acyl Phosphates. Diphosphoglyceric acid is an acyl phosphate which participates in ATP generation steps during anaerobic respiration (conversion of glucose to pyruvate). Its standard free energy of hydrolysis is –11,800 cal/mole at pH 7.0.

Enolic Phosphates. Phosphoenol pyruvate (PEP), an enolic phosphate, is the second energy-rich compound which participates during conversion of glucose to pyruvate and provides energy for the regeneration of ATP from ADP. The total free energy change on hydrolysis of this enolic phosphate is –14,800 cal/mole at pH 7.0.

$$\underset{\text{pyruvate}}{\text{Phosphoenol-}} \xrightarrow[\Delta G' = -6800\ \text{cal}]{} \text{Phosphate} + \underset{\text{(unstable enol form)}}{\text{pyruvate}} \xrightarrow[\Delta G' = -8000\ \text{cal}]{\text{Tautomerization}} \underset{\text{(stable keto form)}}{\text{Pyruvate}}$$

The unstable enolic form, which appears first, instantly isomerises to the stable keto form. This tautomerization is of major importance in making PEP one of the most energy-rich compounds of biological importance.

Thiol Esters. The only energy-rich compound which is not phosphorylated, but nonetheless yields a useful amount of energy on hydrolysis, is the thioester acetyl CoA enzyme. This can in turn be utilized to generate ATP from ADP. The $\Delta G'$ of hydrolysis of this compound is approximately –7500 cal/mole at pH 7.0.

Acetyl coenzyme A ———→ Acetate + Coenzyme A

This compound is also a source of organic carbon and energy for fatty acid synthesis, oxidative respiration and synthesis of some amino acids.

Standard free energy of hydrolysis of some other important metabolites is listed in table 4.1.

Table 4.1 ΔG′ of important metabolites

Compound	*ΔG′ at pH 7.0 (cal/mole)*	*Comments*
Cyclic AMP	– 12,000	Regulation of glucose metabolism
Glucose-1-phoshpate	– 5000	Utilization of polysaccharides
Fructose-6-phosphate	– 3800	An intermediate in glycolytic sequence
Glucose-6-phosphate	– 3300	An activated form of glucose used in respiration
Glycerol-3-phosphate	– 2200	Product of the break down of phosphoglycerides
Sucrose	– 6900	Energy generating mechanism during inulin synthesis

4.3 OXIDATION-REDUCTION REACTIONS

Many biologically important reactions involve gain or loss of electrons. These are called redox reactions or oxidation-reduction reactions. Such reactions involve transfer of electrons from an electron donor (reducing agent) to an electron acceptor (the oxidant). A reducing agent may be defined as a substance that furnishes an electron to be oxidized:

$$Fe^{2+} \xrightarrow{\text{oxidized}} Fe^{3+} + 1 \text{ electron}$$

Similarly, the substance which accepts an electron is said to be reduced:

$$Fe^{3+} + 1 \text{ electron} \xrightarrow{\text{reduced}} Fe^{2+}$$

In some oxidation-reduction reactions, transfer of one or more electrons is made via the transfer of hydrogen. Dehydrogenation is thus equivalent to oxidation.

Oxidizing and reducing agents function as conjugate redox pairs or couplers, consisting of an electron donor and its conjugate electron acceptor:

$$\text{Electron donor} \rightleftharpoons e^- + \text{electron acceptor}$$

Electron Carriers

In cells the movement of an electron is usually accompanied by the movement of a proton (hydrogen ion) so that in most cases reduction means a gain of hydrogen, and oxidation the loss of hydrogen. Redox reactions normally involve electron carriers, such as nicotinamide adenine dinucleotide (NAD^+), nicotinamide adenine dinucleotide phosphate ($NADP^+$), often called pyridine nucleotides, and flavin adenine dinucleotide (FAD) (Fig. 4.5). NAD^+, $NADP^+$ and FAD are electron carriers for the redox reactions catalyzed by a group of enzymes called dehydrogenases. Schematically, the reactions can be illustrated as:

(1) $NAD^+ + 2\,H^+ \underset{\text{oxidation}}{\overset{\text{reduction}}{\rightleftharpoons}} NADH + H^+$

(2) $NADP^+ + 2\,H^+ \underset{\text{oxidation}}{\overset{\text{reduction}}{\rightleftharpoons}} NADPH + H^+$

(3) $FAD + 2\,H^+ \underset{\text{oxidation}}{\overset{\text{reduction}}{\rightleftharpoons}} FADH_2$

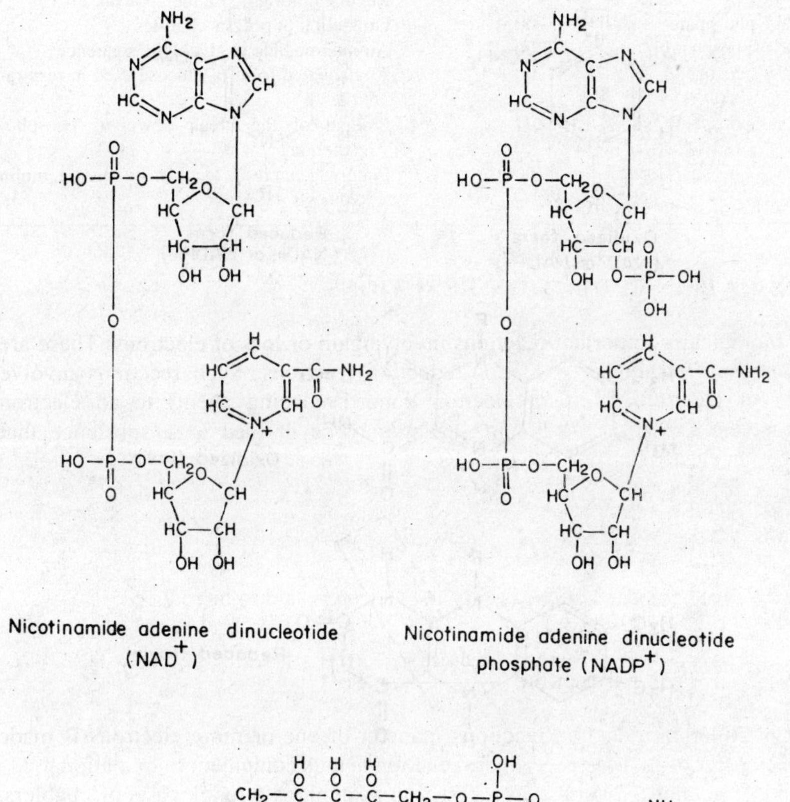

Nicotinamide adenine dinucleotide (NAD^+)

Nicotinamide adenine dinucleotide phosphate ($NADP^+$)

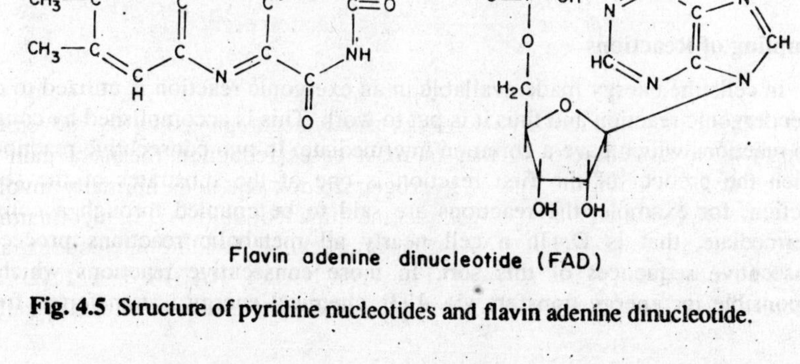

Flavin adenine dinucleotide (FAD)

Fig. 4.5 Structure of pyridine nucleotides and flavin adenine dinucleotide.

The pyridine nucleotides act as electron acceptors or carriers during the enzymatic removal of hydrogen atoms from specific substrate molecules. The hydrogen atom from the substrate is transferred as a hydride ion to the nicotinamide portion of the oxidized forms of these nucleotides, symbolized as NADH and NADPH respectively. The other hydrogen atom from the substrate becomes a hydrogen ion. Flavin nucleotides accept hydride ions from NADH along with a proton from the environment or a pair of hydrogen atoms from any other substrate. Substrates are a wide variety of organic metabolites. The structure of oxidized and reduced forms of these electron carriers have been shown in Fig. 4.6, where R is the remainder of the molecule.

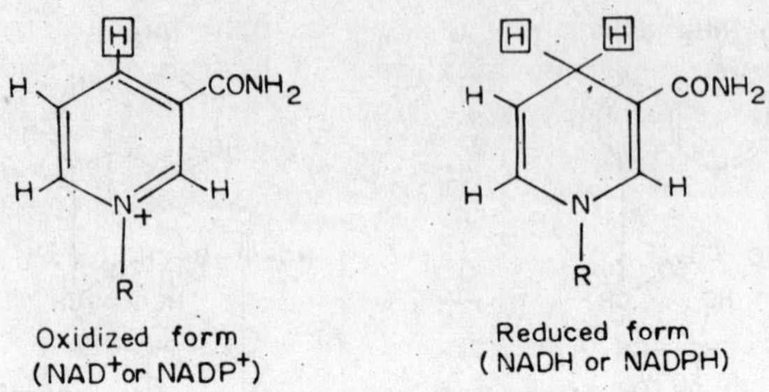

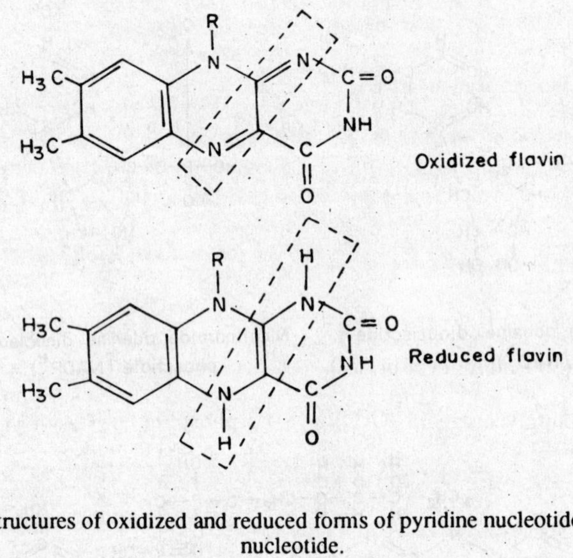

Fig. 4.6 Structures of oxidized and reduced forms of pyridine nucleotides and flavin nucleotide.

Coupling of Reactions

In cells the energy made available in an exergonic reaction is utilized to drive an endergonic reaction and thus it is put to work. This is accomplished by coupling two reactions which have a common intermediate. In two consecutive reactions in which the product of the first reaction is one of the substrates of the second reaction, for example, the reactions are said to be coupled through a common intermediate, that is D. In a cell nearly all metabolic reactions proceed in consecutive sequences of this sort. In those consecutive reactions which are responsible for energy transfers via ATP, chemical energy is transferred from a high energy phosphate group donor to ADP and is conserved in the form of ATP as a reaction product. In succeeding reactions, in which ATP is the substrate, its terminal phosphate group is transferred to an acceptor molecule, causing the latter to be changed to a compound with a higher energy content. ATP is thus a common intermediate that serves to link or couple enzymatic reactions involving the transfer of phosphate groups and is thus a vehicle for the transfer of chemical energy.

$$A + B \rightarrow C + D$$

$$D + E \rightarrow F + G$$

The common intermediate principle can be best illustrated by a specific example. During the conversion of glucose to lactic acid, at one step, the phosphorylated triose glyceraldehyde-3-phosphate is oxidized to 3-phosphoglyceric acid:

$$\text{Glyceraldehyde-3-phosphate} + H_2O \rightarrow \text{3-Phosphoglyceric acid}$$

At pH 7.0 the $\Delta G'$ of the reaction is approximately −12,000 cal/mole, indicating that the reaction is coupled to the generation of ATP, which has a $\Delta G'$ of about 7300 cal/mole at pH 7.0, by the participation of a common intermediate 1,3-diphosphoglyceric acid. The formation of this acyl phosphate represents the expenditure of 11,800 cal/mole:

$$\text{3-Phosphoglyceric acid} + H_3PO_4 \rightarrow \text{1,3-diphosphoglyceric acid} + H_2O$$

But the actual reaction in which 1,3-diphosphoglyceric acid is formed is a combination of oxidation-reduction and phosphorylation reaction with a $\Delta G'$ of 1500 cal/mole:

$$\text{Glyceraldehyde-3-phosphate} + NAD^+ \; H_3PO_4 \rightarrow \text{1,3-diphosphoglyceric acid} + NADH + H^+$$

In a subsequent reaction this acyl phosphate is utilized to convert ADP to ATP with a $\Delta G'$ of − 4500 cal/mole at pH 7.0.

$$\text{1,3-diphosphoglyceric acid} + ADP \rightarrow \text{3-phosphoglyceric acid} + ATP$$

By coupling the above two reactions the sum total may be written as:

$$\text{Glyceraldehyde-3-phosphate} + NAD^+ + H_3PO_4 + ADP \rightarrow \text{3-phosphoglyceric acid} + NADH + H^+ + ATP$$

The $\Delta G'$ of the reaction amounts to approximately −3000 cal/mole. It is clear that during direct oxidation of an aldehyde to a carboxylic acid, a lot of energy was released making the reaction irreversible. But by simple coupling of oxidation-reduction with a phosphorylation reaction, a significant amount of energy is made available which is utilized to drive the formation of ATP rather than being lost to the environment as heat.

4.4 WHAT GOVERNS THE REACTIONS

Reactions are governed by the enzymes which are proteins synthesized in the cell. These are specialized to catalyze or accelerate biological reactions needed to sustain life. They are among the most remarkable biomolecules known because of their extraordinary specificity and catalytic power. Being a protein an enzyme loses its catalytic properties if it is subjected to agents like heat, strong acids or bases, organic solvents or other conditions which denature protein. They work by reducing the amount of energy which reactants must possess before undergoing a chemical change. To understand what governs the speed of a reaction, we must focus our attention on *enzymes* and on the concept of 'activation energy'.

Enzymes—the Biocatalysts

In 1897 E. Buchner succeeded in extracting an enzyme from the juice of yeast cells which catalyzed alcoholic fermentation. He designated the active juice as an enzyme. Soon other extracts with the same ability were discovered and were also termed enzymes. Later the mixture of substances isolated by Buchner was given a new name 'zymase'. However, not many years later an enzyme was isolated for the first time in a pure crystalline form. This was accomplished by J.B. Sumner in 1926 when he first purified and crystallized the enzyme urease, isolated from the extracts of the jack-bean. It was not, however, till the years 1930-1936 that J. Northrop crystallized the enzymes pepsin, trypsin and chymotrypsin to prove their protein nature. To-day nearly 2000 enzymes are known and at least 250 have been purified and have, without exception, been proved to be proteins.

The catalytic function of an enzyme is due to its protein nature, as the highly complex structure of the enzyme protein can provide the environment for a particular reaction mechanism as well as the template function to recognise a limited set of substrates. That is why enzymes are highly specific and catalyze only one reaction or one type of reaction. An enzyme catalyzing a single reaction possesses *absolute specificity*, and an enzyme that attacks one type of chemical bond in a variety of substrates is said to show *group-specificity*. In order to enable it to function as a catalyst, the shape of an enzyme molecule is of critical importance.

The structure and shape of an enzyme molecule depends on various covalent and electrical bonds holding the molecule together. These are hydrogen bonds, disulfide linkages and a variety of weak electrical bonds such as dipole and electrostatic bonds, and also hydrophobic bonds (Fig. 4.7). Factors which affect these bonds will alter the shape of the enzyme molecule and thus its activity. Such factors include heat, hydrogen-ion concentration and several cofactors that are associated with the enzyme molecule.

Enzyme Classification

The number of enzymes known to exist in biological systems is very large, but perhaps not as large as might be expected in view of the complexity of reactions occurring in living matter. On the other hand, the number of reaction types is comparatively small, since there are often many enzymes bringing about the same type of reaction in different molecules.

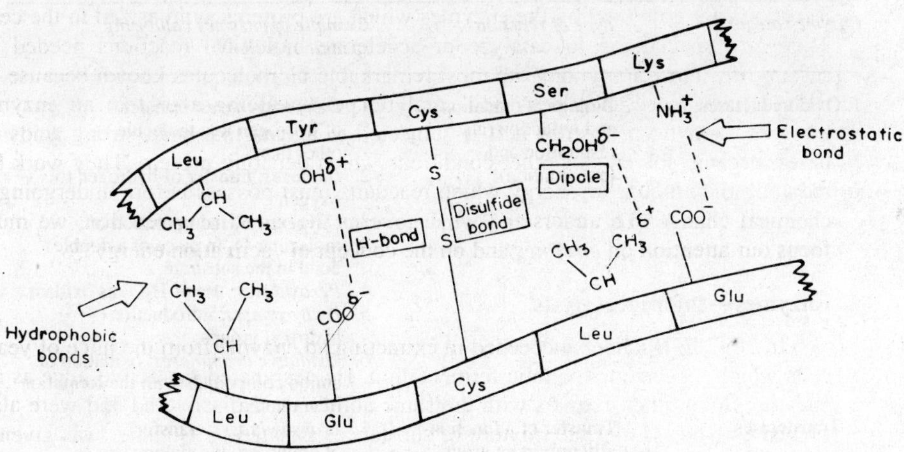

Fig. 4.7 Various bonds responsible for holding the enzyme molecule together.

In view of an increasing number of newly discovered enzymes, a systematic classification has been adopted on the recommendation of the Enzyme Commission of International Union of Biochemistry (1964). In this system, enzymes are divided into six major classes and sets of sub-classes, according to the type of reaction catalyzed. These enzymes are eventually individually numbered, because for most biological purposes classifying enzymes just by their physical characteristics would not be useful. Thus, despite the multiplicity of enzymes, the picture which has emerged from the study of reaction types is one of comparative fundamental simplicity (Table 4.2).

Mechanism of Enzyme Action

Before dealing with the mechanism of enzyme action, we must focus our attention on an important feature, that is, the energy of activation (EA).

Activation Energy. Let us look at the following chemical reaction:

$$\underset{\text{(Initial state)}}{\underset{\text{Substrate}}{A}} \longrightarrow \underset{\text{(Transition state)}}{\underset{\text{Activated substrate}}{A^*}} \longrightarrow \underset{\text{(Final state)}}{\underset{\text{Product}}{B}}$$

In this reaction, conversion of the substrate to the product takes place because a certain fraction of the A molecules at any given time possess much more energy than the rest of the molecules. This energy is normally enough to attain an activated state in which a chemical bond is either broken or made, to form the final product B. However, in every chemical reaction there is a transition state, which is defined as the energy-rich state of the interacting molecules at the top of the activation barrier (Fig. 4.8). The rate of a reaction is proportional to the concentration of the transition state species. Energy of activation (EA) is the amount of energy, in calories, required to bring all the molecules of 1 mole of substrate at a given temperature to its activated, transient state. The rate of reaction is determined by the number of molecules having this 'activation energy' to react together at any one moment. In a comparatively slow reaction only a very small

Table 4.2 Classification of enzymes based on types of reactions catalyzed

Enzyme category	*Type of reaction*	*Example of enzymes catalyzing specific reactions*
1. Oxidoreductases	Biological oxidation and reduction (also respiration and fermentation)	1. *Dehydrogenases*: transfer of hydrogen to a molecule other than O_2. 2. *Oxidases*: transfer of hydrogen to oxygen 3. *Oxygenases:* incorporation of molecular O_2 in place of a double bond in the substrate. 4. *Peroxidases:* use of H_2O_2 as oxidant. 5. *Hydroxylases*: introduction of hydroxyl groups. 6. *Oxidative deaminases*: oxidation of amino compounds with the formation of ammonia.
2. Transferases	Transfer of a functionally important group from one molecule to another	1. *Transaminases*: transfer of amino groups making new amino acids from existing ones. 2. *Kinases:* transfer of phosphate group from ATP to another substrate. 3. *Phosphorylases*: addition of inorganic phosphate without using ATP. 4. *Acyl transferases*: transfer of acetyl/acyl group to a suitable acceptor. 5. *Glycosyl transferases*: transfer glycosyl groups.
3. Hydrolases	Splitting of molecules into two by the action of water	1. *Peptidases*: hydrolysis of peptide bonds (all digestive enzymes fall in this category). 2. *Phosphatases*: removal of PO_4 group from organic molecules by hydrolysis. 3. *Phosphodiesterases*: hydrolysis of phosphate esters. 4. *Glycosidases*: action on glycosidic bonds. 5. *Esterases*: hydrolysis of carboxylic esters. 6. *Deaminases*: hydrolysis of amines. 7. *Deamidases*: hydrolysis of amides.
4. Lyases	Addition or removal of groups without involving water	1. *Carboxylases*: addition of CO_2 2. *Decarboxylases*: removal of CO_2 3. *Aldolases*: cleavage of ketose di and monophosphates. 4. *Dehydratases*: removal of H_2O.
5. Isomerases	conversion of one isomer form of a compound into a different isomer by redistributing the atoms	1. *Mutases* 2. *Racemases* 3. *Epimerases* 4. *cis-trans isomerases*
6. Ligases	Linking together of two molecules at the expense of ATP or a similar triphosphate	*Synthetases*: formation of C – S, C – O, C – N or C – C bonds.

percentage of the molecules involved have the necessary amount of energy. The activation energy, therefore, represents a kind of barrier which must be overcome before a reaction can proceed.

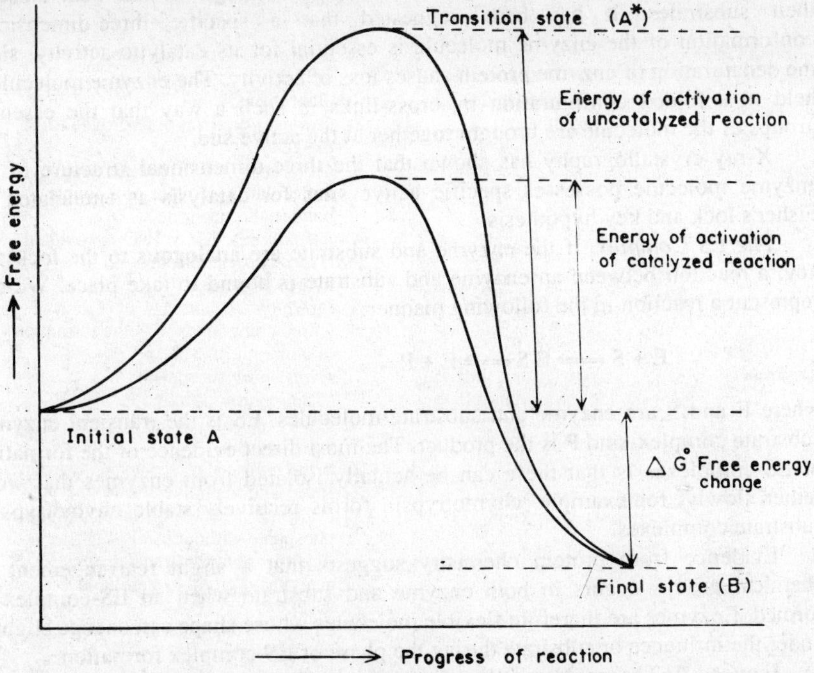

Fig. 4.8 Diagram to show free energy change during uncatalyzed and catalyzed reactions.

In principle, reactions can be speeded up in two ways: Firstly by supplying reactants with more energy e.g., by heating, and secondly by lowering the energy of activation by means of a catalyst. A rise in temperature increases the number of molecules capable of entering the transition state. In most reactions the rate of reaction is almost doubled for a 10°C rise of temperature.

In organisms, there is a limit to the amount of heat that can be supplied because heat progressively denatures proteins. Enzymes, though they are proteins, help in lowering the energy barrier. In our own bodies it would have taken years instead of a few hours for our food to be digested in the absence of the digestive enzymes.

Lock and Key Hypothesis. Enzymes are remarkable catalysts that bring about 100 per cent yield in a biological reaction with amazing speed. The process of enzyme catalysis depends on a combination of the substrate and the enzyme. This may, however, require a consideration of important points such as the nature of enzyme-substrate combination, its effects, and the manner in which catalysis takes place. There is a common idea that the enzyme-substrate combination produces some kind of 'activation' of the substrate molecule. It has long been a working hypothesis as proposed by Emil Fisher in 1894, that the substrate molecule, or a portion of it, fits the active site of the enzyme in a lock and key relationship. For a key to work it must be provided with the right lock and so is the case with enzymes and substrates.

The Active Site Principle. The active site of an enzyme contains specific functional groups that can bind the substrate and subsequently bring about the catalytic event. Normally the enzyme molecules are larger in size than those of their substrates. It has been suggested that a specific three-dimensional conformation of the enzyme molecule is essential for its catalytic activity, since the denaturation of enzyme protein causes loss of activity. The enzyme molecule is held in a folded configuration by cross-links in such a way that the essential groups of the molecule are brought together at the active site.

X-ray crystallography has shown that the three-dimensional structure of the enzyme molecule possesses specific active sites for catalysis as enunciated by Fisher's lock and key hypothesis.

The ES Complex. If the enzyme and substrate are analogous to the lock and key, a reaction between an enzyme and substrate is bound to take place. We can represent a reaction in the following manner:

$$E + S \rightleftharpoons ES \rightleftharpoons E + P$$

where E and S are enzyme and substrate molecules, ES is the transient enzyme-substrate complex, and P is the product. The most direct evidence of the formation of ES-complexes is that these can be actually isolated from enzymes that work rather slowly, for example, chymotrypsin forms relatively stable chymotrypsin-substrate complexes.

Evidence from protein chemistry suggests that a slight rearrangement of chemical groups occurs in both enzyme and substrate when an ES-complex is formed. Enzymes are therefore flexible molecules whose shape can change slightly under the influence of substrate during the phase of ES-complex formation.

Induced-fit Theory. Normally a large enzyme molecule is required to catalyze a reaction at the active site which is only a tiny portion on its surface. It may be argued that if the active site represents only a small part of the molecule, what is the function of the rest of the enzyme molecule? In 1963, Koshland suggested that the enzyme essentially consists of four categories of amino acids (Fig. 4.9).

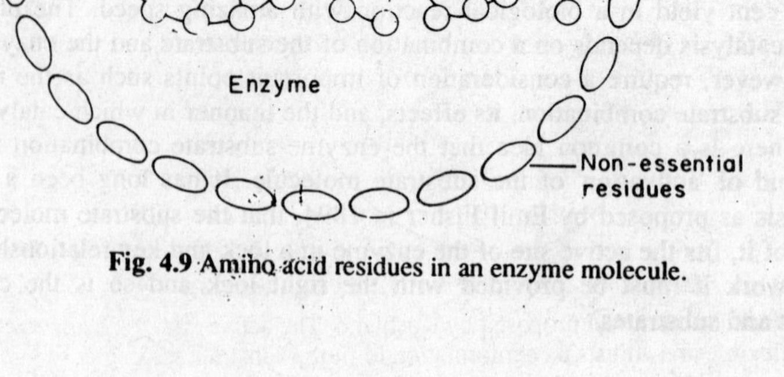

Fig. 4.9 Amino acid residues in an enzyme molecule.

(i) Catalytic residues (catalytic site): These make and break chemical bonds and are required for catalytic activity.
(ii) Binding residues (binding sites): These hold the substrate in place while the reaction (catalysis) is taking place.
(iii) Structural residues: These hold the active site in the correct shape so that it can function properly.
(iv) Non-essential residues: These have no specific function. They are often near the surface of the enzyme molecule and can be removed or replaced without loss of function.

The idea of an enzyme "wrapping-round" a substrate to form a more stable structure is called *induced-fit* hypothesis. These and other considerations led Koshland to postulate that essential functional groups on the active site of the free enzyme are not in their optimal positions for promoting catalysis when the active site is unoccupied, but when the substrate molecule is bound by the enzyme, the binding affinity forces the enzyme molecule into a conformation in which the catalytic groups assume a favourable geometric position to form the transition state. This is the induced-fit model of the enzyme-substrate molecules (Fig. 4.10).

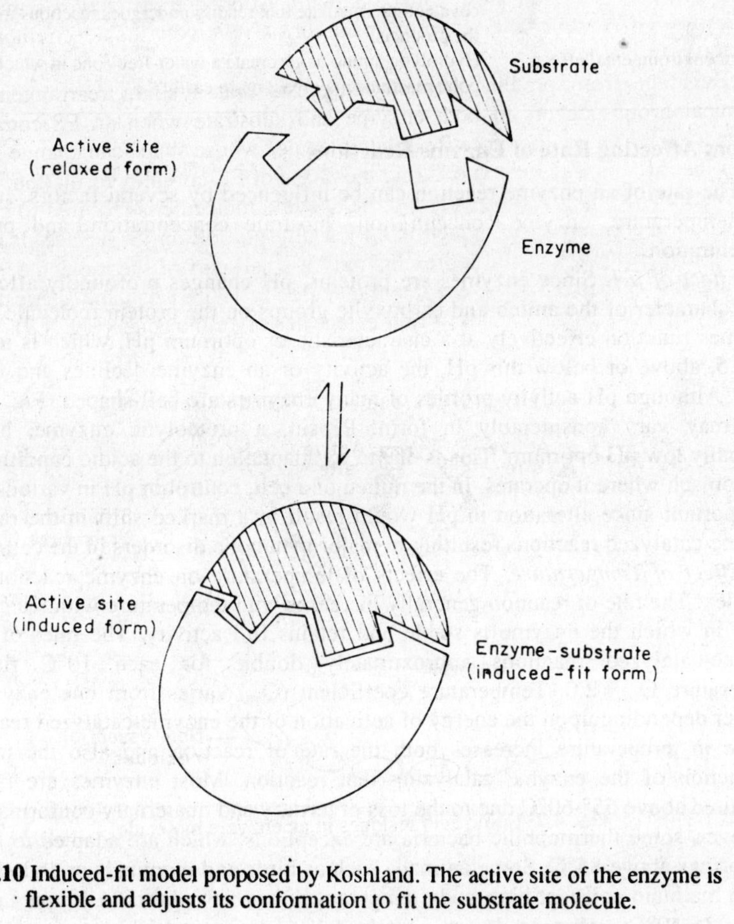

Fig. 4.10 Induced-fit model proposed by Koshland. The active site of the enzyme is flexible and adjusts its conformation to fit the substrate molecule.

Molecular Basis of Enzyme Action. According to thermodynamic considerations, enzymes lower the energy barrier or activation energy through several steps, each step with reduced activation energy, including the formation of ES-complex. On the other hand, in molecular terms, few major factors appear to participate and probably operate simultaneously at the active site, all of which contribute to a lowering of activation energy, and hence to the large rate accelerations produced by the enzymes (Table 4.3).

Table 4.3 Molecular machanisms which contribute to the catalytic efficiency of Enzymes

Mechanisms	*Remarks*
1. Proximity effects	Temporary binding of reactants close to each other on an enzyme increases the chance of a reaction.
2. Orientation effects	Reactants are held by the enzyme in such a way that the bonds are exposed to attack and a transition state is readily achieved.
3. Strain effects	Enzyme may induce strain or distortion in the susceptible bond of the substrate molecule, making the bond easier to break.
4. Acid-base catalysis	Acidic and basic amino acids in the enzyme facilitate transfer of electrons to and from the reactants.
5. Covalent catalysis	Enzyme may combine with the substrate to form an unstable covalent intermediate that readily undergoes reactions to form the products.
6. Microenvironmental effects	Hydrolytic amino acids create a water-free zone in which non-polar reactants may react more easily.

Factors Affecting Rate of Enzyme Reactions

The rate of an enzyme reaction can be influenced by several factors, such as pH, temperature, enzyme concentration, substrate concentration, and product concentration.

Effect of pH. Since enzymes are proteins, pH changes profoundly affect the ionic character of the amino and carboxylic groups on the protein molecule. Most enzymes function effectively at a characteristic or optimum pH, which is usually 7 ± 1.5, above or below this pH, the activity of an enzyme declines and finally stops. Although pH-activity profiles of many enzymes are bell-shaped (Fig. 4.11), they may vary considerably in form. Pepsin, a proteolytic enzyme, has an unusually low pH optimum. This is due to its adaptation to the acidic conditions of the stomach where it operates. In the milieu of a cell, control of pH in various parts is important since alteration in pH would result in a marked shift in the rates of enzyme-catalyzed reactions resulting in major metabolic disorders in the cell.

Effect of Temperature. The effects of temperature on enzyme reactions are complex. The rate of reaction generally increases with temperature within a certain range in which the enzyme is stable and retains full activity. The rates of most enzyme-catalyzed reactions approximately double for each 10°C rise in temperature $Q_{10} = 2.0$. Temperature coefficient (Q_{10}) varies from one enzyme to another depending upon the energy of activation of the enzyme-catalyzed reaction. A rise in temperature increases both the rate of reaction and also the rate of destruction of the enzyme catalyzing that reaction. Most enzymes are rapidly denatured above 55°-60°C due to the loss of tertiary and quaternary conformations. However, some thermophilic bacteria are exceptions, which are adapted to life in hot springs above 85°C. Some animals, such as birds and mammals or endotherms which maintain constant high body temperatures have an optimum temperature of about 35-40°C; whereas lower vertebrates and invertebrates or ectotherms generally have lower optimum temperature between 20-35°C. Temperatures as low as the freezing point do not commonly destroy enzymes. Most enzyme catalyzed reactions appear to have an optimum temperature (Fig. 4.12), showing a typical bell-shaped curve with a peak of activity.

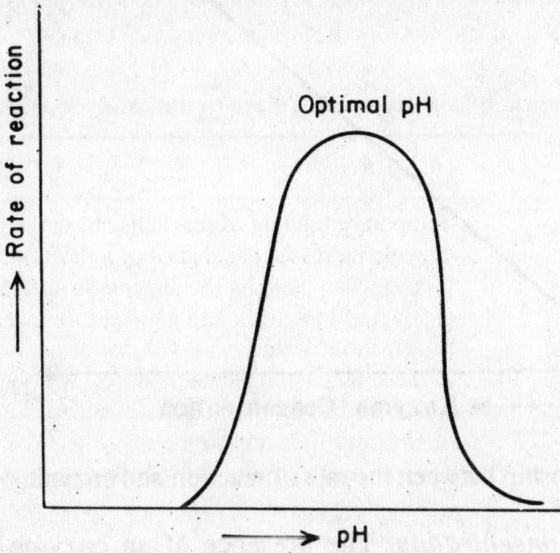

Fig. 4.11 Effect of pH on enzyme activity.

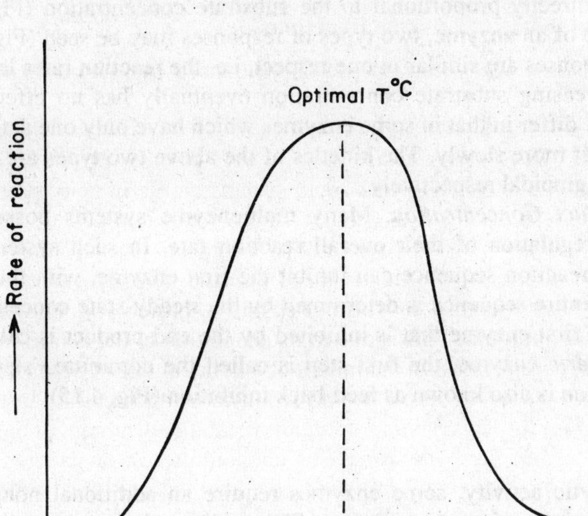

Fig. 4.12 Effect of temperature on the enzyme activity.

Effect of Enzyme Concentration. The rate of an enzyme-catalyzed reaction is directly proportional to the enzyme concentration. Fig. 4.13 shows this relationship between the rate of a reaction and increasing enzyme concentration, assuming that substrate concentration is in saturating amounts.

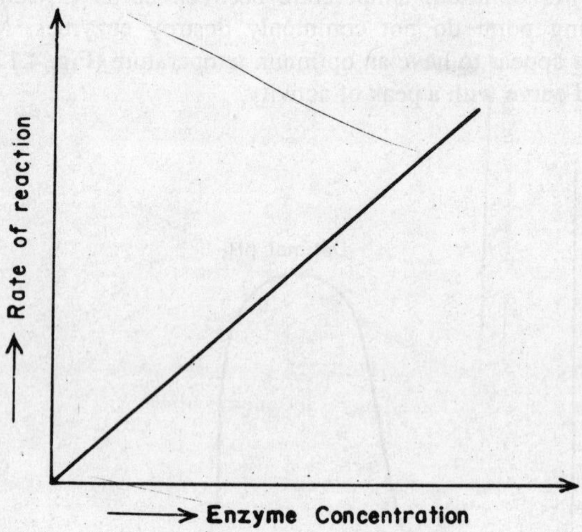

Fig. 4.13 Showing relationship between the rate of reaction and enzyme concentration.

Effect of Substrate Concentration. The presence of an enzyme in a reaction has a marked effect on its kinetics. The general principles of chemical reaction kinetics are also applicable to enzyme-catalyzed reactions. Without an enzyme, the rate of reaction is directly proportional to the substrate concentration (Fig. 4.14 (a)). In the presence of an enzyme, two types of responses may be seen (Fig. 4.14, b and c). These responses are similar in one respect, i.e. the reaction rates level off at a point and increasing substrate concentration eventually has no effect. The responses, however, differ in that in some enzymes which have only one active site the reaction takes off more slowly. The kinetics of the above two types are known as hyperbolic and sigmoidal respectively.

Effect of Product Concentration. Many multienzyme systems possess the capability for self-regulation of their overall reaction rate. In such systems, the end-product of the reaction sequence can inhibit the first enzyme, with the result that the rate of the entire sequence is determined by the steady-state concentration of the product. The first enzyme that is inhibited by the end-product is called the regulatory or *allosteric* enzyme, the first step is called the committed step. This end-product inhibition is also known as feed-back inhibition (Fig. 4.15).

Enzyme Cofactors

For their catalytic activity, some enzymes require an additional nonprotein component which actively promotes reactions. This additional component is called the *cofactor*. Cofactors are generally stable to heat, whereas most enzyme proteins lose their activity on being heated. The enzyme alongwith the cofactor is called a holoenzyme, but when the cofactor is removed, the remaining enzyme protein is called an apoenzyme and is catalytically inactive.

Cofactors are required in small amounts associated with the enzyme protein, but they do not have any catalytic property of their own. They are non-protein components which may undergo a temporary change during a reaction and are normally restored afterwards. There are three types of cofactors: prosthetic groups, coenzymes and metal ions.

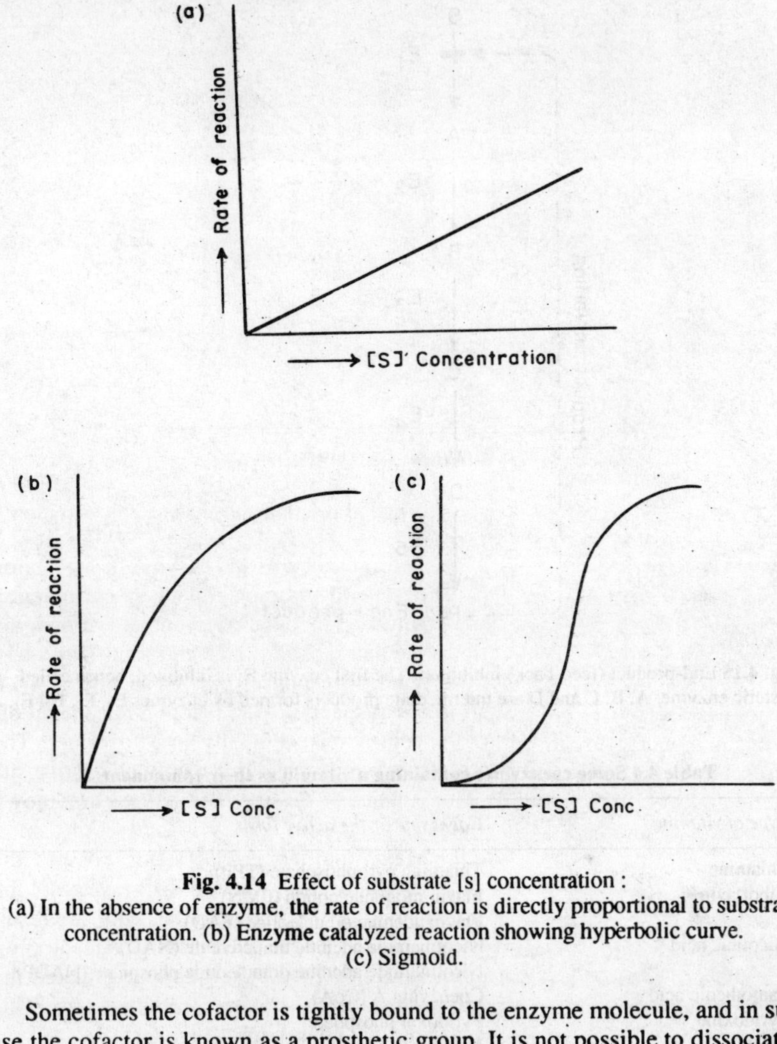

Fig. 4.14 Effect of substrate [s] concentration :
(a) In the absence of enzyme, the rate of reaction is directly proportional to substrate concentration. (b) Enzyme catalyzed reaction showing hyperbolic curve.
(c) Sigmoid.

Sometimes the cofactor is tightly bound to the enzyme molecule, and in such a case the cofactor is known as a prosthetic group. It is not possible to dissociate the cofactor without disrupting the rest of the molecule. Such prosthetic groups include porphyrins, FAD etc. The porphyrin moiety of haem (iron containing prophyrin) is present in enzymes such as catalases, cytochromes and haemoglobin. The cofactor FAD is a component of succinic dehydrogenase.

Coenzymes are small, heat stable organic molecules which bind reversibly and transiently to an enzyme. These are easily dissociated from the enzyme molecule by dialysis. In many respects they are more akin to substrate molecules, but unlike substrate molecules they are regenerated and reused. Important examples include a number of coenzymes which are derivatives of vitamin B group (Table 4.4).

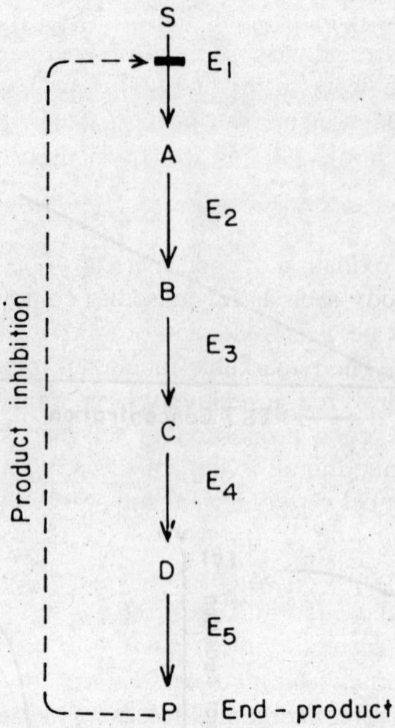

Fig. 4.15 End-product (feed-back) inhibition. The first enzyme E_1 is inhibited, hence called allosteric enzyme. A, B, C and D are intermediate products formed by enzymes E_1, E_2, E_3, E_4.

Table 4.4 Some coenzymes containing a vitamin as their component

Type of vitamin	*Coenzyme or the active form*
Thiamine	Thiamine pyrophosphate (TPP)
Riboflavin	Flavin mononucleotide (FMN)
	Flavin adenine dinucleotide (FAD)
Nicotinic acid	Nicotinamide adenine dinucleotide (NAD)
	Nicotinamide adenine dinucleotide phosphate (NADP)
Pantothenic acid	Coenzyme A (CoA)
Pyridoxine	Pyridoxal phosphate
Biotin	Biocytin
Folic acid	Tetrahydrofolic acid
Vitamin B_{12}	Coenzyme B 12

4.5 GENERATION OF ENERGY

In order to maintain life, the organisms need a source of energy. For most organisms this is provided by oxidation of food stuffs by molecular oxygen. In animals and certain other organisms food is obtained from exogenous sources in organic form (heterotropic nutrition). Green plants, however, can synthesize their own food from inorganic sources (autotrophic nutrition). An organism's problems do not end after the food and oxygen have been taken into the body, because these have to be transported to the constituent cells by active and passive transport mechanisms. Energy is released from these food stuffs when they are broken down to simpler and thermodynamically more stable molecules such as CO_2 and H_2O.

Carbohydrates and fats are primary sources of energy, while proteins are required for growth and repair work in the body. However, if carbohydrates and fats are in short supply, tissue proteins can be converted into carbohydrates to supply the energy needs of the body.

ATP—The Vital Link

The energy released from the oxidation of food materials does not directly supply the immediate needs of the body such as for muscular contraction, absorption of food stuffs and so on. The link between oxidation of food and the energy requiring activities is provided by ATP. The remarkable functional importance of ATP lies in the fact that when it is hydrolyzed, a relatively large amount of free energy is released (–7300 cal/mole). A good proportion of this can be used directly for biological activities. ATP is found in all living cells and is generally the universal supplier of energy in all biological systems; plant and animal alike.

Energy from Glucose

How exactly does the oxidation of glucose provide energy for the synthesis of ATP? This question can be answered without going into details of the metabolic pathways involved. Energy comes from two sources: first, some of the individual steps in the overall process are exergonic, releasing sufficient energy for a molecule of ATP to be synthesized from ADP and inorganic phosphate (Pi). Comparatively few steps are capable of doing this, however, and the total number of ATP molecules built up in this way is relatively small. The second and the most important source of energy is oxidation-reduction reactions involving removal of hydrogen atoms from the substrates to be oxidized. Such reactions are dehydrogenations which are linked with ATP synthesis through oxidative phosphorylations. The hydrogen atoms pass through a number of hydrogen carriers.

Fermentation and Respiration

All heterotropic organisms ultimately obtain their energy from oxidation-reduction reactions. Aerobic organisms obtain most of their energy from respiration, defined as the oxidation of organic fuels by molecular oxygen. In anaerobic respiration, however, there is no net oxidation of the fuel. The fuel molecules are partially broken down in the absence of oxygen (anaerobiosis). Yeasts can respire anaerobically, as can various other fungi and bacteria. The most common fuels for anaerobic respiration are sugars, particularly glucose.

The main difference between aerobic and anaerobic respiration is that in the latter case glucose is only partially broken down and converted into either alcohol as in yeasts and some bacteria or lactic acid as in the case of animals, whereas in the former case glucose is completely oxidized to CO_2 and H_2O (Fig. 4.16).

Anaerobiosis (alcoholic fermentation):

$$\underset{\text{Glucose}}{C_6H_{12}O_6} \longrightarrow \underset{\text{Ethanol}}{2\,CH_3CH_2OH} + 2\,CO_2$$

Glycolysis:

$$C_6H_{12}O_6 \longrightarrow 2CH_3CH\,(OH)\,COOH$$
Lactic acid

Aerobic respiration:

$$C_6H_{12}O_6 + 6O_2 \longrightarrow 6\,H_2O + 6\,CO_2$$

In anaerobic respiration considerable amount of energy remains locked up in the ethanol or lactic acid molecules. In animals this energy can be liberated by conversion of lactic acid into pyruvic acid, which is then oxidized in the usual manner. As a short term measure this is extremely useful. In humans, for example, lactic acid which accumulates during muscular exercise can be subsequently broken down or reconverted to carbohydrate when oxygen becomes available.

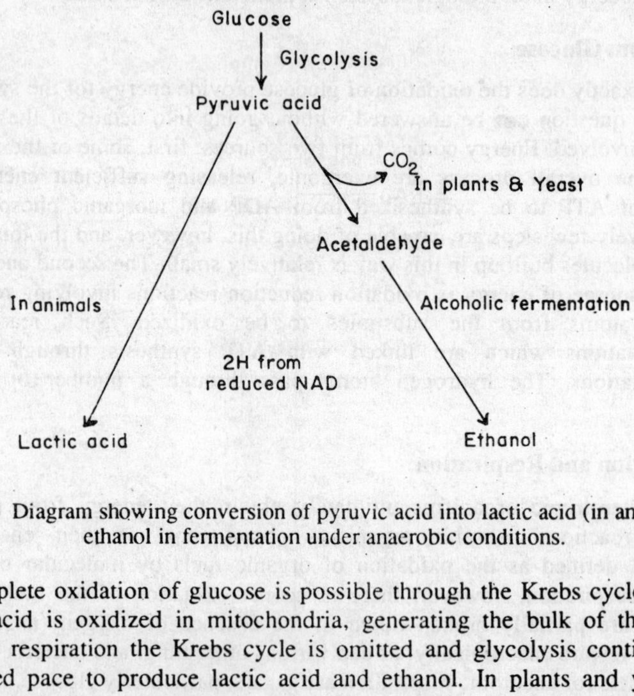

Fig. 4.16 Diagram showing conversion of pyruvic acid into lactic acid (in animals) and ethanol in fermentation under anaerobic conditions.

Complete oxidation of glucose is possible through the Krebs cycle, in which pyruvic acid is oxidized in mitochondria, generating the bulk of the ATP. In anaerobic respiration the Krebs cycle is omitted and glycolysis continues at an accelerated pace to produce lactic acid and ethanol. In plants and yeast cells, pyruvic acid is first converted to acetaldehyde, which is subsequently reduced by hydrogen atoms to ethanol. Omission of Krebs cycle ensures incomplete oxidation of glucose and fewer ATP molecules. In fact anaerobic breakdown of one molecule of glucose yields 2 molecules of ATP as compared to 38 moles of ATP in aerobic conditions. The ability of organisms to utilize oxygen is of vital importance for energy production.

Glycolysis: Non-oxidative Respiration

Glucose and its polymers, starch and glycogen, provide a substantial amount of the energy required by organisms. The sequence of reactions by which glucose is degraded anaerobically is called glycolysis, which literally means *splitting glucose*. Monosaccharides other than glucose can also be broken down by glycolysis, provided they are first converted into an intermediate of that sequence.

The glycolytic sequence which is catalyzed by the sequential action of eleven enzymes, if glucose is considered as the starting point, takes place in the cytoplasm. This can be conveniently divided into two stages. In the first stage, glucose is phosphorylated and cleaved to form two moles of 3-carbon sugar, glyceraldehyde-3-phosphate, which are then cleaved into two moles of lactic acid in the second stage (Fig. 4.17).

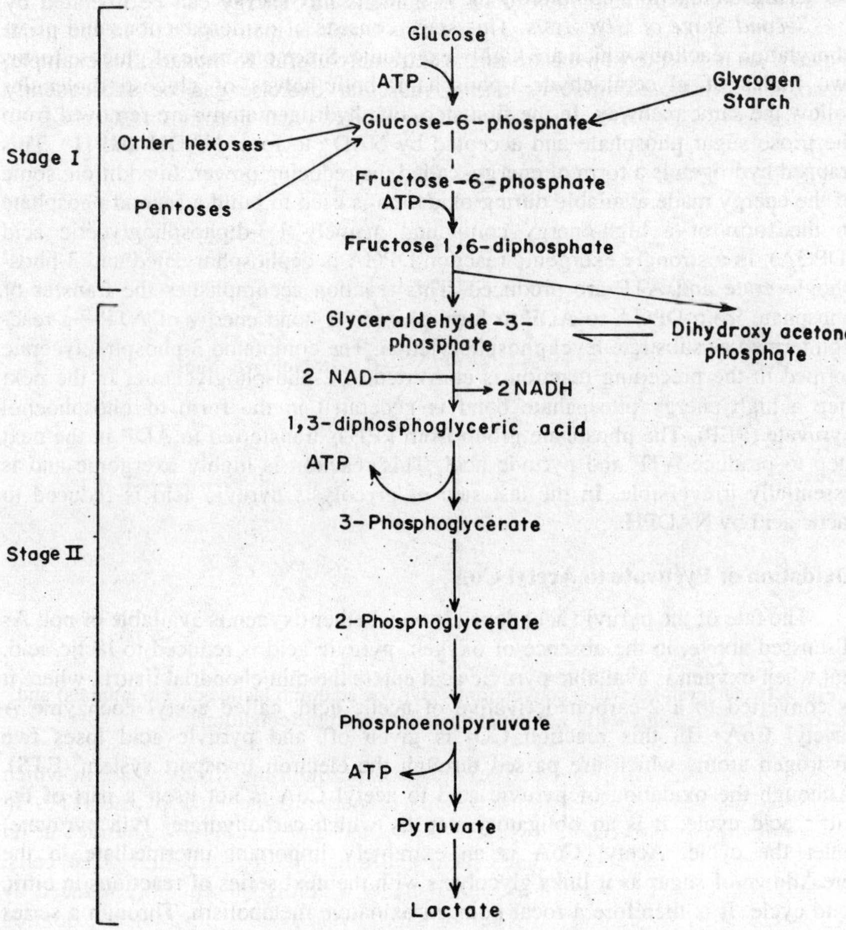

Fig. 4.17 The two stages of glycolysis.

First Stage of Glycolysis. The first stage of glycolysis serves as a preparatory or collection process in which glucose is phosphorylated at the expense of ATP. resulting in a common product, i.e. glyceraldehyde-3-phosphate. In the initial step glucose is phosphorylated at the expense of ATP to produce gluccse-6-phosphate. The reaction is endergonic and the requisite energy to prime the glucose comes from hydrolysis of terminal phosphate of ATP. In the next reaction, glucose-6-

phosphate is isomerized to fructose-6-phosphate. The third step involves phosphorylation of fructose-6-phosphate, resulting in the synthesis of fructose-1,6-diphosphate, a low energy compound containing an esterified phosphate bond. In the next step this compound splits into two triose phosphate sugars, dihydroxyacetone phosphate and glyceraldehyde-3-phosphate. Production of the latter compound in the above reaction technically completes the preparatory phase of glycolysis. It may be noted that so far only half of the glucose molecule has been converted to glyceraldehyde-3-phosphate, while the other 3-C compound dihydroxyacetone phosphate (DHAP) can yield glyceraldehyde-3-phosphate by an isomerization reaction, completing the first stage of glycolysis.

Second Stage of Glycolysis. This stage consists of oxidoreductions and phosphorylation reactions which are highly exergonic. Since one mole of glucose forms two moles of glyceraldehyde-3-phosphate, both halves of glucose molecules follow the same pathway. In the first step, two hydrogen atoms are removed from the triose sugar phosphate and accepted by NAD^+ to form NADH and H^+. This trapped hydrogen is a form of energy, called the reducing power. In addition, some of the energy made available during oxidation is used to build a second phosphate in the form of a high-energy compound, namely 1,3-diphosphoglyceric acid (DPGA). In a strongly exergonic reaction DPGA is dephosphorylated and 3-phosphoglycerate and ATP are produced. This reaction accomplishes the transfer of phosphate from DPGA to ADP to form phosphate bond energy of ATP—a reaction termed as substrate level phosphorylation. The compound 3-phosphoglycerate formed in the preceding reaction is converted to 2-phosphoglycerate. In the next step a high energy phospahate bond is generated in the form of phosphoenol pyruvate (PEP). The phosphate group from PEP is transferred to ADP in the next step to produce ATP and pyruvic acid. This reaction is highly exergonic and is essentially irreversible. In the last step of glycolysis pyruvic acid is reduced to lactic acid by NADPH.

Oxidation of Pyruvate to Acetyl CoA

The fate of the pyruvic acid depends on whether oxygen is available or not. As discussed above, in the absence of oxygen, pyruvic acid is reduced to lactic acid, but when oxygen is available pyruvic acid enters the mitochondrial matrix where it is converted to a 2-carbon derivative of acetic acid, called acetyl coenzyme A (acetyl CoA). In this reaction CO_2 is given off and pyruvic acid loses two hydrogen atoms which are passed through the electron transport system (ETS). Although the oxidation of pyruvic acid to acetyl CoA is not itself a part of the citric acid cycle, it is an obligatory step by which carbohydrates (via pyruvate) enter the cycle. Acetyl CoA is an extremely important intermediate in the breakdown of sugar as it links glycolysis with the next series of reactions in citric acid cycle. It is therefore a focal point in oxidative metabolism. Through a series of reactions involving a multienzyme complex, pyruvate combine with coenzyme A to form acetyl CoA:

$$\text{Pyruvic acid} + \text{Co A} \xrightarrow[\text{NAD} \;\to\; \text{NADH} + \text{H}^+]{} \text{Acetyl CoA} + CO_2$$

Citric Acid Cycle

In eukaryotes, aerobic respiration takes place exclusively in the mitochondria. It has also been discovered that mitochondria isolated from yeasts, protozoa and other microorganisms can carry out oxidation of tricarboxylic acids, hence the oxidative cycle is also called a tricarboxylic acid cycle (TCA cycle). The reactions of the cycle take place in the inner compartment of the mitochondria as most of the enzymes are found in the matrix.

In the process of oxidative respiration, simple organic (carboxylic) acids derived from carbohydrates, fats and proteins are metabolized to CO_2 and H_2O, releasing a substantial amount of energy for ATP synthesis. The cyclical series of reactions were worked out by Sir Hans Krebs and his associates in 1937. The fundamental mechanism of the series of eight reactions is therefore known as the Krebs cycle (Fig. 4.18). The cycle is also known as the citric acid cycle since citric acid is the first compound to be synthesized, or it is sometimes also known as the tricarboxylic acid cycle because most of the intermediates in the cycle have three carboxyl groups.

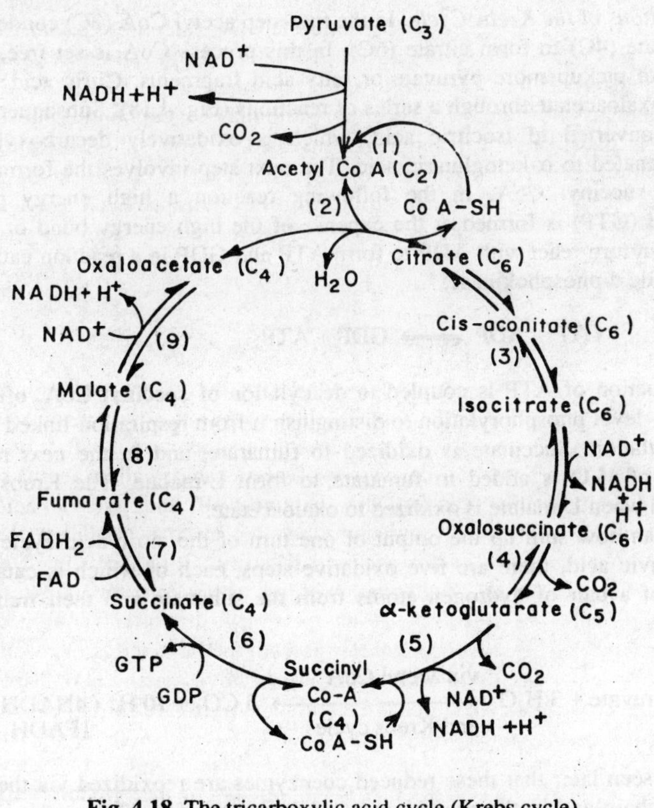

Fig. 4.18 The tricarboxylic acid cycle (Krebs cycle).

Carbondioxide is released by decarboxylation reactions and H_2O is formed as a result of the oxidation reactions. The hydrogen component in the water molecule originates from organic acids, whereas oxygen originates from the atmosphere. It is during the transport of hydrogen to oxygen that most of the energy for ATP synthesis becomes available through the electron transport chain.

Entry Points in Krebs Cycle. Acetyl CoA and α-ketoglutaric acid are the principal points of entry into the Krebs cycle. Acetyl CoA is generated in two ways:

(i) via glycolysis
(ii) via β-oxidation of fatty acids

Glycolysis generates two moles of pyruvate per glucose molecule in the cytosol. Pyruvate diffuses into the mitochondrial matrix and is converted to acetyl CoA. A small amount of pyruvate may also be formed directly in the mitochondria by deamination of the amino acid alanine. Whatever be the source, pyruvate is converted to acetyl CoA before complete oxidation.

β-oxidation of fatty acids takes place in the mitochondria where fatty acids are chopped into fragments of 2-carbon compound acetyl CoA. Proteins enter the Krebs cycle via α-ketoglutarate. α-ketoglutarate is produced by deamination of glutamate, which in turn is formed by transamination from a variety of amino acids.

Reactions of the Krebs Cycle. In the first step acetyl CoA (2C) condenses with oxaloacetate (4C) to form citrate (6C). In this process CoA is set free, which is recycled to pickup more pyruvate or fatty acid fragments. Citric acid is broken down to oxaloacetate through a series of reactions (Fig. 4.18). Subsequently, citric acid is converted to isocitric acid, which is oxidatively decarboxylated and dehydrogenated to α-ketoglutaric acid. The next step involves the formation of a thioester, succinyl CoA. In the following reaction a high energy phosphate compound (GTP) is formed at the expense of the high energy bond of thioester. GTP can in turn react with ADP to form ATP and GDP in a reaction catalyzed by a nucleoside diphosphokinase.

$$\text{GTP} + \text{ADP} \rightleftharpoons \text{GDP} + \text{ATP}$$

The production of ATP is coupled to deacylation of succinyl CoA, often called substrate - level phosphorylation to distinguish it from respiration-linked oxidative phosphorylation. Succinate is oxidized to fumarate, and in the next reaction a molecule of H_2O is added to fumarate to form L-malate. The Krebs cycle is completed when L-malate is oxidized to oxaloacetate.

We can now sum up the output of one turn of the citric acid cycle. Starting from pyruvic acid, there are five oxidative steps, each of which is catalyzed by removal of a pair of hydrogen atoms from the substrate and their transfer to a coenzyme.

$$\text{Pyruvate} + 3\,H_2O \xrightarrow[\text{and Krebs cycle}]{\text{via acetyl CoA}} 3\,CO_2 + 10\,H^+ \begin{pmatrix} 4\text{NADH} + H^+ \\ 1\text{FADH}_2 \end{pmatrix}$$

It will be seen later that these reduced coenzymes are reoxidized via the electron transport chain and in the process most of the energy for ATP synthesis is made available. The citric acid cycle cannot be made to proceed in the reverse direction, although individual steps are reversible except for the oxidative decarboxylation of α-ketoglutarate with a large decrease in the standard free energy i.e., $\Delta G = -8000$ cal, which renders the complete cycle essentially irreversible.

Energy from Protein

Tissue proteins are used as a source of energy in abnormal circumstances only, as normally a certain amount of energy is always derived from the excess dietary proteins. Initially, dietary proteins are hydrolyzed into their constituent amino acids, which undergo deamination and the residual carbon skeleton then enters into carbohydrate metabolism with subsequent release of energy. The amino acid residue may be converted into acetyl CoA, pyruvic acid or a Krebs cycle intermediate (α–ketoglutarate, succinate or oxaloacetate). The reactions are reversible so that certain amino acids can be resynthesized from carbohydrates and carbohydrates can be resynthesized from amino acids. The reactions by which the intermediates of the cycle are dranied off and formed remain in a state of dynamic equilibrium so that the concentrations of the intermediates in the mitochondria remain constant.

It is thus clear that the Krebs cycle not only supplies energy but also provides the opportunity for metabolic interconversions between carbohydrates, fats and proteins (Fig. 4.19). In mammals the liver plays a prominent role in controlling these metabolic interconversions.

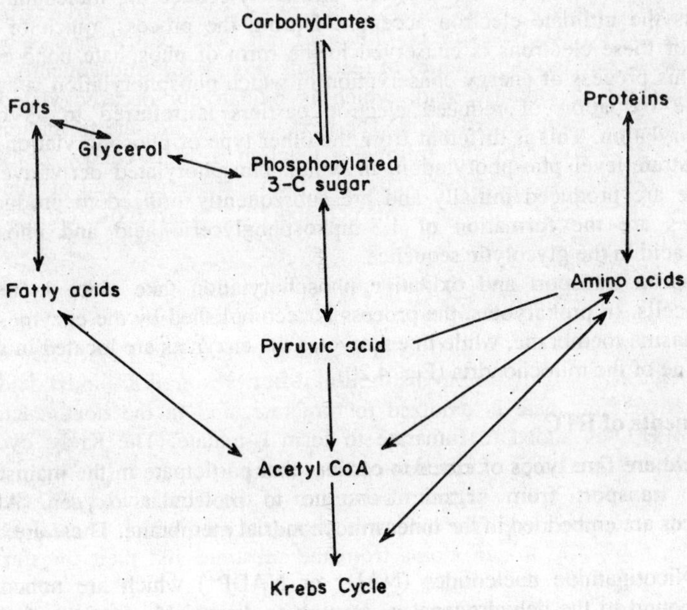

Fig. 4.19 Simplified scheme showing interconversions between carbohydrate, fat and protein metabolism.

Energy from Fats

As discussed earlier, fats can be utilized in releasing energy for the synthesis of ATP. When demands of energy are greater, or carbohydrate is in short supply, fats are hydrolyzed into glycerol and fatty acids. The former is phosphorylated and converted into a 3-carbon sugar, which is then converted into pyruvic acid and recycled into the Krebs cycle. Meanwhile, the fatty acid component goes through a series of reactions in which carbon atoms are lost, two at a time. Each of the two carbon units forms a molecule of acetyl CoA which enters the Krebs cycle. The oxidation of fatty acids takes place in the mitochondrial matrix. The successive oxidative removal of acetyl CoA units from the chain of fatty acids is called β-oxidation. A complete cycle involves a number of steps, at the end of which one acetyl CoA unit is released. For complete oxidation of a 16-carbon palmitic acid, seven oxidative cycles will be required, yielding 8 molecules of acetyl CoA units. About 48% of the standard free energy is recovered from the palmitic acid molecule in the form of phosphate bond energy (ATP). The number of ATP molecules produced by the complete oxidation of a fatty acid depends on the number of carbon atoms it contains.

4.6 ELECTRON TRANSPORT

Oxidative Phosphorylation

A major objective of catabolism of carbon compounds by a living organism is the production of energy for growth and development. The electrons derived from the intermediates of the TCA cycle flow down through the sequence of carriers in the electron transport chain (ETC) and ultimately reduce the molecular oxygen, which is the ultimate electron acceptor. During the process, much of the free energy of these electrons is conserved in the form of phosphate bond-energy of ATP. This process of energy conservation in which phosphorylation is associated with the oxidation of reduced electron carriers is referred to as oxidative phosphorylation. This is different from the other type of phosphorylation process, the substrate-level phosphorylation, in which phosphorylated derivatives of the substrate are produced initially and are subsequently utilized to produce ATP. Examples are the formation of 1,3-diphosphoglyceric acid and phosphoenol pyruvic acid in the glycolytic sequence.

Electron transport and oxidative phosphorylation take place in nearly all aerobic cells. In prokaryotes, the process is accomplished by the enzymes located in the plasma membrane, while in eukaryotes the enzymes are located in the inner membrane of the mitochondria (Fig. 4.20).

Components of ETC

There are five types of electron carriers that participate in the mainstream of electron transport from organic substrate to molecular oxygen. All these substances are embedded in the inner mitochondrial membrane. These are:

(i) Nicotinamide nucleotides (NAD^+ or $NADP^+$) which are noncovalently bound to the dehydrogenases, serving as dissociable carriers of electrons. In animal cells NAD^+ is usually present in much greater amounts than $NADP^+$. In oxidative respiration, primarily NAD^+ - linked dehydrogenases serve to transfer electrons from substrates towards oxygen via a number of other carrier proteins.

(ii) Flavoproteins (FAD or FMN) are tightly bound prosthetic groups of flavin-linked dehydrogenases. These do not leave the enzyme during or after the catalytic cycle. Out of a number of flavinlinked dehydrogenases, the two important ones are NADH dehydrogenase (Fp1) containing FMN and succinate dehydrogenase (Fp2) containing FAD as the prosthetic group.

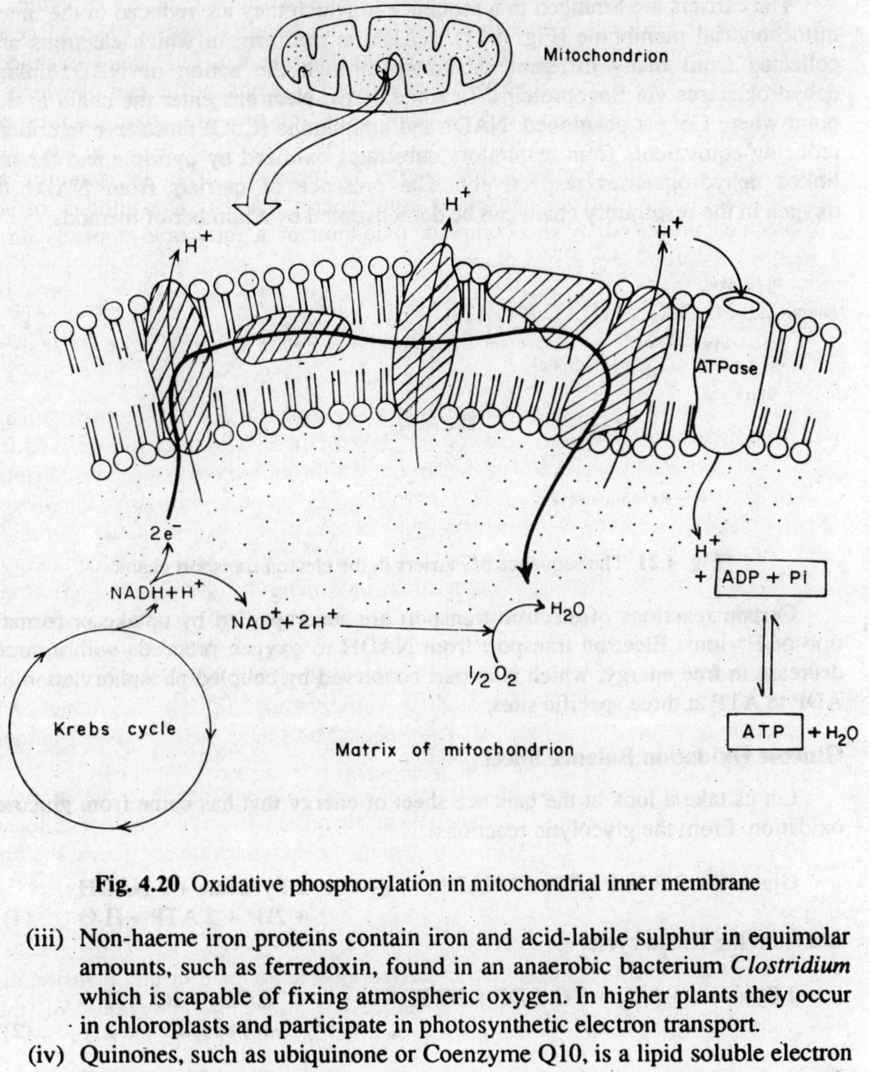

Fig. 4.20 Oxidative phosphorylation in mitochondrial inner membrane

(iii) Non-haeme iron proteins contain iron and acid-labile sulphur in equimolar amounts, such as ferredoxin, found in an anaerobic bacterium *Clostridium* which is capable of fixing atmospheric oxygen. In higher plants they occur in chloroplasts and participate in photosynthetic electron transport.

(iv) Quinones, such as ubiquinone or Coenzyme Q10, is a lipid soluble electron carrier, which participates in the transport of electrons from organic substrates to molecular oxygen.

(v) Cytochromes are electron transferring proteins containing an iron porphyrin ring system. These are of various types and are found only in aerobic cells. At least five types of cytochromes have been identified in the mitochondrial membrane, and these are cytochromes *b*, C_1, *c*, *a* and a_3. With the exception of cytochrome *c*, the rest are very tightly bound to the mitochondrial membrane. Cytochromes *a* and a_3 are collectively called the respiratory enzyme or cytochrome oxidase.

The Respiratory Chain

The carriers are arranged in a sequence in which they are reduced in the inner mitochondrial membrane (Fig. 4.21). NADH is the form in which electrons are collected from many different substrates through the action of NAD^+-linked dehydrogenases via flavoproteins. In some cases, electrons enter the chain at the point where CoQ is positioned. NAD^+ and ubiquinone (CoQ) thus serve to collect reducing equivalents from respiratory substrates oxidized by pyridine and flavin-linked dehydrogenases respectively. The presence of carriers from NAD^+ to oxygen in the respiratory chain can be demonstrated by a number of methods.

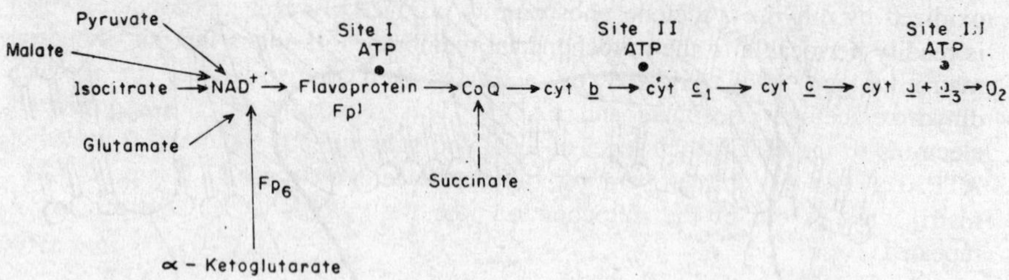

Fig. 4.21 The sequence of carriers in the elecon transport chain.

Certain reactions of electron transport are accompanied by uptake or formation of H^+ ions. Electron transport from NADH to oxygen proceeds with a large decrease in free energy, which is in part conserved by coupled phosphorylation of ADP to ATP at three specific sites.

Glucose Oxidation Balance Sheet

Let us take a look at the banance sheet of energy that has come from glucose oxidation. From the glycolytic reactions:

$$\text{Glucose} + 2\ \text{Pi} + 2\ \text{ADP} + 2\ \text{NAD}^+ \longrightarrow 2\ \text{Pyruvate} + 2\ \text{NADH} + 2\text{H}^+ + 2\ \text{ATP} + \text{H}_2\text{O} \quad ...(1)$$

and from the Krebs cycle:

$$2\ \text{Pyruvate} + 5\ \text{O}_2 + 3\text{O}\ \text{ADP} + 3\text{O}\ \text{Pi} \longrightarrow 6\ \text{CO}_2 + 3\text{O}\ \text{ATP} + 34\ \text{H}_2\text{O} \quad ...(2)$$

For the complete equation, to this should be added the oxidation of 2 moles of extramitochondrial NADH formed during glycolysis. Oxidation of extramitochondrial NADH in porkaryotes generates three molecules of ATP per pair of electrons. Thus we have the following equation:

$$2\ \text{NADH} + 2\text{H}^+ + \text{O}_2 + 6\ \text{Pi} + 6\ \text{ADP} \longrightarrow 2\text{NAD}^+ + 6\text{ATP} + 8\ \text{H}_2\text{O} \quad ...(3)$$

The sum of equations 1, 2 and 3 is:

$$\text{Glucose} + 6\text{O}_2 + 38\ \text{Pi} + 38\ \text{ADP} \longrightarrow 6\text{CO}_2 + 38\ \text{ATP} + 44\ \text{H}_2\text{O}$$

The ΔG' for the oxidation of glucose by O_2 to CO_2 and H_2O as estimated by colorimetric data is

$$C_6H_{12}O_6 + 6\,O_2 \longrightarrow 6\,CO_2 + 6H_2O$$
$$\Delta G' = -686{,}000 \text{ calories at pH7.0.}$$

The cell can conserve a large amount of this energy by coupling the energy released, to the synthesis of the energy-rich ATP from ADP and Pi. If 38 moles of ATP are formed from complete oxidation of one mole of glucose, it represents a total of –277,000 calories. The overall efficiency of energy recovery is about 40 per cent.

In eukaryotes the inner mitochondrial membrane is not permeable to NADH, and thus a shuttle process involving dihydroxyacetone phosphate and glycerol 3-phosphate is utilized. In this process, NADH produced in glycolysis is first oxidized by dihydroxyacetone phosphate to produce glycerol 3-phosphate, which is readily permeable to the mitochondrial membranes. It enters the mitochondrial matrix, where it is oxidized by a FAD-linked dehydrogenase to produce dihydroxyacetone phosphate and $FADH_2$. This reduced FAD now contributes electrons to the ETC at the level of CoQ and subsequently generates 2 moles of ATP. To keep the shuttle operating, dihydroxyacetone phosphate produced in the matrix cleaves out of the mitochondria into the cytoplasm and the process is repeated.

Chemiosmosis in Mitochondria

The mitochondria are the main site for ATP production. But how? It has been seen that the reactions of the Krebs cycle take place in the matrix while the ATP producing elctron transfer reactions occur in the inner membrane and cristae. The currently held theory is that the hydrogen carriers span the inner mitochondrial membrane from one side to the other. The hydrogen is picked up on one side and is moved across to the other side, resulting in an electrochemical gradient of H^+ ions being set up across the membrane. Energy from this gradient is used for synthesis of ATP (Fig. 4.22). This theory is known as chemiosmotic theory, proposed by P. Mitchell, who got a Nobel Prize for this discovery in 1978. Chemiosmosis involves four independent postulates:

1. The inner membrane of the mitochondria is generally impermeable to H^+ ions.
2. The electron transport chain pumps H^+ ions out of the matrix into the inter-membranal space.
3. Accumulation of H^+ ions on the cytosolic side of the inner membrane is a source of potential energy for ATP synthesis.
4. The H^+ ions diffuse back into the mitochondrial matrix under their own electrochemical gradient through special membrane-bound ATPases. ATP synthesis occurs where high concentration of H^+ ions is located.

4.7 AUTOTROPHY

All animals depend on plants directly or indirectly, for their food. Plants manufacture their food by a process called photosynthesis in which the solar energy is trapped and converted into chemical bond-energy. Solar energy is not only the immediate source of energy for plants and other photosynthetic autotrophs, but is the ultimate source of energy for nearly all heterotrophic organisms because the complex organic matter upon which the animals feed comes from plants only. Plants are, therefore, called autotrophs, a condition shared by certain protists and bacteria as well, which are capable of utilizing external sources of energy for the manufacture of new organic materials. In other words, the world's stockpile of organic materials is being continuously replenished by these autotrophs.

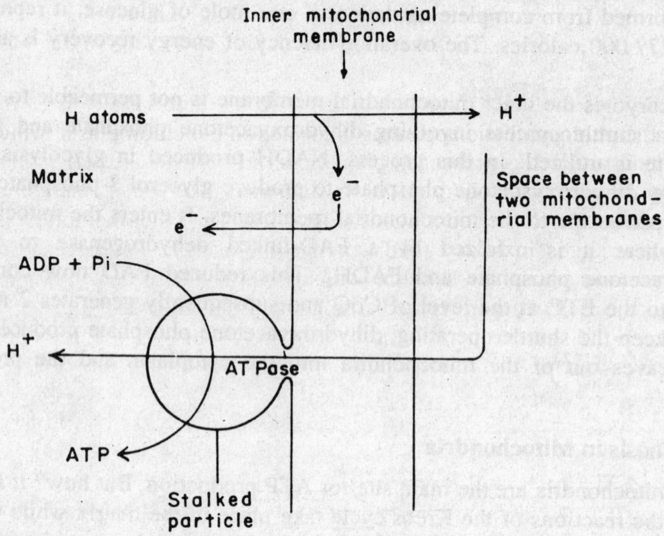

Fig. 4.22 A diagrammatic representation of Mitchell's chemiosmotic theory.

Autotrophic organisms employ two methods of synthesizing organic materials from inorganic precursors.

(i) *Phototrophy or Photosynthesis.* This method is employed by all green plants and some bacteria in which synthesis of organic compounds is carried out from CO_2 and H_2O using sunlight as the energy source and chlorophyll for trapping the energy.

(ii) *Chemotrophy or Chemosynthesis.* This method is employed by certain bacteria, which synthesize organic compounds from CO_2 and H_2O, but the energy comes from the oxidation of inorganic materials such as hydrogen sulphide, ammonia and iron.

The process of oxidation or respiration is exergonic while autotrophy is endergonic. Synthesis of glucose from CO_2 and H_2O, requires a considerable amount of energy, at least equal to that released during its oxidation to CO_2 and H_2O.

$$C_6H_{12}O_6 + 6O_2 \underset{\text{reduction (fixation)} \; \nwarrow \text{solar energy}}{\overset{\text{(respiration) oxidation} \; \nearrow \text{ATP}}{\rightleftharpoons}} 6CO_2 + 6H_2O$$

Photosynthetic organisms incl de green plants, unicellular and multicellular green, brown and red algae, euglenoids, dinoflagellates and diatoms. Among photosynthetic prokaryotes, blue-green algae, green sulphur bacteria and purple bacteria are important examples. Some blue-green algae live on CO_2 for carbon source, but there are otheres which can fix atmospheric nitrogen. The anaerobic green sulphur bacteria live in ponds and lakes rich in sulphur containing organic matter. The purple-sulphur bacteria are strict anaerobes and require H_2S, sulphur or thiosulphate. These are found in sulphur springs. More than half of the earth's photosynthetic activity is carried out in the oceans by algae, diatoms and dinoflagellates.

Photosynthesis in Green Plants

In plants the photosynthetic apparatus is located in the chloroplasts, which are membrane bound organelles present in plant cells only. These are self-replicating units, varying in size from 1 to 10 μ in diameter. The outer membrane of chloroplast is rather fragile and continuous (Fig. 4.23). The inner membrane is also continuous but arranged in folds called lamellae enclosing a compartment called stroma. At regular intervals, the lamellae widen to form flattened membrane sacs, called thylakoids which occur in stacked arrangements called grana.

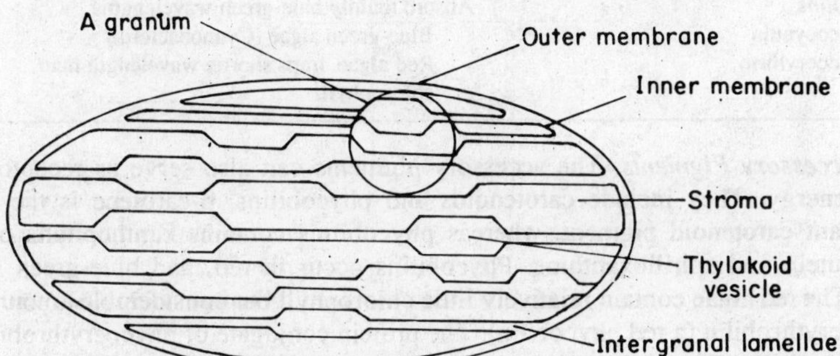

Fig. 4.23 Schematic drawing of a chloroplast showing internal structure.

Organization of Photosynthetic Systems

The thylakoid membranes and the intergranal lamellae contain photosynthetic pigments as well as the enzymes required for primary light-dependent reactions. There are several photosynthetic pigments involved in photosynthesis. Each pigment absorbs light at particular wavelengths (Table 4.5).

Chlorophylls. The main pigments in eukaryotes are the chlorophylls which are normally associated with specific proteins. These are indispensable components of the photosynthetic apparatus since whatever energy is absorbed by accessory pigments must be transferred as excitation energy to the chlorophyll molecule before it can be used for photosynthesis. In higher plants these fall mainly into two categories, chlorophyll *a* and chlorophyll *b*. Prokaryotic photosynthetic cells which do not produce oxygen lack chlorophyll *a*. Instead, they contain bacteriochlorophyll *a* or bacteriochlorophyll *b*. All chlorophylls absorb light in the visible range very efficiently, because they have many conjugated double bonds in their structures.

Table 4.5 Photosynthetic pigments and their characteristics

Pigment	*Occurrence and comments*
Chorophylls	Absorb mainly blue and red wavelengths
Chl *a* (blue-green)	All photosynthetic eukaryotes, more abundant in PS I
P 700	A variety of chlorophyll *a* which absorbs at 700 nm and is associated with PS I
P680	A variety of chlorophyll *a* which absorbs at 680 nm and is associated with PS II
Chl *b* (yellow-green)	All green plants, more common in PS II
Chl *c,d*	Various non-green algae
Carotenoids	Absorb mainly blue-green wavelengths
β-carotene (orange-yellow)	All photosynthetic eukaryotes except some algae, probably passes energy on to chlorophyll
α-carotene, fucoxanthine	Various algae and to some extent green plants
Phycobilins	Absorb mainly blue-green wavelengths
Phycocyanin	Blue-green algae (Cyanobacteria)
Phycoerythrin	Red algae, traps shorter wavelength than chlorophyll.

Accessory Pigments. The accessory pigments can also serve as receptors of light energy. They include carotenoids and phycobilins. β-carotene is the most abundant carotenoid pigment, whereas phycobilins contains xanthophylls of the type lutein and spirilloxanthine. Phycobilins occur in red, and blue-green algae only. The red algae contain relatively little chlorophyll but considerable amounts of phycoerythrobilin (a red phycobilin). The protein conjugate of phycoerythrobilin is phycoerythrin. Blue-green algae contain the blue phycocyanin as the protein conjugate.

The Photosystems

Photosynthetic pigments are organized into two functional sets or assemblies, which in turn are interconnected through an electron transport system. These functional units are called photosystem I (PS I) and photosystem II (PS II).

It has been observed that two different light reactions, each having a distinctive optimum wavelength, are required for maximum photosynthetic efficiency. L.N.M. Duysena has postulated that the long wavelength system with a maxima near 700 nm, designated as PS I, is associated with forms of chlorophyll *a* absorbing longer wavelengths. This system is not responsible for the evolution of oxygen. PS II, on the other hand, is activated by shorter wavelengths of 680 nm or below and is required for oxygen evolution. Oxygen evolving photosynthetic cells contain both PS I and PS II systems. Although both photosystems contain chlorophyll *a* and chlorophyll *b*, the ratio of *a* to *b* is higher in PS I than in PS II.

The pigment assembly of PS I in higher plants contains some 200 molecules of chlorophyll *a*, 50 molecules of chlorophyll *b*, 50 – 200 molecules of carotene, all forming the *antenna complex*, and a single molecule of P 700 constituting the reaction centre. A quantum of energy trapped anywhere in this photosynthetic unit, whether by *a* carotenoid or *a* chlorophyll molecule, migrates through a set of pigment molecules in the form of an *exciton*, until it is funnelled into a single molecule of P 700 which accepts the exciton. As a consequence, it loses an electron having a large amount of energy. Photon capture and exciton transfer within the photosystem are extremely rapid processes. A similar process of photon capture and exciton transfer occurs in the other photosystem until the exciton reaches P 680, the reaction centre of PS II which then loses an electron. The pigment assemblies of photosystems I and II are embedded in the membranes of thylakoids, predominantly in the regions where adjacent thylakoids meet. The distribution of photosystems and electron transport system in the thylakoid membrane is not symmetrical (Fig. 4.24).

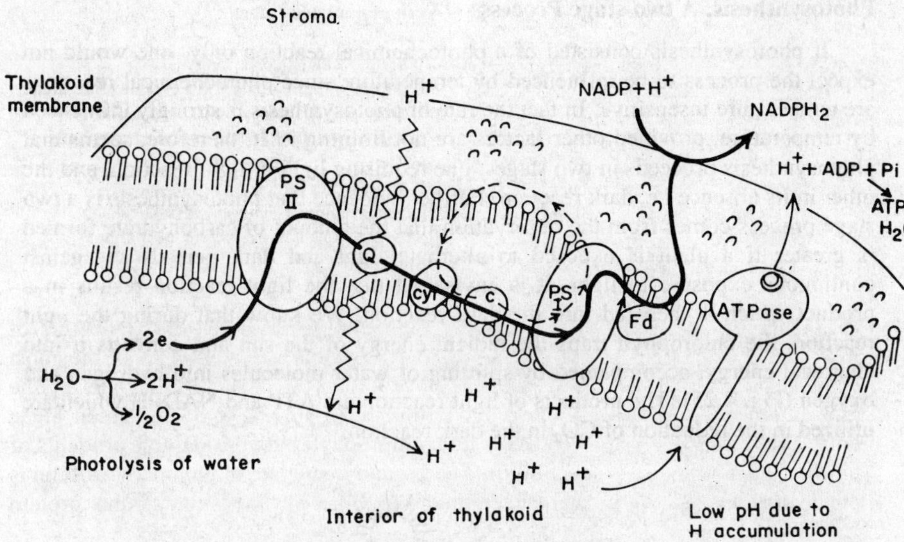

Fig. 4.24 Light reaction in the chloroplast.

——→ path of electrons during non-cyclic phosphorylation.

- - -→ path of electrons during cyclic phosphorylation.

——→ path of protons resulting from pumping which occurs during electon transport.

Q – Plastoquinone, cyt – cytochromes.

C – Plastoquinone, Fd – ferredoxin.

The Hill Reaction

In 1937, Robert Hill of Cambridge University found that isolated chloroplasts, when illuminated in the presence of artificial electron acceptors such as ferricyanide, produced oxygen and acquired reducing properties accompanied by reduction of electron acceptors.

$$H_2O + A \xrightarrow{\text{light}} AH_2 + 1/2\ O_2$$

According to the general equation, CO_2 was not apparently required nor was it reduced to a stable form that accumulated, suggesting that the photoreduction of CO_2 to hexose sugar is a later step in photosynthesis. Hill concluded that water is the source of oxygen. The reduction of acceptor molecule (hydrogen/electron acceptor) did not take place until the chloroplast extracts were illuminated. This is now called the Hill reaction, and the acceptor molecule (A) is known as the Hill reagent. That the light reaction involves photolysis of water has been confirmed by the use of isotope labelling. In an experiment, the green protist, *Chlorella* was placed in water whose oxygen atom was labelled with heavy oxygen (^{18}O), it was found that the oxygen which was evolved contained heavy oxygen. However, when *Chlorella* was grown in ordinary water without labelled oxygen but CO_2 was labelled with ^{18}O, the oxygen liberated did not contain any ^{18}O. This experiment confirmed that the oxygen evolved during photosynthesis comes from water only.

Photosynthesis: A two-stage Process

If photosynthesis consisted of a photochemical reaction only, one would not expect the process to be influenced by temperature since photochemical reactions are temperature insensitive. In fact the rate of photosynthesis is strongly influenced by temperature, provided other factors are not limiting it. It, therefore, seems that photosynthesis proceeds in two stages: one requiring light or light reaction, and the other in its absence i.e. dark reaction. Further evidence that photosynthesis is a two stage process comes from the observation that the amount of carbohydrate formed is greater if a plant is exposed to alternate light and dark periods as against continuous exposure to light. It is envisaged that the light reaction results in a product which is then fed into the dark reaction. We know that during the light reaction, the chlorophyll traps the radiant energy of the sun and converts it into chemical energy, accompanied by splitting of water molecules into hydrogen and oxygen (Fig. 4.25). The products of light reaction are ATP and NADPH which are utilized in the reduction of CO_2 in the dark reaction.

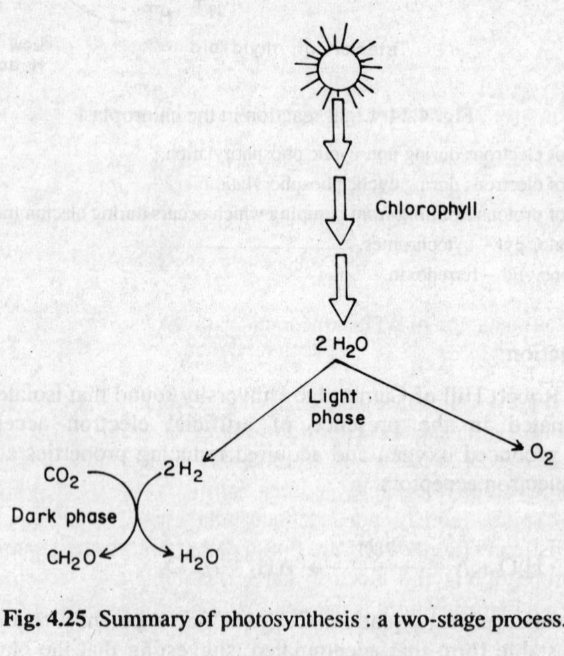

Fig. 4.25 Summary of photosynthesis : a two-stage process.

Light Reaction

The function of light reaction is twofold:

(i) Photochemical splitting of water molecule produces hydrogen atoms for reduction of CO_2, and
(ii) Production of ATP and its utilization for carbohydrate synthesis.

When light strikes the chlorophyll molecule, energy level of an electron is raised. The chlorophyll becomes excited and the electron is emitted and taken up by an electron acceptor. The electron is then passed through a series of carriers, the last one handing it back to the chlorophyll molecule. After losing the electron, the chlorophyll molecule becomes unstable and is restored to normalcy when the electron is returned. During the transfer of the electron from one carrier to the next, energy is removed from it for synthesis of ATP. The carrier system includes the cytochromes. This process is called cyclic photophosphorylation (Fig. 4.26).

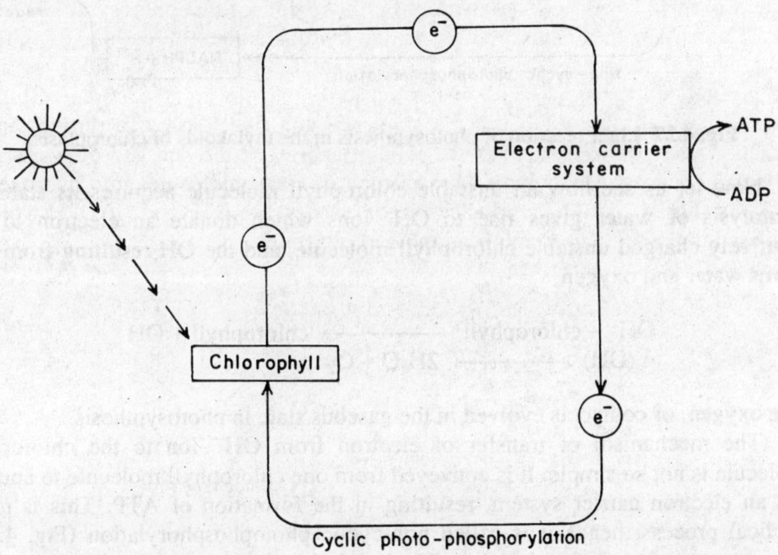

Fig. 4.26 Production of ATP in the light phase of photosynthesis through cyclic phosphorylation.

Like oxidative phosphorylation during respiration, photophosphorylation is concerned with synthesis of ATP but the source of energy is different. In oxidative phosphorylation energy is derived from oxidation of organic compounds, while in photophosphorylation the energy comes from the sunlight. In the light reaction, water splits spontaneously into H^+ and OH^- ions. We have seen that when chlorophyll is excited, electrons are emitted, but not all the electrons are returned to the chlorophyll via the electron carrier system, and sometimes, an electron together with a hydrogen ion, is taken up by a hydrogen acceptor (NADP) which is ultimately reduced to NADPH (Fig. 4.27). This NADPH is the reductant in the dark reaction.

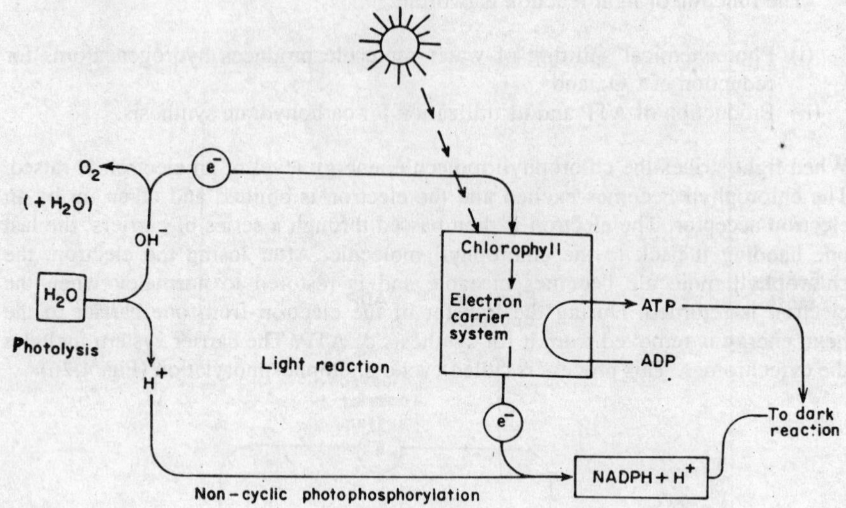

Fig. 4.27 Light reaction of photosynthesis in the thylakoids of chloroplasts.

Now let us see how an unstable chlorophyll molecule acquires its stability. Photolysis of water gives rise to OH^- ions which donate an electron to the positively charged unstable chlorophyll molecule, and the OH resulting from this forms water and oxygen.

$$OH^- + \text{chlorophyll}^+ \longrightarrow \text{chlorophyll} + OH$$
$$4\,(OH) \longrightarrow 2H_2O + O_2$$

The oxygen, of course, is evolved in the gaseous state in photosynthesis.

The mechanism of transfer of electron from OH^- ion to the chlorophyll molecule is not so simple. It is conveyed from one chlorophyll molecule to another via an electron carrier system, resulting in the formation of ATP. This is not a cyclical process, hence it is called non-cyclic photophosphorylation (Fig. 4.28). The passage of electron in non-cyclic phosphorylation is complex, involving two distinct groups of chlorophyll molecules called PS I and PS II. These are at different energy levels, and under the influence of light the energy state of an electron can be raised, enabling it to be transferred from one system to the other. In the course of its journey, the electron passes through a series of electron carriers, generating ATP. This energy conversion process of photosynthesis proposed by R. Hill and F. Bendall, is popularly known as the Z-scheme.

When a chlorophyll molecule in PS I is excited, its energy level is raised and the excitation energy is then transferred to the reaction centre, P 700 from which electrons are expelled for the reduction of NADP to yield NADPH. The PS II with its own set of pigments and reaction centre P 680, is excited by shorter wavelengths and is responsible for oxygen evolution. In green plants PS I and PS II are linked in series. Boosting an electron by excitation of PS I leads to the reduction of NADP via a chain of carriers that include ferredoxin and ferredoxin NADP-oxidoreductase. The electrons required to fill the electron holes in PS I come from the excited PS II via a central electron transport chain, which includes

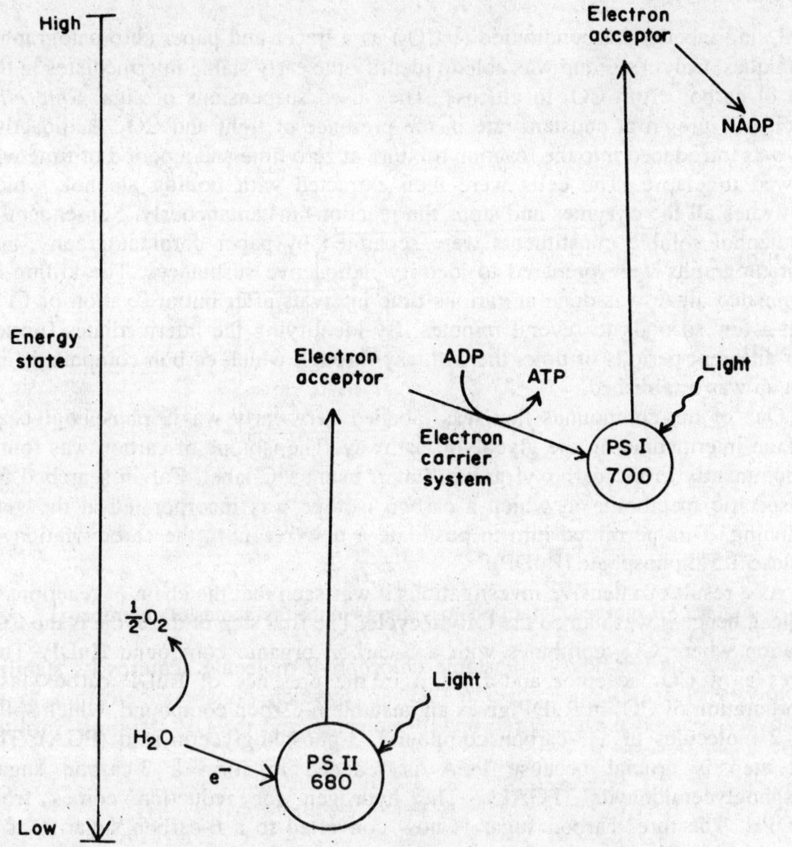

Fig. 4.28 Non-cyclic photophosphorylation (Z-scheme).

plastoquinone, cytochrome *b* 559, cytochrome *f* and plastocyanin. Finally, the electrons required to fill in the electron holes of PS II come from photolysis of water. Phosphorylation of ADP is coupled to the photoinduced electron transport chain between PS I and PS II. At least one molecule of ATP is generated per pair of electrons transferred from H_2O to NADP.

The Dark Reaction

The enzymes and other materials for the reduction of CO_2 to the level of carbohydrate $(CH_2O)_n$ are found in the stroma of the chloroplast where carbohydrate synthesis occurs. Synthesis is an endergonic process requiring energy in the form of ATP and the reducing power from NADPH. Experiments have been carried out to show that isolated chloroplasts can fix CO_2 in the dark and synthesize carbohydrate, provided ATP and NADPH are available.

Reduction of CO_2 and subsequent synthesis of carbohydrate takes place in a number of steps, each of which is catalyzed by an enzyme. These steps were worked out by M. Calvin and his associates in 1947 at the university of Calfornia.

The Calvin Cycle

Using labelled carbondioxide ($^{14}CO_2$) as a tracer and paper chromatographic techniques, Calvin's group was able to identify the early stable intermediates in the path of carbon from CO_2 to glucose. They used suspensions of alga, *Chlorella*, which was grown at constant rate in the presence of light and CO_2. Radioactive CO_2 was introduced into the reaction mixture at zero time and a period of time was allowed to elapse. The cells were then extracted with boiling alcohol, which inactivates all the enzymes and stops the reactions instantaneously. Subsequently, the alcohol soluble constituents were separated by paper chromatography, and autoradiographs were prepared to identify radioactive susbtances. The killing of illuminated algae was done at various time intervals after initial fixation of CO_2, from a few seconds to several minutes. By identifying the intermediates formed after different periods of time, the pathway through which carbon compounds are built up was established.

One of the compounds that was labelled very early was 3-phosphoglyceric acid, an intermediate in the glycolytic pathway. The isotope of carbon was found predominantly in the carboxyl group. Later, using ^{14}C-label, Calvin searched for the specific metabolite in which a carbon isotope was incorporated at the very beginning. This permitted him to postulate a new reaction, the carboxylation of ribulose-1,5-diphosphate (RuDP).

As a result of intensive investigations it was seen that the chain of reactions is cyclical, hence it was named the Calvin cycle. The first step of the cycle is the CO_2 -fixation where CO_2 combines with a 5-carbon organic compound RuDP. This serves as a CO_2 acceptor and fixes it in the presence of RuDP carboxylase. Incorporation of CO_2 in RuDP gives an unstable 6-carbon compound, which splits into 2 molecules of a 3-carbon compound, 3-phosphoglyceric acid (PGA). The next step is crucial because PGA is reduced to form a 3-carbon sugar, phosphoglyceraldehyde (PGAL). The hydrogen for reduction comes from NADPH. The three-carbon sugar is now converted to a 6-carbon sugar, that is glucose, through a number of steps. Although the sequence of reaction shows how one of the carbon atoms of the glucose molecule arises from CO_2, it does not account for the fact that all six carbon atoms of hexose are ultimately formed from CO_2 during photosynthesis. In order to elucidate such a pathway, Calvin and his colleagues proposed a cyclic mechanism of glucose synthesis where one molecule of RuDP is regenerated for each molecule of CO_2 reduced. This implies that not all the 3-carbon sugar is converted into a 6-carbon sugar. In fact, the majority of it enters a series of reactions resulting in the regeneration of RuDP. This is important because by ensuring a supply of RuDP, continued fixation of CO_2 is possible. However, it may be noted that not a single carbon atom is lost in the process.

Fig. 4.29 shows the outline of the Calvin cycle in which one mole of glucose is produced from 6 moles of CO_2 and RuDP. Steps 1 to 7 are concerned with the formation of glucose, while steps 8 to 16 are involved in the regeneration of RuDP. The overall equation for one turn of the cycle, expresssed in terms of CO_2 equivalents is:

$$\text{RuDP} + CO_2 + 3\ \text{ATP} + 2\ \text{NADPH} + 2H^+ + 2\ H_2O \longrightarrow \text{RuDP} + (CH_2O) + 3\ \text{ADP} + 3\ \text{Pi} + 2\ \text{NADP}^+$$

RuDP is written on both sides to show that it is a necessary component which is regenerated at the end of each cycle. After summing up the reactions through six turns of the Calvin cycle (Fig. 4.29) and subtracting 6 molecules of RuDP from both sides of the equation, we have the following:

$$6CO_2 + 18\ ATP + 12\ NADPH + 12H^+ + 12H_2O \longrightarrow C_6H_{12}O_6 + 18\ ADP + 18\ Pi + 12\ NADP^+$$

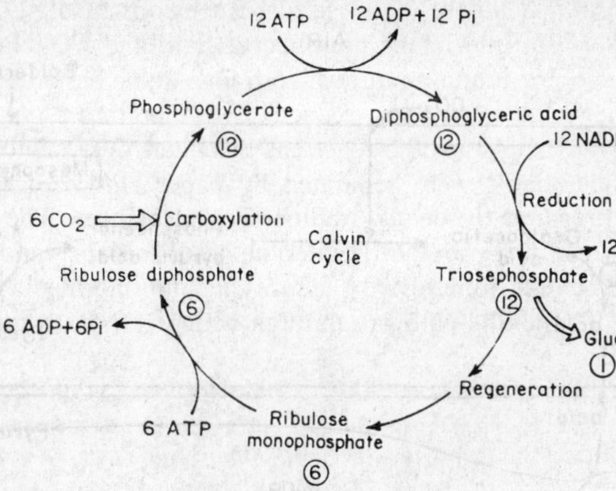

Fig. 4.29 The Calvin cycle.

The Calvin cycle described above explains the mode of photosynthesis in most plants of temperate zones. Since a 3-carbon 3-phosphoglyceric acid is the first metabolite in which the carbon atom of CO_2 is incorporated, the cycle is also known as the C 3 pathway, and such plants are known as C 3 plants.

C4 Pathway

C4 pathway, also called the Hatch-Slack pathway, is a photosynthetic mechanism in which the mesophyll cells of some tropical plants fix CO_2 into 4-carbon acids which are transported into bundle-sheath cells where the acids are decarboxylated and CO_2 is refixed via the C3 pathway. There are about 100 species of C4 plants which include sugar cane, maize, desert plants, crab grass, bermuda grass, and sorghum, in which 4-carbon dicarbxylic acids i.e. oxaloacetic acid and malic acid appear to be the earliest products of CO_2 fixation. Instead of RuDP these plants use phosphoenol pyruvate (PEP) as the substrate for CO_2 fixation in mesophyll cells. The leaves of C4 plants contain two types of photosynthetic cells which have quite different biochemical and structural organization: (1) the bundle sheath cells surrounding the veins, (2) the mesophyll cells loosely arranged around the bundle-sheath cells.

Experiments have been performed in which leaves of such plants were supplied with radioactive $^{14}CO_2$ and it was observed that the immediate product of CO_2 fixation was a 4-carbon compound, oxaloacetic acid instead of PGA. This fixation takes place in the mesophyll cells and oxaloacetic acid is reduced to malic acid (Fig. 4.30). The malic acid so produced is transported to bundle sheath cells where it is decarboxylated to pyruvic acid, and CO_2 reacts with RuDP to yield 3-PGA which is converted to glucose by the Calvin cycle. The pyruvic acid formed in the bundle sheath cells is returned to the mesophyll cells where it is converted to phosphoenol pyruvate at the expense of two high-energy phosphate groups to start another cycle. The overall equation of photosynthesis in C4 plants may be written as:

$$6CO_2 + 30\ ATP + 12\ NADPH + 12\ H^+ + 24\ H_2O \longrightarrow Hexose + 30\ ADP + 30\ Pi + 12\ NADP^+$$

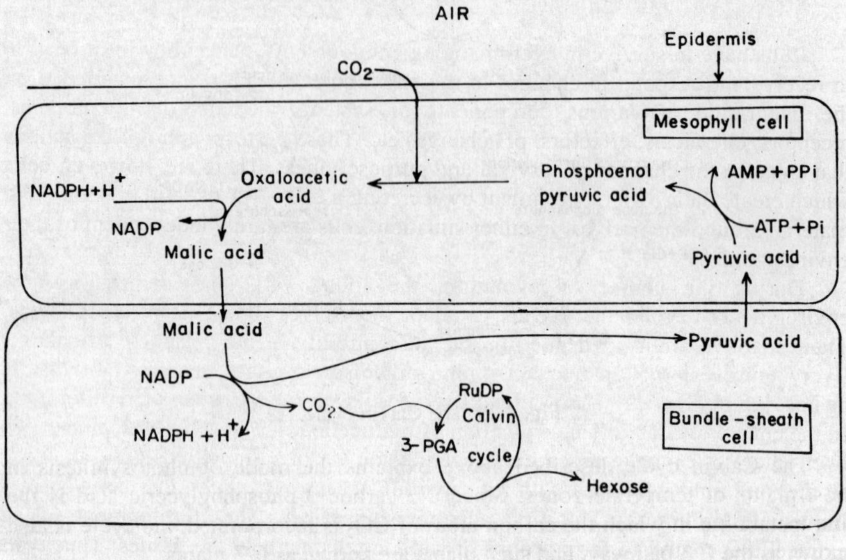

Fig. 4.30 The Hatch-Slack pathway : fixation of CO_2 in C 4 plants.

The C4 plants require more ATP to synthesize a hexose unit as compared to C3 plants, and they also fix CO_2 more efficiently even in low concentrations. The process therfore, provides the plant with a means of building up a store of fixed CO_2 which is subsequently converted into carbohydrate. This is valuable to plants living in an environment of low CO_2 concentration, as in dense tropical vegetation. It is also useful to plants which live in very dry conditions since it permits them to close their stomata during the day time. The C4 system operates efficiently at high temperatures prevailing in tropical regions.

CELL ENVIRONMENT INTERACTIONS

Cells have to survive in ever-changing environments, hence they must be able to receive and respond to changes in the environment. Such responses depend on the integration of various components or systems mediated by membranes, receptors, chemicals, effectors, pH changes etc. These perform appropriate actions that enhance the chances of survival and purposefulness. There are, however, cells which create their own environment by secreting a thin layer of solid or semi-solid material around themselves. In either situation, cells are rarely independent of their environment.

During the course of evolution, the living cells started probing new environments beyond the oceans to adapt themselves to the new surroundings. Their adaptive ability depended on the susceptibility of their genome to change. Every thing external to the cell or the organism is either beneficial or hostile against which the cell has to defend itself for its survival,-growth or proliferation. In the course of adaptation, acquisition of multicellularity was found advantageous over unicellularity, hence cellular specialization was the natural outcome. In doing so they not only interacted with the environment but also with other cells in their immediate proximity. The first accomplishment was the development of an extracellular matrix to form groups of similar cells, namely the tissues. Thus cells created their own environment to become dependent on each other.

Though cells remain embedded in their environment, they retain their identity because of the limiting plasma membrane. We have earlier said that a cell is thermodynamically an open system, it exchanges both matter and energy with its environment. Therefore, every cell is responsive to stimuli or changes in the environment, chiefly mediated through the plasma membrane.

5.1 EXTRACELLULAR ENVIRONMENT

Cells are infact the product of the environment or *milieu exterior* in which they live, but an environment perfectly normal for one set of cells may be lethal to another group of cells. Let's examine the components of this environment.

Natural Environment

Biological components of the natural environment include water, salinity, pressure, temperature, air and radiation.

Water is an absolute requirement of life, and about 80 % of a cell is water. The Major functions of water are :

1. It provides a solvent system for nutrients and metabolites,
2. It acts as an electron donor in a variety of oxidation-reduction reactions such as photosynthesis, and
3. It maintains the structural organization of the plasma membrane by keeping its lipid molecules in a minimum free energy state.

Thus all vital life activities of the cell depend on the availability of water, the absence of which would result in cessation of metabolism, growth and proliferation. Every cell exhibits a sort of *water activity* which is influenced by the degree of absorption of water by its surface, and the extent to which the solute molecules remain hydrated.

Salinity: Most organisms grow at higher water activity (0.9), but many are osmotolerant and osmophilic. For example, *Halobacterium* grows in high Na^+ concentration. The osmotic pressure created by aquatic environments has been solved by cells in two ways, firstly by selecting an environment having salt concentration equal to that of the cell interior (*milieu interior*), and secondly by creating their own environments by changing the composition of body fluids and blood plasma.

The pH: Water molecules dissociate into H^+ and OH^- ions, and any changes in the hydrogen ion concentration will have a drastic effect on the aqueous environment of the cell or the organism. Proteins are highly sensitive to pH changes, and a slight change may result in structural deformity and consequently their inactivation. Most natural environments are acidic. For example, fresh water bodies like lakes, rivers and ponds have a pH range between 5 and 6, which is comfortable for most organisms. The sea water is slighly alkaline (pH 8.0).

Pressure: All aquatic organisms live under pressure exerted by the column of water in which they live. The hydrostatic pressure increases with the increase in depth; for example, an increase of a depth of 10 metres would enhance the pressure by one atmosphere. Animals living in the deep seas have adapted themselves by developing certain counter-measures. They have become *barotolerant* and can withstand high pressures without showing structural deformities. Some enzymes such as lactate degydrogenase and pyruvate kinase are pressure dependent for their activities. There is evidence to indicate that under high pressure, cells of protozoans and multicellular forms lose their characteristic shapes and become spherical.

Air: Air is not a natural environment for unicellular organisms or individual cells. The effect of air on living cells is harsh, it causes desiccation and lowers the water content to a level at which the metabolic activities become almost impossible. This has resulted in various cellular modifications such as evolution of a cell wall as protection against desiccation. In *Mycobacterium* a waxy cell coat is produced to enable these bacteria to survive for months in a dry environment. Many bacteria, protozoans and plant germ cells survive by producing protective cysts. Another adaptation to avoid unfavourable aerial environment is the evolution of multicellularity. Cells aggregate and enclose themselves in a protective cellular coat to avoid desiccation.

Radiation: Radiation forms an extremely lethal environment, especially for micro-organisms. It destroys them. The hazardous component of radiation is the ultraviolet light which photodimerizes the thymine base in the DNA to form a cyclobutane dimer (see Chapter 11). Many micro-organisms develop a low level of a photosensitive enzyme, *photolase*, to overcome radiation hazards. This enzyme is activated in sunlight and catalyzes the reversal of cyclobutane dimer back to thymine. Some lower multicellular and unicellular forms develop specific pigments which act as an anti-radiation screen. Higher forms develop keratinization of epidermal cells, a hairy or feathery covering for protection against radiation. Plants have developed photoabsorbing pigments like chlorophyll *a* and *b* which convert radiant energy into chemical form by synthesizing photosynthetic products.

Temperature: Temperature is an important component of the natural environment which influences cell functions to a great extent. Cells and organisms can comfortably survive in the 10 to 45C range which may be called the *biokinetic* zone. Certain organisms, especially bacteria can however survive extremes of temperature. Bacteria living in high temperature zones are known as *thermophilic*, while those which thrive in subzero temperatures are called the *crinophilics*. The cells that survive at higher temperatures have developed heat stable membranes, enzymes and nucleic acids. Heat stable membranes contain lipids with high melting points. *Thermoplasma*, a bacterium found in hot springs with a temperature range of 70-80C has a membrane with the unusual lipid diglyceroltetraether, which is highly thermostable. Thermophilics synthesize heat stable enzymes which can be put to many industrial uses.

Cell-Created Environment

Living cells inhabiting harsh surroundings create their own environment to protect themselves from the deleterious effects of the natural environment. This is accomplished by development of an extracellular matrix around a group of cells in the form of a collagen and non-collagen substance or a rigid cell wall. The procaryotes and plant cells have highly specialized cell-walls. In Gram-positive and Gram-negative bacteria the cell wall is composed of peptidoglycans forming a complex network of macromolecules, which maintain the shape of the cell and also prevent cell-lysis in hypotonic media. Plant cell-walls are made up of rigid cellulosic material and perform the same functions as suggested for the bacteria.

The formation of an extracellular matrix around multicellular systems has led to the creation of an aqueous environment around cells to protect them from desiccation and radiation damage. In animals, the extracellular matrix is thin and chiefly composed of collagen and non-collagen material. Collagen material is abundant and common in tendons and skin. The collagens are fibrous proteins, formed by a combination of five types of α - helical chains, each of which contains about 33% glycine accompanied by hydroxyproline and hydroxylysine. These are formed by pro-collagen molecules secreted by the cells. Presently five types of collagens have been identified as I, II, III, IV, and V, depending upon the combination of α chains. Collagen I is found in bones, skin and tendons, collagen II in extracellular matrix of cartilage, collagen III in foetal skin and blood vessels, collagen IV in basement membranes, and collagen V in the walls of blood vessels and smooth muscles. Synthesis of all these types of collagens is under genetic control. Collagen increases the ability of cellular attachment, growth and differentiation.

The non-collagen materials are mainly proteoglycans composed of proteins and glucosaminoglycans. Proteoglycans interact with collagen molecules and regulate the process of cytodifferentiation and formation of hard tissues such as bones and cartilage. The extracellular matrix is capable of trapping water molecules in the fibrous network which provide a congenial environment to the cells of a tissue.

The Plasma Membrane

The contents of a cell are enclosed in a limiting membrane, called the plasma membrane. The plasma membrane is a dynamic barrier which regulates exchange of materials between the cell interior and its environment through selective permeability, receives signals, coordinates molecular interactions at the surface such as cell to cell recognition, adhesion and communication.

Chemical Nature of Plasma Membrane: The plasma membrane is composed of 37-40% lipids, 55-60% proteins, and 1-10% carbohydrates. These percentages may however vary from one cell type to another. Membrane lipids are predominantly phospholipids which include phosphatidylcholine, phosphatidyl serine, phosphatidylethanolamine, phosphatidylinositol and sphingolipids. These lipids are amphipathic in nature since they have a hydrophilic or polar head represented by a phosphate group linked to a nitrogenous residue, and a non-polar or hydrophobic hydrocarbon chains. The hydrocarbon chains are two in number, one of which may be unsaturated. The unsaturated chain bends at the site of the double bond forming an unstable lipid aggregate (Fig. 5.1).

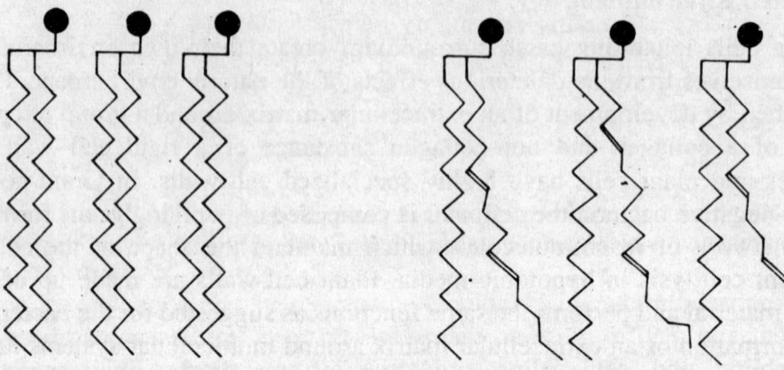

Fig. 5.1 Arrangement of phospholipids : (a) closely packed aggregates; (b) loosely packed aggregates

The other type of membrane lipids are glycolipids which are also amphipathic, but differ from phospholipids in having one or more sugars linked with the polar head. These include cerebrosides and gangliosides. A third kind of lipid present in the membrane is cholesterol, which consists of a polar head formed by a hydroxyl group, a non-polar sterol ring and a non-polar hydrocarbon tail as shown below (Fig. 5.2).

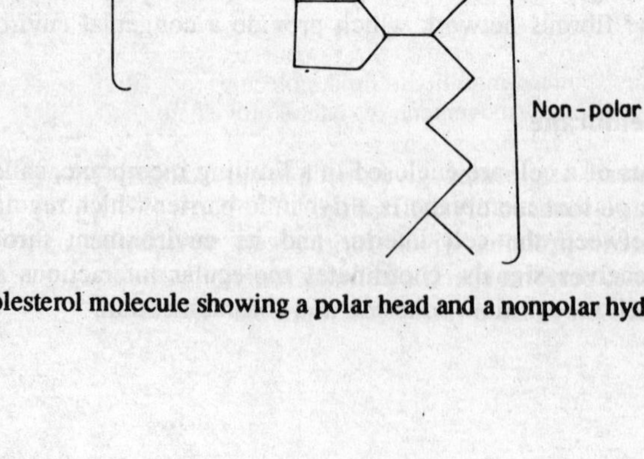

Fig. 5.2 Cholesterol molecule showing a polar head and a nonpolar hydrocarbon tail

Physical Nature of the Membrane

In the aqueous phase the amphipathic lipid molecules tend to orient in such a way that the polar heads are directed towards water and the hydrophobic hydrocarbon chains move away from the water molecules, giving rise to either *spherical micelles* or *lipid bilayers* (Fig. 5.3).

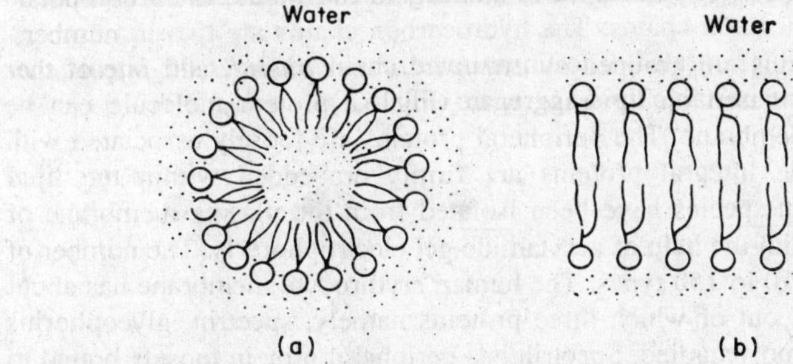

Fig. 5.3 Amphipathic phospholipid molecules in aqueous phase. (a) micelles showing arrangement in which polar heads are directed towards water, (b) lipid bilayer

Most phospholipids and glycolipids spontaneously form bilayers in aqueous environment and tend to close on themselves to form sealed compartments. The self-assembling and self-sealing properties of membrane lipids were first demonstrated by Langmuir in 1917. In 1971, Kornberg and McConnel created artificial lipid micelles, called *liposomes* and suggested that the lipid bilayer is fluid in nature and the lipid molecules exchange places with the adjacent molecules within the monolayer. Besides, the lipid molecules were occasionally shown to migrate from one monolayer to another—a process called the flip-flop phenomenon. The study further concluded that each lipid molecule rotates freely along its axis and its unsaturated hydrocarbon chain shows maximum flexibility at the bilayer centre (Fig. 5.4).

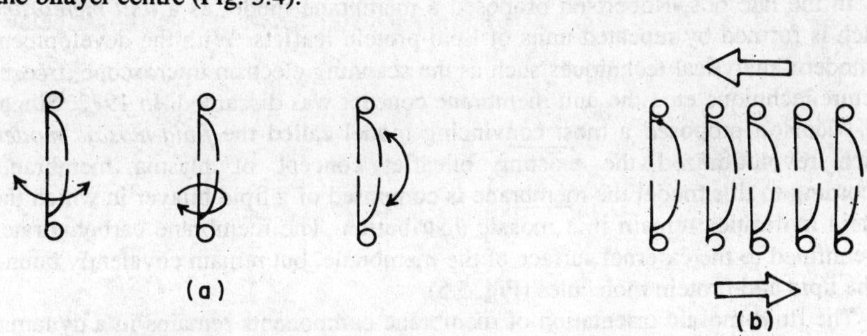

Fig. 5.4 Figure showing mobility of lipid molecules in lipid bilayer. (a) rotational movement, (b) lateral movement.

The cholesterol molecules give mechanical stability to lipid bilayers and inhibit temperature-induced crystallization.

The membrane proteins[1] which form about 50% of the dry weight of the plasma membrane, perform specific cellular functions. Membrane proteins are also amphipathic and they have both hydrophobic and hydrophilic regions. The hydrophilic regions (NH_2^+ and COO^- terminals) remain exposed to the water phase on one or both the surfaces of the membrane, while hydrophobic regions interact with the hydrophobic hydrocarbon chains of lipid molecules in the membrane interior. Like lipid molecules, proteins also exhibit mobility within the lipid bilayer. They show both lateral and rotational movements. Various studies have shown that membrane proteins move laterally to form clusters in response to various stimuli.

Membrane proteins are grouped as *extrinsic or peripheral*, and *intrinsic or integral*, a grouping based on the ease with which a protein molecule can be extracted from the membrane. The peripheral proteins are loosely associated with the membrane, while integral proteins are firmly embedded within the lipid bilayer. Many protein species have been isolated from the plasma membrane of different cell types with the help of acrylamide gel electrophoresis. The number of proteins varies from 10 to 150 types. The human erythrocyte membrane has about 15 types of proteins, out of which three proteins namely, spectrin, glycophorins and band-III are the most studied. Spectrin is a peripheral protein loosely bound to the cytoplasmic face of the membrane. Glycophorin and band-III proteins are transmembrane glycoproteins and hence constitute the integral proteins.

Membrane carbohydrates make up about 2 to 10 % dry weight of the plasma membrane. There are about nine monosaccharides and their derivatives that are commonly found in the plasma membrane and include glucose, galactose, mannose, fucose, glucosamine, galactosamine and sialic acid. These sugars are found as oligosaccharide side chains covalently bound to the membrane proteins and lipids. These oligosaccharide side chains contain 3 to 5 monosaccharides and project from the outer surface of the plasma membrane forming structures called *glycocalyx*. Besides, the surface carbohydrates also participate in cell to cell recognition and immune responses.

Molecular Structure of Plasma Membrane

In the late 60s, Robertson proposed a membrane model as a *unit membrane* which is formed by repeated units of lipid-protein leaflets. With the development of modern analytical techniques such as the scanning electron microscope, freeze-fracture technique etc., the unit membrane concept was discarded. In 1972, Singer and Nicolson proposed a most convincing model called the *fluid-mosaic model*, which revolutionized the existing bileaflet concept of plasma membrane. According to this model the membrane is composed of a lipid bilayer in which the protein molecules remain in a mosaic distribution. The membrane carbohydrates are confined to the external surface of the membrane, but remain covalently bound to the lipid and protein molecules (Fig. 5.5).

The fluid-mosaic orientation of membrane components remains in a dynamic state.

The orientation of phospholipids, sphingolipids and cholesterol molecules in the lipid bilayer imparts an asymmetrical character to it. The phosphatidylcholine and sphingolipids are generally confined to the outer lipid layer, whereas phosphatidylserine and ethanolamine are located in the inner layer facing the cell interior. The membrane asymmetry is perhaps helpful in the flip-flop movement of lipid molecules to maintain proper distribution of proteins. The mosaic distribution of proteins in the membrane is explained by the freeze-fracture technique applied for the erythrocyte membrane. In this technique, the cells are frozen in a block of

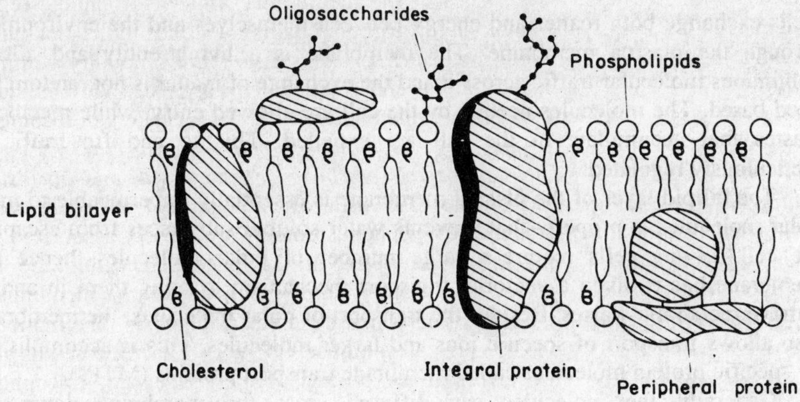

Fig. 5.5 Fluid-mosaic model of plasma membrane showing orientation of phospholipids, cholesterol, proteins and carbohydrates.

ice at about - 196°C and then the ice is fractured in the plane of the membrane. The fracture passes through the middle of the lipid bilayer separating into two monolayers. The fractured face is then shadowed with platinum or carbon, and the organic substances are digested to obtain the fractured faces of the two monolayers. During the fracturing technique, the integral proteins are either pulled out from the bilayer or split from the middle. However, proteins that do not traverse both the monolayers are exposed (Fig. 5.6).

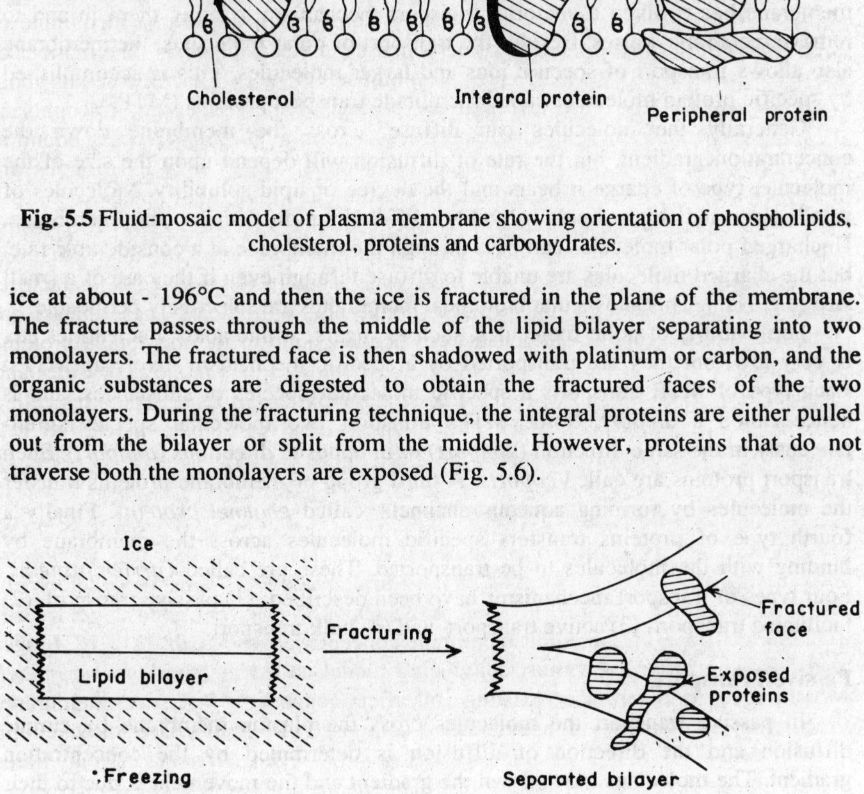

Fig. 5.6 Freeze-fracture technique showing separation of bilipid layer with protein moleeules protruding on fractured surface.

The asymmetric distribution of proteins has been demonstrated by vectorial labelling techniques also using radioactive and fluoroscent tags which cannot penetrate the membrane. The proteins exposed to the outer face of the membrane take the label, which may be identified with the help of polyacrylamide-gel electrophoresis (PAGE).

5.2 MEMBRANE TRANSPORT

Cells exchange both matter and energy between themselves and the environment through the plasma membrane. The membrane is a living entity and allows continuous molecular traffic across it and the exchange of matter is not random but need based. The molecules needed by the cell are allowed entry, while metabolic wastes that accumulate in the cell are expelled. This to and fro traffic is meticulously regulated.

The bilipid layer of the plasma membrane is essentially impermeable to most polar molecules, a property that prevents water soluble substances from escaping the cell. Living cells require a large number of polar molecules, hence the membrane has evolved a vectorial transport mechanism to bring them in and to remove metabolic wastes. Besides the transport of polar molecules, the membrane also allows transport of specific ions and larger molecules. This is accomplished by specific protein molecules called membrane transport proteins (MTPs).

Generally the molecules can diffuse across the membrane down the concentration gradient, but the rate of diffusion will depend upon the size of the molecule, type of charge it bears and the degree of lipid solubility. Molecules of small size and high lipid solubility diffuse rapidly across the membrane. Uncharged polar molecules can pass through the membrane at a considerable rate, but the charged molecules are unable to diffuse through even if they are of a small size (e.g. Na^+). This shows that biological membranes are selectively permeable.

Permeability of polar molecules, such as sugars, amino acids, nucleotides etc. is very low, but they are transported by a specific mechanism involving MTPs. Each type of MTP transports a specific molecular species of substances, and is hence called a *uniport.* Some MTPs transport two molecular species simultaneously in the same direction (*symport*) or in opposite directions (*antiport*). Such transport proteins are called *coports.* A third group of membrane proteins transfer the molecules by forming aqueous channels, called *channel proteins.* Finally a fourth type of proteins transfers specific molecules across the membrane by binding with the molecules to be transported. These are called *carrier proteins.* Four types of transport mechanisms have been described: (1) passive transport, (2) facilitated transport, (3) active transport, and (4) bulk transport.

Passive Transport

In passive transport the molecules cross the plasma membrane by simple diffusion and the direction of diffusion is determined by the concentration gradient. The molecules move down the gradient and the movement is due to their own kinetic energy. The rate of diffusion is, however, limited by the size of the particles and their solubility in lipids. The process does not require any expenditure of metabolic energy.

The water molecules can pass through the membranes by simple diffusion. In fact water molecules continuously move in and out without affecting the shape and size of the cell. When the concentration of solutes on the two sides of the plasma membrane differs, water moves from the region of low solute concentration to that of higher concentration until both sides attain equal concentration. This is called *osmosis.*

Small molecules of non-polar substances rapidly diffuse through the plasma membrane (e.g. CO_2 and ethanol), whereas the charged molecules can not cross the membrane barrier irrespective of their size. However, some essential charged particles may be transported into the cell through specific aquatic channels or pores formed temporarily by channel proteins. The channel may be formed by a single protein or by a cluster of proteins. It has been suggested that temporary channels may be formed by the lipid bilayer due to thermal mobility of phospholipid tails. These channels are referred to as *statistical channels* ((Fig 5.7).

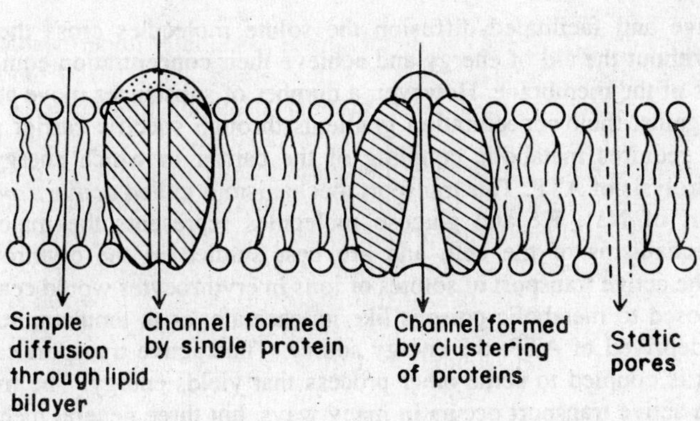

Fig. 5.7 Passive transport by simple diffusion through channel formation

Facilitated Transport

A variety of molecules, such as glucose, amino acids, nucleotides, are transported across the membrane with the help of a carrier molecule, which is invariably a protein. In order to ferry a specific molecule, a special type of carrier protein is involved. The carrier protein is always embedded in the membrane which picks up the substance to be transported and releases it across the other side. The transport of sugars in the erythrocytes has been explained by postulating mobile carriers which can circulate in the membrane (Fig. 5.8).

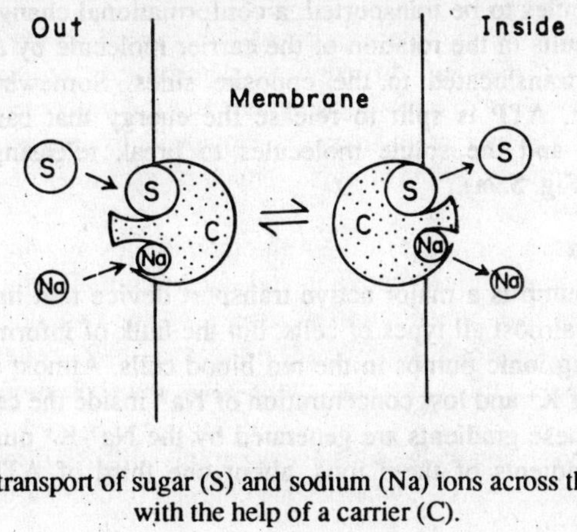

Fig. 5.8 Cotransport of sugar (S) and sodium (Na) ions across the membrane with the help of a carrier (C).

As indicated in the figure, the carrier is loaded with the sugar (S) and then unloads it on the other side of the membrane. Mitchell considers a mobile carrier analogous to an enzyme molecule which binds to the solute at the active site. If the carrier is saturated with solute molecules, the rate of solute transport is maximum, Further, like enzymes, carriers also have conventional equilibrium constants as K_1 and K_3 i.e., dissociation constants, and K_2 and K_4 or distribution constants.

Active Transport

In passive and facilitated diffusion the solute molecules cross the plasma membrane without the aid of energy and achieve their concentration equilibria on the two sides of the membrane. However, a number of substances move across the membrane against their concentration gradients through specific carrier proteins. The process requires metabolic coupling of the carrier in which energy comes from the hydrolysis of ATP. This transport mechanism is called *active transport.*

Transport of Na^+, K^+ and glucose molecules represents the major active transport mechanisms of the cell, and are best studied in the erythrocytes of mammals. The active transport of solutes or ions in erythrocytes would cease when they are exposed to metabolic poisons like iodoacetate, or to temperatures below 4°C, or are deprived of ATP, the energy source. Thus active transport can occur only when it is coupled to some other process that yields energy. The molecular movement in active transport occurs in many ways, but three general mechanisms have been demonstrated to be common. These are: uniport, coport and antiport. In the uniport mechanism, the molecular movement is unidirectional against the concentration gradient at the expense of ATP. Such transports are independent of ionic gradients across the membrane involving only one molecular species. In the coport type, the carrier binding with one molecular species facilitates binding with another molecular species on the same side. Subsequently, the carrier is translocated or rotated to transfer two types of molecules to the other side of the membrane. Transport of glucose is linked with the transport of Na^+ ions, which is like one person hiring a taxi to travel from one place to another, and another person getting free ride without any liability.

In the antiport type of transport, the carrier has binding sites for two different types of molecules on opposite surfaces. For example, if the part of the carrier facing the cytoplasm of the cell has a binding site for molecule A, then the opposite face can bind with the molecule B. As a result of this simultaneous binding of the molecules to be transported, a conformational change in the carrier is induced which results in the rotation of the carrier molecule by about 180°C so that A and B are translocated to the opposite sides. Somewhere during the rotational movement, ATP is split to release the energy that causes the bonds between the carrier and the solute molecules to break releasing A and B in opposite directions (Fig. 5.9a).

The Na^+ - K^+ Pump

Sodium-potassium pump is a major active transport device that maintains proper Na^+-K^+ gradients in almost all types of cells, but the bulk of information has been generated by studying ionic pumps in the red blood cells. Almost all cells have a high concentration of K^+ and low concentration of Na^+ inside the cell as compared to the outside, and these gradients are generated by the Na^+-K^+ pump. In order to maintain relative gradients of these ions, about one third of ATP is consumed while the cell is at rest. In 1957, Jens Skou made a startling discovery that an enzyme ATPase in the nerve cells of the crab is most active in the presence of Na^+ and K^+ ions, and that this enzyme hydrolyzes ATP to form ADP and phosphate to release energy which actively transports these ions. The enzyme was named as Na^+-K^+-ATPase, which is an integral part of the Na^+-K^+ pump. The enzyme always remains inside the cell along with Na^+ and K^+ ions, which continually energize the enzyme. For each molecule of ATP hydrolyzed, three Na^+ ions are expelled from the cell and two K^+ ions are moved in to the cell from outside (Fig. 5.9a). This represents an antiport mechanism.

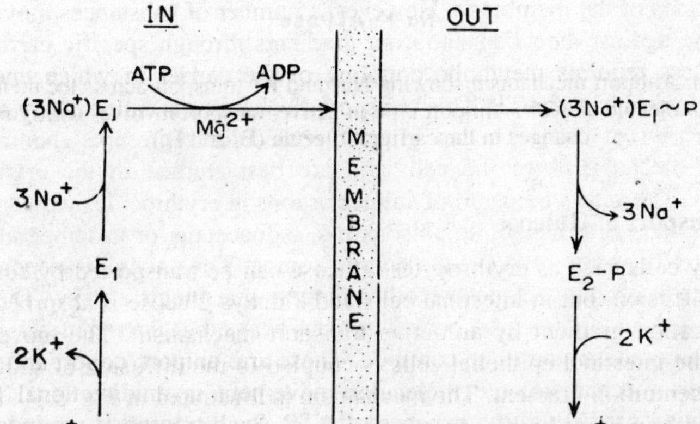

Fig. 5.9(a) Active transport of Na^+ and K^+ through ATPase pump

The mechanism by which sodium and potassium are pumped is quite complicated and is believed to occur in several steps. In 1966,0. Jardetzky proposed a model to explain the Na^+-K^+-pump. Ion transport and ATP hydrolysis are tightly coupled, and there is a transport protein involved which undergoes conformational changes to fulfil different requirements. There are two conformational states E_1 and E_2 which show different affinities for binding the ions. The E_1 state that has the sodium binding site has a cavity facing the inside of the cell, whereas the E_2 conformation has a high affinity for potassium faces outside of the cell. When the carrier protein binds 3 Na^+ ions, it triggers phosphorylation and shifts from the E_1 to the E_2 conformation (Fig. 5.9b).

The binding of potassium to the E_2 conformation triggers dephosphorylation, releasing the sodium ions into the interior of the cell. Thus Na^+ triggers phosphorylation and stabilizes the E_2 conformation, whereas K^+ triggers dephosphorylation and stabilizes the E_1 conformation. After releasing the potassium ions to the inside of the cell, the protein comes back to the E_1 state to repeat the cycle. The Na^+-K^+-ATPase is specifically inhibited by cardiotonic steroids such as ***digitoxigenin*** and ***ouabain***, making the pump inactive.

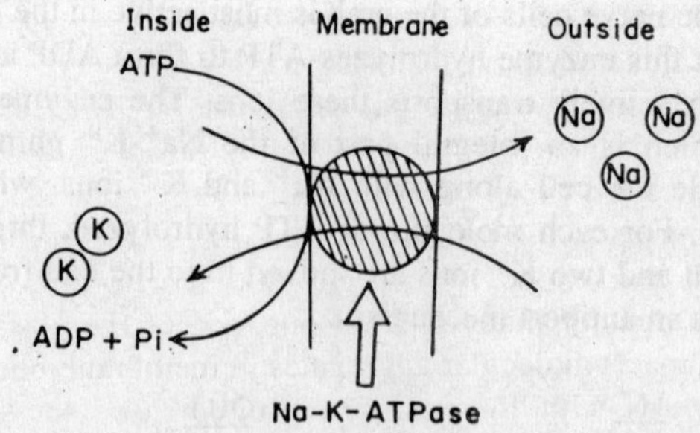

Fig. 5.9(b). Antiport mechanism showing Na^+ and K^+ transport across the membrane utilizing a molecule of ATP. Antiport type of active transport involves configurational changes in the carrier molecule (E_1 and E_2).

Active Transport of Glucose

In many cells such as erythrocytes, glucose can be transported by simple and facilitated diffusion, but in intestinal cells and kidneys glucose is absorbed against the concentration gradient by an active transport mechanism. The movement of glucose in the intestinal epithelial cells is coupled to the diffusion of sodium ions down a concentration gradient. The mechanism is illustrated in Fig. 5.10, in which the carrier protein binds both sodium and glucose. Since sodium ion concentration inside the cell remains very low, the carrier releases the sodium inside thus triggering a conformational change in the protein so that it loses its affinity for glucose. Thus, glucose is released into the cell against the concentration gradient. Once glucose is released, the protein again picks up sodium ions and pumps them back across the membrane to adjust the ionic balance. Glucose is subsequently transported into the blood stream by diffusion.

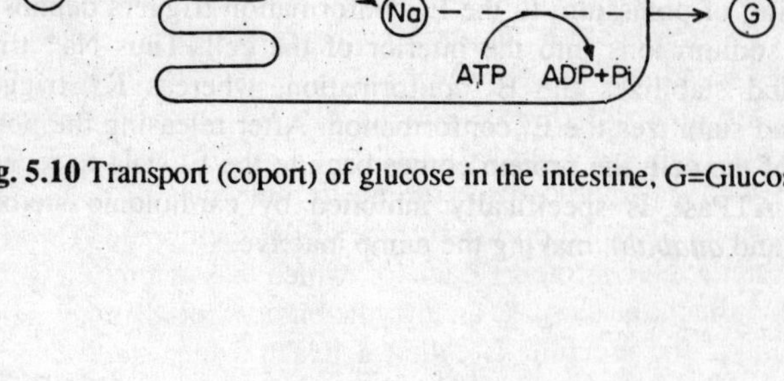

Fig. 5.10 Transport (coport) of glucose in the intestine, G=Glucose

Bulk Transport

Cells have also developed mechanisms to transport across the membranes, larger molecules which otherwise cannot diffuse through the membrane barrier. Generally secretory molecules like proteins and hormones are transported out of the cell by a process called *exocytosis*. Similarly, particulate matter from the cell surface can be transported to the inside of the cell by a process called *endocytosis*. Although exocytosis and endocytosis occur in opposite directions, both these processes essentially transfer material from one side of the plasma membrane to the other by packing macromolecular aggregates in membrane-bound vesicles. The vesicles ultimately fuse with the membrane to release their contents on the opposite surface (Fig. 5.11). Endocytosis is essentially an energy dependent process, whereas exocytosis does not involve any energy consumption.

Practically all eukaryotic cells secrete macromolecules by exocytosis. For example, insulin, secreting pancreatic cells pack insulin molecules in membrane-bound vesicles formed by Golgi complex. The insulin-containing vesicles fuse with the plasma membrane, then open to the cell exterior and release the hormone in the extracellular fluid.

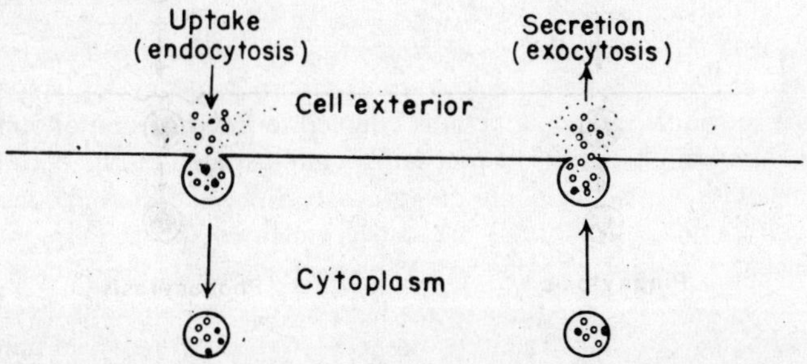

Fig. 5.11 Bulk transport across the plasma membrane, exocytosis and endocytosis.

Endocytosis occurs in two ways: (1) *pinocytosis* in which macromolecules are taken in solution, and (2) *phagocytosis* where particulate matter is taken up without fluid. Sometimes endocytosis involves intake of macromolecules by binding of specific membrane proteins, termed as adsorptive or receptor-mediated pinocytosis. The pinocytic vesicles are of various shapes and sizes, ranging from 50 to 400 nm. These vesicles are formed by invagination of the plasma membrane in the underlying cytoplasm and become pinched off internally. The vesicles then fuse with each other or with other intracellular vesicles, and ultimately fuse with lysosomes leading to enzymic digestion of their contents (Fig. 5.12).

Phagocytosis is a major mechanism for ingestion of large of particulate matter, such as viruses, bacteria, diatoms etc. by protozoans, and in multicellular organisms it is carried out by specific cells such as macrophages, polymorphic leucocytes and lymphocytes. Phagocytes have a variety of membrane receptors which bind the material to be ingested.

In receptor-mediated pinocytosis, the macromolecules to be transported bind with specific receptors located on the surface of the membrane. The receptors are generally membrane glycoproteins each of which has a binding affinity for a specific molecule much the same as an enzyme binds, the substrate molecules. The molecule that binds the receptor is called a ligand which gets internalized in a vesicle and carried through the cytoplasm (Fig. 5.13).

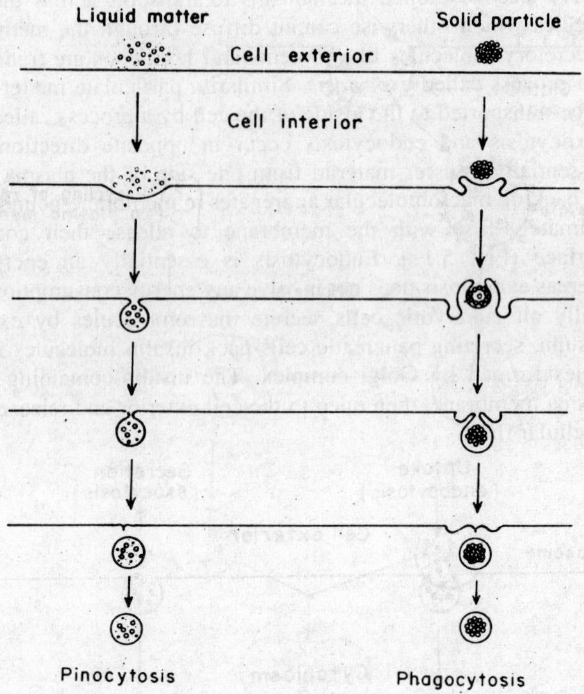

Fig. 5.12 Phagocytosis and endocytosis across the membrane

Generally vesicles are formed in specific regions called coated pits. In coated pits, the plasma membrane is slightly depressed with the receptors projecting into the extracellular fluid. The underlying cytoplasmic surface is coated with fibrous proteins called *coat proteins*, which may vary from one type of pit to another in the same cell or in different cells.

One of the best studied examples of receptor-mediated transport is that of cholesterol. The cholesterol molecules are esterified with long chain fatty acids and then are surrounded by a lipid bilayer with only one protein. The spheres of cholesterol-lipid-protein complex are called low-density lipoproteins (LDL). An LDL-receptor in plasma membrane is a glycoprotein with a terminal galactose unit and is termed a sialoglycoprotein. The LDL-receptor molecules are localized in coated pits and when LDL binds the receptor, a vesicle is formed which is pinched off in the cytoplasm. Subsequently, the vesicle sheds its coat and the receptor is released. The LDL vesicles now fuse with lysosomes and cholesterol molecules are released.

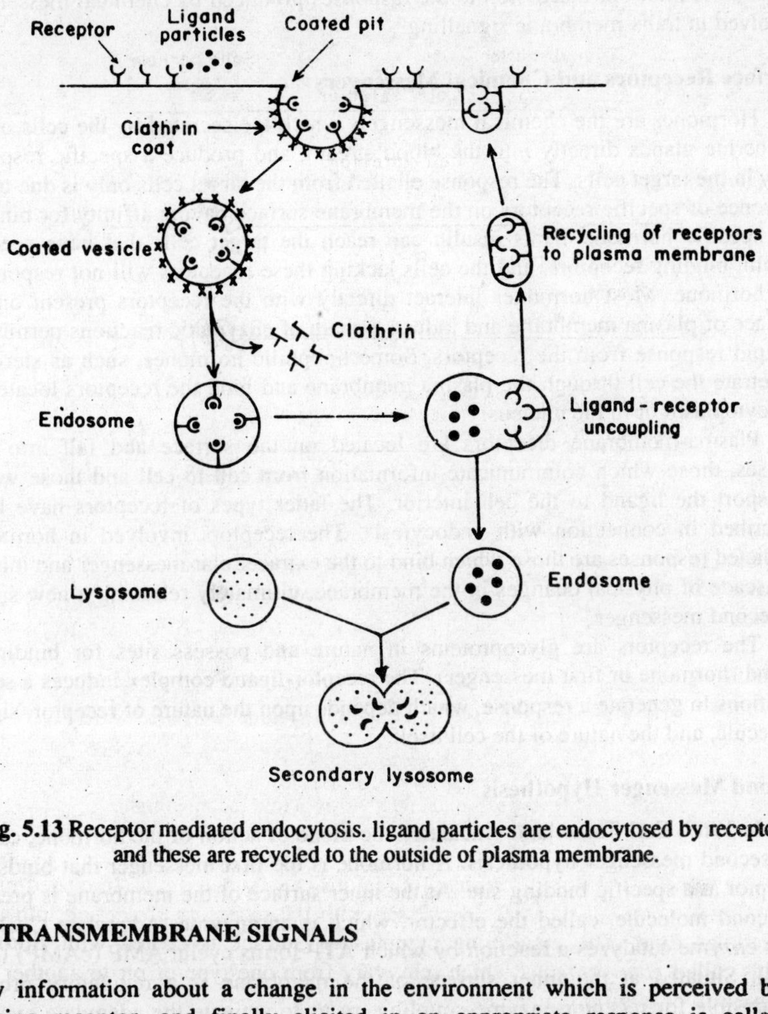

Fig. 5.13 Receptor mediated endocytosis. ligand particles are endocytosed by receptors and these are recycled to the outside of plasma membrane.

5.3 TRANSMEMBRANE SIGNALS

Any information about a change in the environment which is perceived by a sensing receptor, and finally elicited in an appropriate response is called a stimulus. In animals, two systems are responsible for generating responses, one is the system of chemical messengers or hormones, and the other is the nervous system. In the hormonal system, the chemical messages are internally transmitted through diffusion from cell to cell and are ultimately picked up by the target cells or organs, whereas in the nervous system the information is received and communicated over chains of specialized cells, called neurons or nerve cells.

The external environment of an organism consists of many variables, such as temperature, salinity, pH, humidity, pressure, air, light and radiation etc., and all of these factors act as stimuli. These are all, in fact, external or physical stimuli which generate direct responses. For example, the epithelial cells of the skin have undergone a physiological change so as to synthesize keratin in the skin in response to direct exposure to air. Similarly, a change in temperature brings about biochemical and behavioural changes in enzymes and their catalytic properties.

Numerous examples may be cited to show the impact of envirorment on cells, but we shall restrict our discussion to the responses produced by chemical messengers involved in trans-membrane signalling.

Surface Receptors and Chemical Messengers

Hormones are the chemical messengers which are secreted by the cells of the endocrine glands directly into the blood stream, and produce a specific response only in the target cells. The response elicited from the target cells only is due to the presence of specific receptors on the membrane surface having affinity for binding the specific hormone. Thus insulin can reach the target cells that have specific insulin binding receptors, and the cells lacking these receptors will not respond to the hormone. Most hormones interact directly with the receptors present on the surface of plasma membrane and induce a chain of enzymatic reactions permitting a rapid response from the receptors. Some lipophilic hormones, such as steroids, penetrate the cell through the plasma membrane and bind the receptors located in the cytoplasm or in the nucleus.

Plasma-membrane receptors are located on the surface and fall into two classes, those which communicate information from cell to cell and those which transport the ligand to the cell interior. The latter types of receptors have been described in connection with endocytosis. The receptors involved in hormone-mediated responses are those which bind to the extracellular messenger and initiate a cascade of physical changes in the membrane, ultimately releasing a new signal or second messenger.

The receptors are glycoproteins in nature and possess sites for binding a ligand (hormone or first messenger). The receptor-ligand complex induces a set of reactions to generate a response, which depends upon the nature of receptor, signal molecule, and the nature of the cell itself.

Second Messenger Hypothesis

In 1967, Earl Sutherland elucidated the mode of action of the hormone, called the second messenger hypothesis. A hormone is the first messenger that binds the receptor at a specific binding site. At the inner surface of the membrane is present a second molecule, called the effector, which is an enzyme, Adenylate Cyclase. This enzyme catalyzes a reaction by which ATP forms cyclic AMP (*c*AMP) (Fig. 5.14). Situated at the inner surface of the membrane is a regulatory protein responsible for receptor-enzyme coupling, so as to activate the adenylate cyclase molecules. The molecule of GTP is an essential requirement for activation of adenylate cyclase.

Cyclic AMP is the second messenger which diffuses into the cytoplasm and activates a specific enzyme, protein kinase. The activated protein kinase transports a phosphate from a donor to an acceptor molecule, which is also a protein. For example, in liver cells the glucagon induced *c*AMP-dependent protein kinase, when activated, controls the activity of glycogen synthetase and phosphorylase kinase. Phosphorylation of glycogen synthetase prevents the conversion of glucose to glycogen, and the phosphorylation of phosphorylase kinase stimulates glycogenolysis. Cyclic AMP is the second messenger in the action of many hormones and acts inside the cell to control a number of processes. The second messenger can be destroyed by the action of phosphodiesterase.

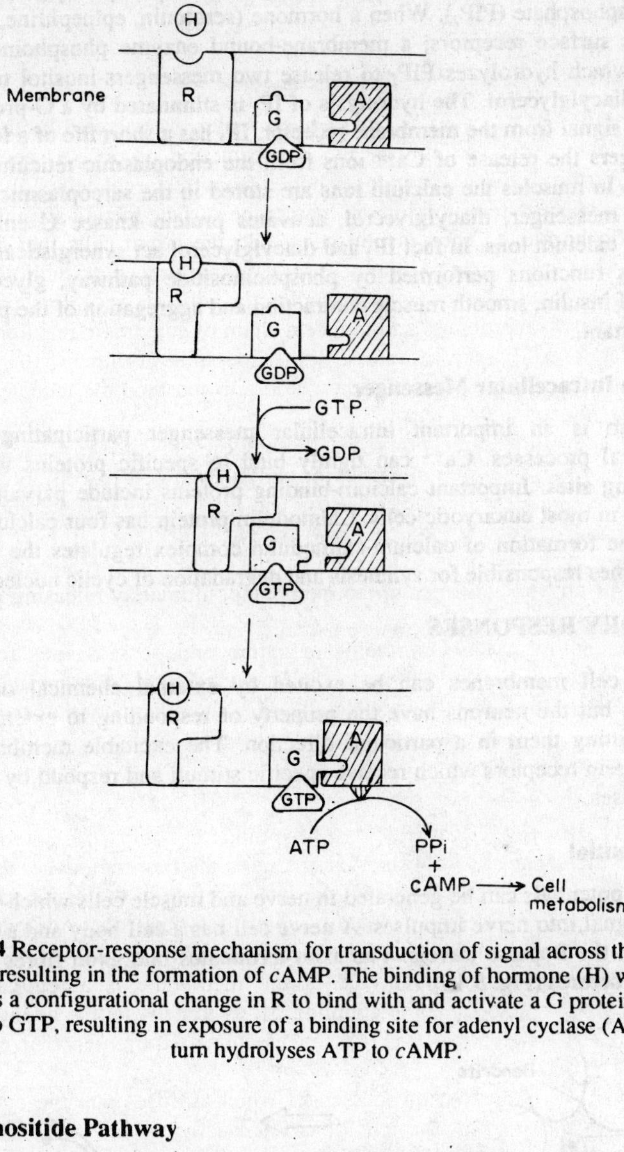

Fig. 5.14 Receptor-response mechanism for transduction of signal across the plasma membrane resulting in the formation of *c*AMP. The binding of hormone (H) with receptor (R) induces a configurational change in R to bind with and activate a G protein. Activated G binds to GTP, resulting in exposure of a binding site for adenyl cyclase (A), which in turn hydrolyses ATP to *c*AMP.

Phosphoinositide Pathway

The adenylate cyclase pathway is a major transducing pathway by which many hormones exert their actions. However, there are other second messengers, such as *c*GMP, Ca^{++}, inositol triphosphate (IP_3) and diacylglycerol, which are involved in the action of many hormones. The two messengers, IP_3 and diacylglycerol are components of phosphoinositide pathway through which the effects of many hormones are mediated. We shall consider this pathway in some detail.

The plasma membrane of many cells contains a phospholipid, phosphatidyl inositol bisphosphate (PIP_2). When a hormone (serotonin, epinephrine, etc.) binds to specific surface receptors, a membrane-bound enzyme phosphoinositidase is activated which hydrolyzes PIP_2 to release two messengers-inositol triphosphate (IP_3) and diacylglycerol. The hydrolysis of IP_3 is stimulated by a G-protein which carries the signal from the membrane receptor. IP_3 has a short life of a few seconds but it triggers the release of Ca^{++} ions from the endoplasmic reticulum into the cytoplasm. In muscles the calcium ions are stored in the sarcoplasmic reticulum. The other messenger, diacylglycerol, activates protein knasee C enhancing its affinity for calcium ions. In fact IP_3 and diacylglycerol act synergistically. Among the various functions performed by phosphoinositide pathway, glycogenolysis, secertion of insulin, smooth muscle contraction and aggregation of the platelets are most important.

Ca^{++} as an Intracellular Messenger

Calcium is an important intracellular messenger participating in many physiological processes. Ca^{++} can tightly bind to specific proteins which have Ca^{++} binding sites. Important calcium-binding proteins include parvalbumin and calmodulin in most eukaryotic cells. Calmodulin protein has four calcium binding sites and the formation of calcium-calmodulin complex regulates the activity of many enzymes responsible for synthesis and degradation of cyclic nucleotides.

5.4 SENSORY RESPONSES

Nearly all cell membranes can be excited by external chemical or physical stimulation, but the neurons have the property of responding to external signals and transmitting them in a particular direction. The excitable membranes have specific protein receptors which receive specific stimuli and respond by producing nerve impulses.

Action Potential

Action potentials can be generated in nerve and muscle cells which modify an electrical signal into nerve impulses. A nerve cell has a cell body and a long axon along which the impulse travels. The axon terminates into axon fibres ending in bulb-like structures (Fig. 5.15).

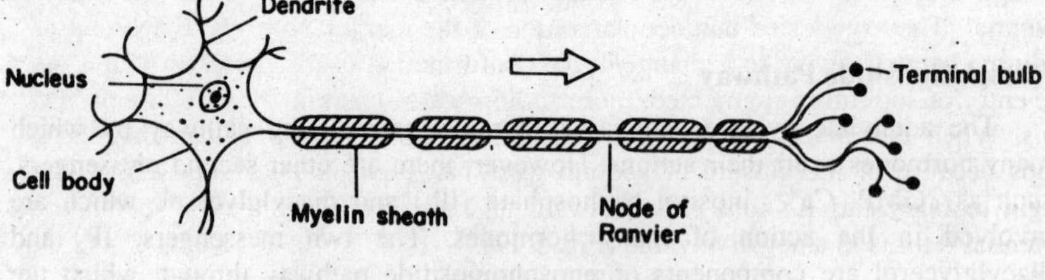

Fig. 5.15. A Neuron with myelinated axon

The terminal bulbs are contiguous with the main cell body of another neuron, forming a junction called synapse (Fig. 5.16).

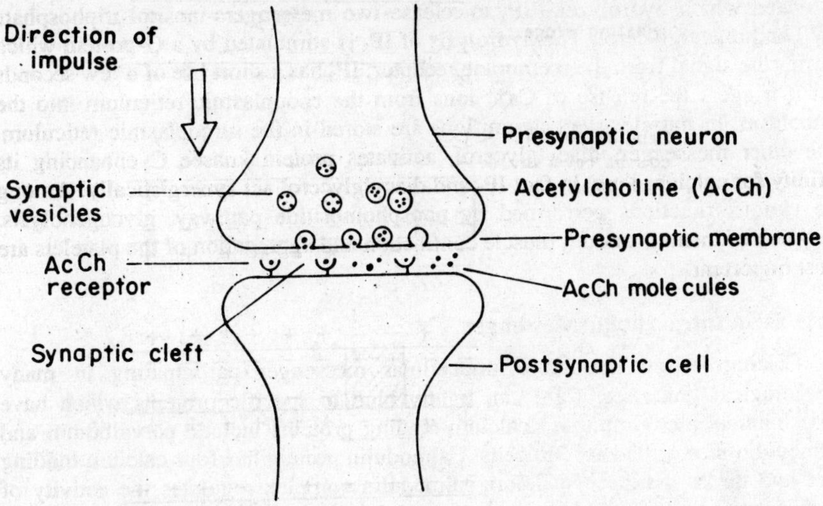

Fig. 5.16 Diagram of a synapse showing release of acetylcholine during transmission of a nerve impulse.

A nerve cell, like any other cells, has a high concentration of K^+ ions inside it than Na^+ ions. Outside the cell, Na^+ ions are in a higher concentration. In a resting state the axon membrane is more permeable to K^+ ions. These ionic gradients are maintained by the Na^+-K^+- ATPase pump. In the resting state the membrane potential is estimated to be about -75 mV, and the inside of the cell membrane is negatively charged with respect to outside. When the nerve cell is stimulated, the membrane gets depolarized and generates an action potential. At this stage the inside of the membrane is positively charged with respect to outside and the membrane becomes permeable to Na^+ which start moving in from outside until the membrane potential reaches the level of about a +30 mV, i.e. the threshold level. The reversal of membrane polarity due to rapid influx of Na^+ ions generates an action potential which has to be propagated down the axon to another nerve cell through a synapse. Hodgkin and Huxley explained the mechanism of the action potential. They suggested that depolarization of the membrane leads to opening of sodium channels owing to a change in the conformation of membrane proteins. As the entry of sodium is augmented, more sodium gates open to allow inflow of Na^+ ions. When the membrane potential reaches a level of +30 mV, sodium inflow stops because of the closure of sodium gates. At this juncture, potassium gates begin to open and K^+ ions flow outward and as a consequence the membrane potential returns to a negative value (Fig. 5.17).

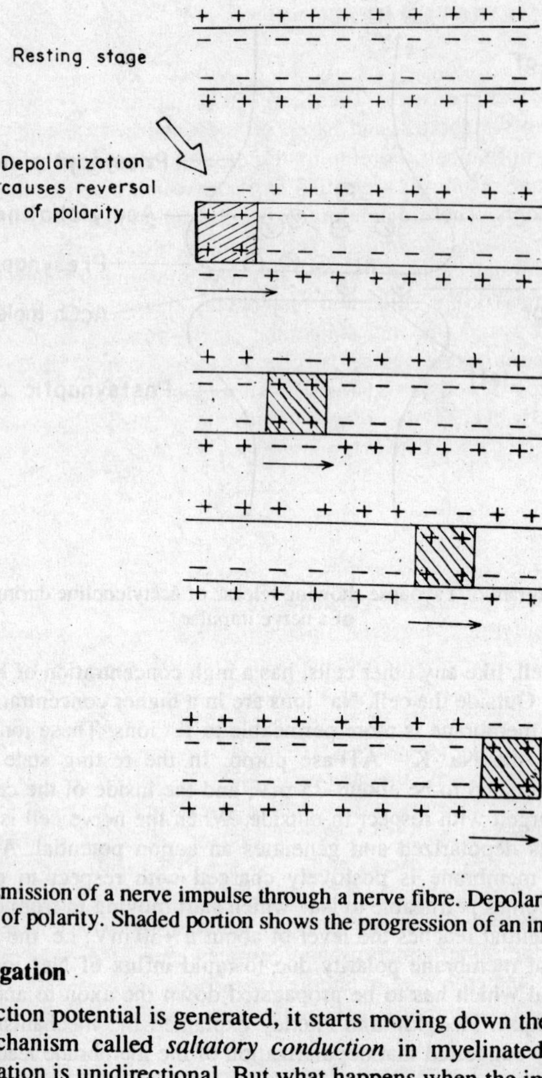

Fig. 5.17 Transmission of a nerve impulse through a nerve fibre. Depolarization causes reversal of polarity. Shaded portion shows the progression of an impulse.

Impulse Propagation

Once the action potential is generated, it starts moving down the length of the axon by a mechanism called *saltatory conduction* in myelinated axons. Thus impulse propagation is unidirectional. But what happens when the impulse reaches the axon terminal? The nerve cells communicate with other nerve cells through specialized junctions called *synapses*. Each axon terminates into a number of fibres, each of which ends up in a bulb-like structure which is apposed to the membrane of another nerve cell body. A synapse is in fact a gap junction between two nerve cells with a narrow synaptic cleft, about 500Å wide. When the impulse arrives at the end of the presynaptic axon, a neurotransmitter acetylcholine is released in the synaptic cleft from the synaptic vesicles. The acetylcholine molecules diffuse to the postsynaptic membrane and bind the receptors located in the membrane to allow the wave of depolarization to the postsynaptic cell. In this way the action potential is propagated from one nerve cell to the other. Once the wave of depolarization has crossed the synapse, the neurotransmitter acetylcholine is hydrolyzed and the resting potential of the membrane is restored.

5.5 ANTIGEN MEDIATED RESPONSES

Antigen mediated responses are meant to protect animals from invasions of microorganisms and parasites. These are called immune responses which are highly evolved in vertebrates, and based on these the body is able to recognize a foreign material or non-self. Foreign materials, cellular or non-cellular, capable of generating immune responses are called *antigens* which give rise to molecules that destroy the invaders. These defending molecules are called *antibodies*, which are glycoproteins or *immunoglobulins*. The immune responses are generated by a special category of cells, called *lymphocytes* which fall under four types: B cells, T cells, NK or natural killer cells, and phagocytes. These cells respond to foreign invaders in two ways: in one mechanism the B cells produce immunoglobulins that remain in circulation and recognize specific antigens, in the second mechanism the antigen stimulates T cells to synthesize antibodies that remain attached to the surface of the cells. The former mechanism is known as humoral response and the latter one is called cell-mediated response.

Antigen

An antigen is a substance that can elicit an immune response, and its antigenic potency depends upon its molar mass, complexity with reactivity, degradability and foreignness. In general, large molecules are potent antigens which include complex heteropolymers, bacterial lipopolysaccharides, and serum proteins. All immune responses are antigen mediated, but if the invading molecule is inert e.g. plastic particles an immune response cannot be generated. The antigen molecule has specific areas on its surface through which it binds the antibody molecule. These antibody binding sites are called antigenic determinants or *epitopes*.

Many small molecules, though foreign, cannot initiate an immune response, but once a response is generated they actively react with antibodies. These small molecules are called *haptens*. For example, a drug molecule, such as paracetamol may not be immunogenic but can be made so if it is chemically linked to a large antigenic molecule, which is generally a protein. A hapten-linked carrier protein will generate a primary immune response, and a secondary response will be generated when the same hapten-protein complex enters the body after a lapse of time. If the same hapten bound to another carrier protein enters the second time in the same body, it will not evoke a secondary response, but will be recognized as a different antigen.

Some common antigens are bacterial antigens, viral antigens and cell-surface antigens and include glycoproteins, polysaccharides, lipopolysaccharides, peptidoglycans etc. Bacterial flagellin antigens, also called *H-antigens*, are potent antigens. Usually all viral capsid proteins behave as antigens.

Antibodies

Antibodies are specific glycoproteins called immunoglobulins (Ig) that are produced in response to an immunogenic reaction. These are highly antigen-specific. An immunoglobulin consists of two identical heavy chains (H) and two identical smaller or light chains (L) organized in a Y - shaped molecule (Fig. 5.18).

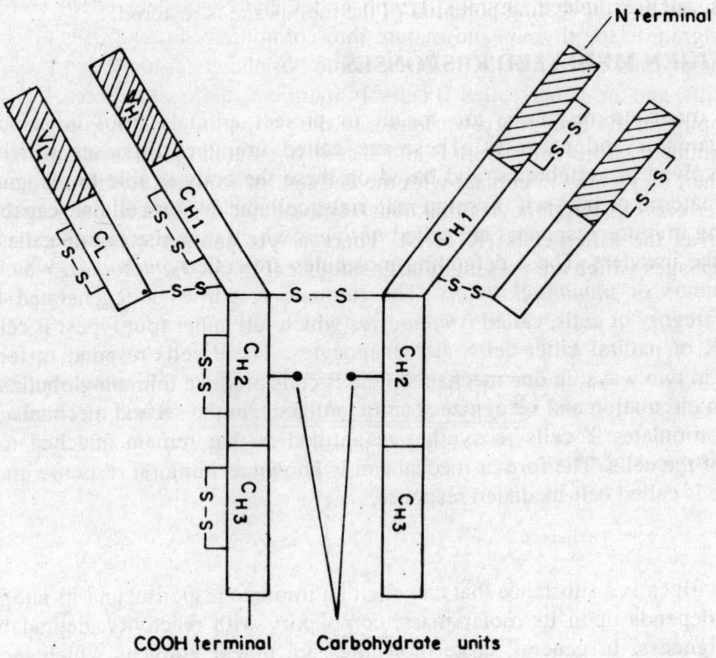

Fig. 5.18 Structure of immunoglobulin, IgG type

Further, each H-chain is held with the other H-chain by disulphide bonds. The region of H-chain pairing with the L-chain forms the forked arm of the Y. Both H and L chains have a constant (C) and a variable (V) region. Five types of immunoglobulins have been recognized : IgA, IgD, IgE, IgG and IgM, out of which IgG forms the major blood antibodies. These antibodies are produced in large numbers during a secondary immune response.

The synthesis of Ig molecules is under genetic control. There are about 10^6 types of Ig molecules, each of which can bind to a specific antigen. This specific antigen-antibody binding is due to conformational fitness between antigen surface epitopes and the binding site on the antibody surface. When an epitope and an antibody combine, they interact through chemical groups on the surface of the epitope and on variable L and H chains of the Ig molecule.

The antibodies themselves are not capable of destroying the antigen, but they bind with the antigen to block its entry or preclude its binding with the receptors on its target cells. However, the antigen-antibody complex activates the complement complex, phagocytes and natural killer cells to destroy the antigen. The complement complex, or also simply called the complement, is formed by 20 different plasma proteins, factors B and D, and a variety of regulatory proteins.

Cellular Components of Immune System

The immune system in a ver. brate body is composed of lymphocytes, a class of white blood corpuscles, which are produced from precursor stem cells in the bone marrow. From there the lymphocytes migrate to the peripheral lymphoid tissues, such as spleen, thymus, lymph nodes and liver sinusoids. Lymphocytes that migrate to the thymus and mature into committed immune cells are called T cells or thymus derived cells. In birds, some lymphocytes migrate to the Bursa of Fabricius, and are hence called B cells. In mammals there is no bursa, but the cells developing in the bone marrow migrate to the peripheral lymphoid organs and get differentiated into B cells. Both B and T cells circulate in the blood and the lymph, and when they come in contact with the antigen, an immune reaction is generated. Another class of lymphocytes that attack and kill the invader by phagocytosis, are known as the killer cells (K cells). There is yet another class of cells, called macrophages which are present in large number (Fig. 5.19).

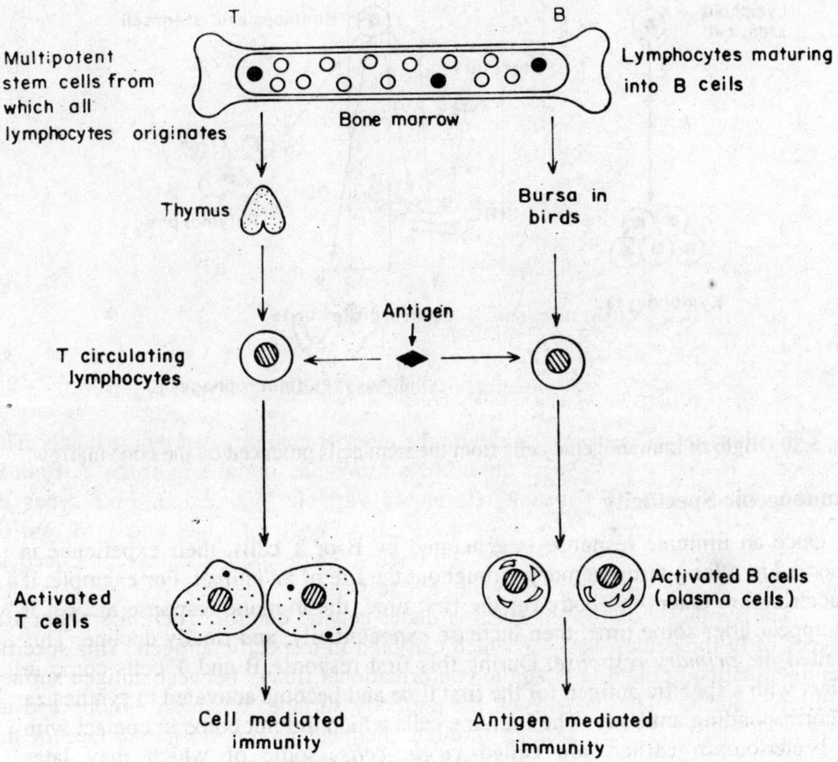

Fig. 5.19 Development of T and B cells from multipotent stem cells

Macrophages are mononuclear lymphocytes originating from the bone marrow and enter the blood stream in an immature form called monocytes. These however mature into macrophages in various tissues and lymph nodes, and are attracted towards the antigen-antibody complex on a chemical basis. All macrophages do not react with the antigen-antibody complex, but those with histocompatible antigens on their surfaces can only react with antigens. On activation, macrophages secrete a host of molecules including lysozyme, collagenase, elastase etc. They also secrete a special class of proteins called *interleukins*, which are responsible for stimulating response against injury. The origin of various types of immune cells is given in Fig. 5.20.

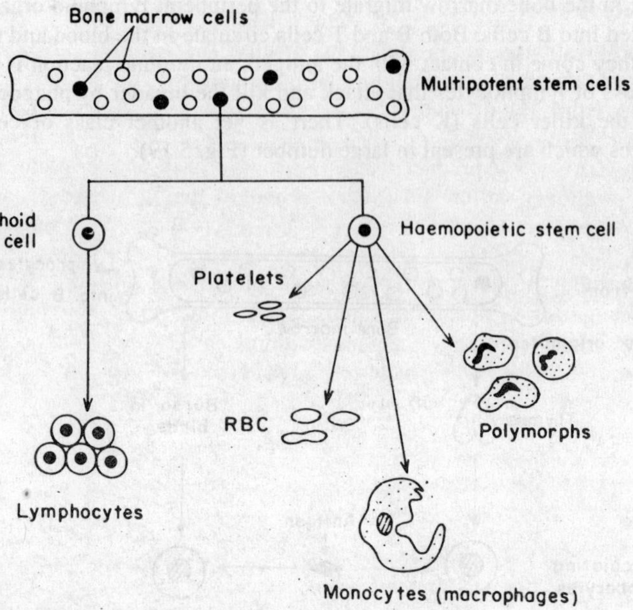

Fig. 5.20 Origin of immunogenic cells from the stem cells produced by the bone marrow.

Immunogenic Specificity

Once an immune response is generated by B or T cells, their experience is preserved as a long-term memory throughout the life of an animal. For example, if a bacteriotoxin enters the body for the first time, the immune response against it will appear after some time, then increase exponentially, and finally decline. This is called the *primary response*. During this first response B and T cells come in contact with a specific antigen for the first time and become activated to synthesize the corresponding antibody. The immune cells which did not come in contact with the bacteriotoxin earlier, are called *virgin cells*, some of which may later participate in the destruction of the bacteriotoxin. However, some virgin cells that dc not participate in an immune reaction at all are retained as memory cells, and do so when confronted with a bacteriotoxin.

The antibodies are antigen specific, i.e. for every antigen a specific antibody would be available. This high antigen specifcity is explained on the basis of the *clonal selection theory*. According to this theory, each lymphocyte in the bone marrow is committed to react with a specific type of antigen even before it is exposed to it. This commitment is due to specific surface receptor proteins having antigen complimentary binding sites. Thus, when a virgin cell encounters an antigen, it undergoes division a number of times to produce a clone of differentiated memory cells. The antigen selectively activates the virgin cells which possess complimentary receptors. The formation of a large number of memory cells from a single virgin cell is called clone expansion. In a vertebrate there are millions of different antigen-specific clones.

Humoral Responses

Humoral responses are produced by B cells. The virgin B cells have antigen-specific receptors which after binding with the antigens differentiate into effector and memory cells. The effector B cells synthesize large numbers of antibodies and release them in the blood stream. Free antibody molecules bind the antigens against which they are produced and form an antigen-antibody complex in the plasma. The antigen-antibody complex provokes the formation of complement complexes and activates NK cells, T cells and macrophages to destroy the antigen (Fig. 5.21).

The antibody itself is not capable of destroying the antigen. In fact an antibody recognizes its specific antigen, traps it to prevent its binding with target cells and ultimately presents it to other components of the immune system (i.e. complement, NK cells, macrophages etc.) for destruction. The antibody molecules have antigen trapping fragment or Fa, and another fragment Fc which binds the carbohydrate moiety. In an antigen-antibody complex, the Fc fragment is free and recognized by the receptors on NK cells, macrophages and complement etc. for destroying the antigen.

Cell Mediated Responses

Cell mediated immune responses are generated by T cells which are of three types : (1) cytotoxic T cells (Tc cells), helper cells (Th cells), and suppressor T cells (Ts cells) Tc cells are directly involved in trapping and destruction of antigens and may themselves be destroyed in the process. The T-helper cells activate other components of the immune system, especially macrophages, B cells, and NK cells to destroy the antigens. The T-suppressor cells are involved in the suppression of Tc and Th cells and also of macrophages as well as B cells. Thus Th and Ts cells perform a regulatory function and form the regulatory control mechanism for both humoral and cell mediated responses. The macrophages are phagocytic and engulf the antigenic materials.

The T cells and B cells cannot be easily differentiated on the basis of their size and form. They are, however, differentiated by their surface proteins. B cells have antibodies on the surface as well as a specific protein called the SB protein. T cells have surface proteins called Thy-I. The T cells recognize antigen receptor proteins on the surface of other cells.

The cytotoxic cells (Tc) defend the body from viral and bacterial infections. They react against viral or bacterial antigens that are expressed on the infected cell-surface before viral multiplication starts. Tc cells kill the virus infected cells to prevent viral multiplication. T-helper cells do not secrete antibodies but cooperate with B cells, Tc cells and Ts cells. Helper cells respond to an antigen by secreting substances such as *lymphokines* and *interleukins*. Lymphokines do not recognize or react with antigens but prepare B and T cells for immunologic reactions. According to a working hypothesis for B cell activation, the Th cells on recognizing a foreign antigen on the surface of an infected cell, secrete an interleukin to interact with B and T cells. The B cells recognize the hapten on the

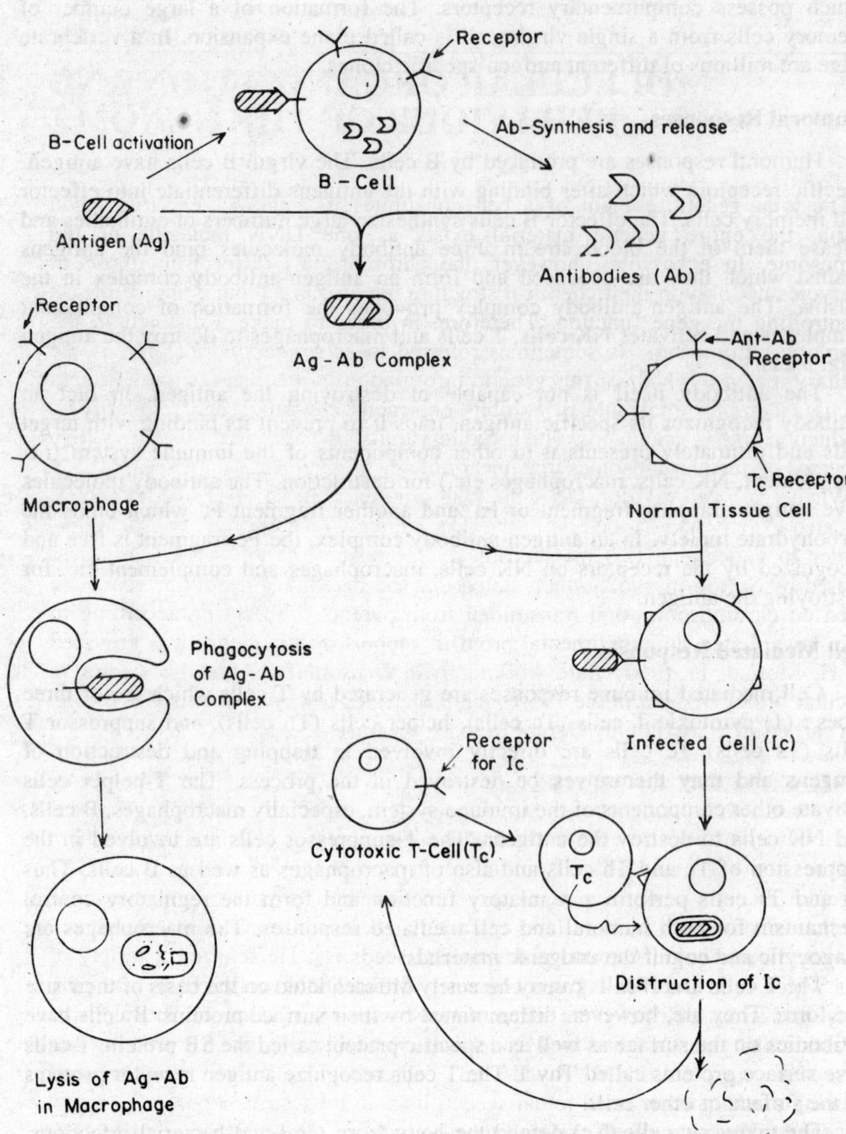

Fig. 5.21 Antigen provokes B cell to synthesize antibodies, which actively participate in the immune respones.

carrier molecule, whereas a Th cell recognizes the carrier on the B cell receptor-heptencarrier complex and stimulates B cells to synthesize and secrete antibodies.

The T-suppressor cells suppress the response of both B and T cells to the antigens. They do so only when they are continuously provoked by Th cells. In the process the Ts cells themselves are suppressed.

CHAPTER 6

MOLECULAR ORGANIZATION AND BEHAVIOUR OF THE GENOME

The total genetic content of a cell constitutes its genome. In viruses and bacteria, the entire genetic material exists as one DNA molecule or the chromosome. In eukaryotes, however, the chromosomes differ from those of prokaryotes not only in their chemical structure but also in having a unique means of controlling the gene function. Therefore, in order to make a distinction, the prokaryotic chromosome is sometimes referred to as a *genophore*. DNA is the hereditary material which carries genetic information in a coded fashion. Though DNA is a complex molecule, it exhibits enormous constancy and regularity in reproduction among organisms. The hereditary traits are controlled by genes which are located on the chromosomes in a linear order.

Whenever a cell divides, DNA-the blue print carrying a set of instructions, is duplicated exactly and handed on to the daughter cells without any modifications. In order to understand this mechanism, a chromosome theory of heredity was proposed by Sutton in 1902, which stated that the hereditary factors or genes are located on chromosomes and transmitted from parents to offspring according to certain laws. Later, an experimental proof in support of this theory was provided by T.H. Morgan in 1910 while working with *Drosophila*. Although studies in molecular biology began in the 1950s, the basic rules of heredity were discovered much earlier in 1856-1866 by Gergor Mendel, an Austrian monk.

6.1 THE CONCEPT OF MENDEL

Mendel began his experiments in 1856 with the garden pea, *Pisum sativum*, which could be easily grown and self-pollinated. It is possible to cross-pollinate the flowers manually without damaging the seed forming ability of the flowers. He concentrated on a few, well defined characters, such as tall and dwarf plants, red and white flower colours, round and wrinkled seeds etc. He selected a variety of plants which bred true for a given character. Mendel took pollen from one pure variety and cross-pollinated an other variety with them. One of the pairs of contrasting characters studied by Mendel was round versus wrinkled-seed bearing plants. He cross-pollinated round-seed bearing plants with wrinkled-seed producing plants and obtained round-seeded lines in the first generation (F1). The first generation plants, when self-pollinated, produced the second generation (F2) in which original parental types reappeared. His count of the two types of plants in F2 generation was 5474 round and 1850 wrinkled, representing a ratio of 3 to 1. Out of these, plants bearing wrinkled-seeds are pure homozygous for the given character, while the round-seeded lines were either homozygous or heterozygous (hybrids). In order to determine whether the plant is pure or hybrid, Mendel crossed round-seeded lines with wrinkled-seeded plants. If the plants are heterozygous, two types of progeny would be produced in equal numbers (round and wrinkled seed producing plants). This type of cross is termed a *test cross* or back cross. In this way Mendel studied seven pairs of alternative characters and obtained remarkably uniform results.

Monohybrid Cross. In his breeding experiments, Mendel found that certain traits, if present in one of the parents, appear in all offspring in the next generation. These traits were called *dominant* and indicated in capital letters. On the contrary, some traits were expressed in the offspring only when present in both the parents were marked in small letters and designated as *recessive*. For example, when a pea plant with red flowers was crossed with another white flowered pea plant, all plants in the F1 generation were red-flowered. Further, when F1 plants were interbred, the resulting plants of the second generation (F2) bore red and white flowers in the ratio of 3:1. In this case red flower colour is dominant over white colour (Fig. 6.1).

From the above results Mendel concluded that each plant carried a pair of factors that governed the inheritance of each trait. He suggested that during gamete formation the paired factors separate in such a way that each gamete receives only one of the factors. A zygote is formed by the fusion of male and female gametes to restorc the pair of factors. Mendel called this as the *Law of Segregation*. When the zygote develops into an adult, the trait may be expressed in two ways: (1) if one of the factors is stronger, it will express itself by masking the expression of the other factor, and (2) if the two factors are identical, they will express themselves as one trait. Based on these observations, Mendel postulated another law, called the *Law of Dominance*.

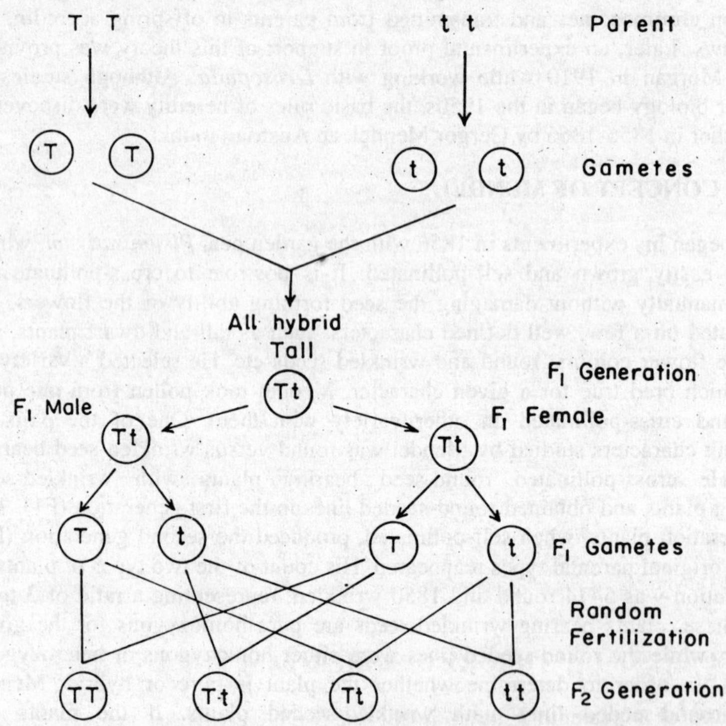

Fig. 6.1 The results of Mendel's monohybrid cross between tall (T) and dwarf (t) plants with tall (TT or Tt) and dwarf (tt) plants in F_2 generation.

Mendel further experimented with two or more than two traits. For example, he took pure breeding red-flowered tall plants (RRTT) and crossed them with white-flowered dwarf plants (rr tt). The results of F_1 and F_2 generations revealed that all F_1 plants were red-flowered and tall, but F_2 generations plants were of four types: red-flowered, tall; white-flowered and tall; red-flowered dwarf; and white-flowered dwarf in the ratio of 9 : 3 : 3 : 1 (Fig. 6.2). These results suggested that the flower colour did not remain associated with the height of plants. These observations led Mendel to suggest another law called *Law of Independent Assortment* which states that during gamete formation factors of two different traits segregate independently without interference of other factors.

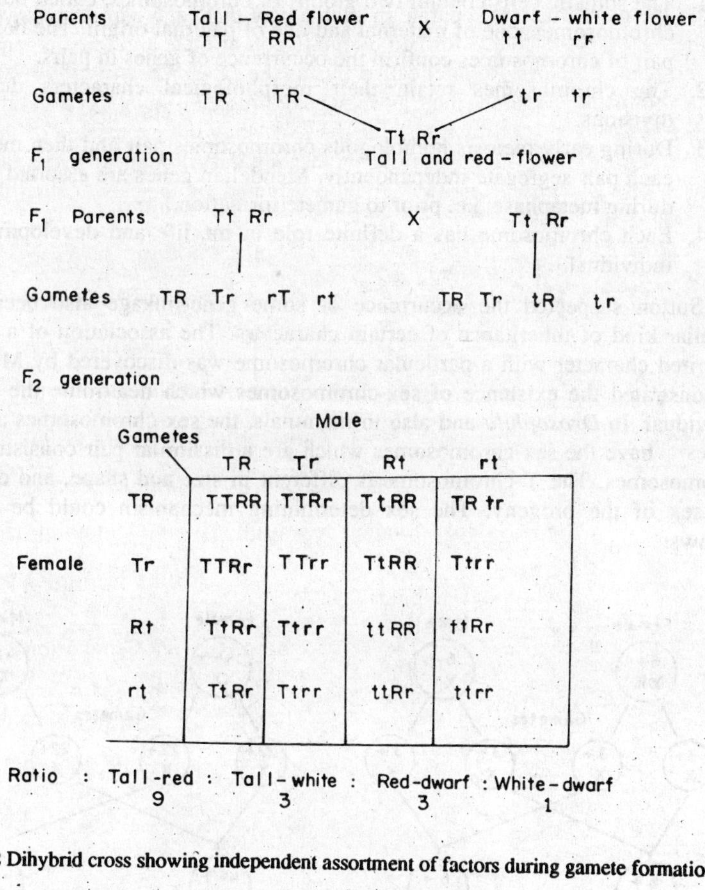

Fig. 6.2 Dihybrid cross showing independent assortment of factors during gamete formation.

6.2 CHROMOSOMAL THEORY OF INHERITANCE

By the end of nineteenth century it became well known that every cell in an organism is formed from the division of pre-existing cells. Walter Fleming in 1880 described *mitosis*, a somatic cell division, during which process the chromosomes of the parental cell splits longitudinally into two to produce two daughter chromosomes, each of which passes on to the daughter cells. Subsequently, it was discovered in sexually reproducing multicellular organisms that a special kind of cell division, know as *meiosis*, occurs to achieve a reduction in the number of genes and chromosomes to one half. This division occurs in the sex cells to produce gametes. In meiosis the sex cells first undergo a reduction division to form two daughter cells with haploid number of chromosomes, followed by another division of chromosomes resulting in four haploid daughter cells.

The relationship between chromosomes and genes was suspected when Mendel's laws were rediscovered. Walter Sutton, a graduate student at Columbia university, began to study the relationship between meiosis and Mendelian factors, and in 1930 advanced a hypothesis correlating gene and chromosome transmission known as the Sutton-Boveri Hypothesis. The observations are as follows:

1. The somatic cells contain two groups of chromosomes, called homologous chromosomes, one of maternal and one of paternal origin. The homologous pair of chromosomes confirm the occurrence of genes in pairs.
2. The chromosomes retain their morphological characters during cell divisions.
3. During early meiosis homologous chromosomes pair and then members of each pair segregate independently. Mendelian genes are assorted randomly during metaphase, i.e. prior to gamete formation.
4. Each chromosome has a definite role in the life and development of an individual.

Sutton suspected the occurrence of some gene-linkage also because of a peculiar kind of inheritance of certain characters. The association of a particular inherited character with a particular chromosome was discovered by Morgan. He demonstrated the existence of sex-chromosomes which determine the sex of an individual. In *Drosophila* and also in mammals, the sex-chromosomes are paired. Males have the sex-chromosomes which are a dissimilar pair consisting of XY chromosomes. The Y-chromosome is different in size and shape, and determines the sex of the progeny. The sex determining mechanism could be shown as follows:

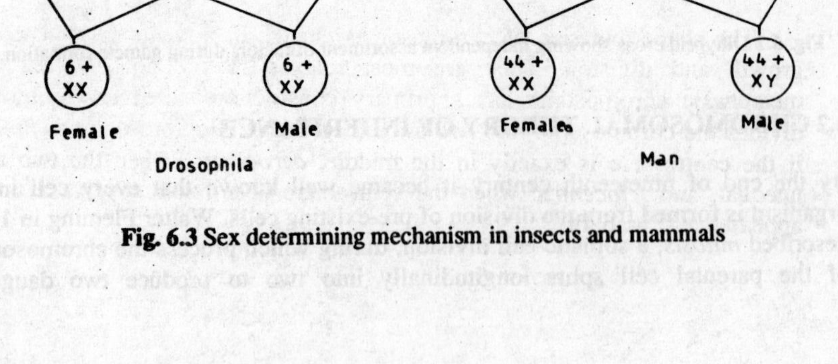

Fig. 6.3 Sex determining mechanism in insects and mammals

Sutton also pointed out that the number of genes far exceed the number of chromosomes, hence many genes would not show free recombinations. He concluded that some kind of linkage of genes on chromosomes existed, a fact confirmed by Bateson and Punnet on the basis of experiments conducted on sweet peas. Subsequently, it was shown that genes have a linear arrangement on chromosomes whose loci could be mapped by three-point crosses. If the genes lie on the same chromosome, they recombine through a process of crossing-over involving actual breakage and rejoining of chromatids during meiosis.

6.3 STRUCTURE OF CHROMOSOMES

Thread-like structures present in the nucleus of eukaryotic cells were named as chromosomes, a term coined by Waldeyer in 1888. In 1924, by specific staining technique, Feulgen and Rossenbeck demonstrated that deoxyribonucleic acid (DNA) is localized in the chromosomes. It is now firmly established that the chromosomes are the carriers of genes and the total number of genes present in a haploid or single set of chromosomes in a cell is called *genome*. In other words, all the chromosomes of the cell nucleus put together constitute the genome of an organism. In order to understand the behaviour of chromosomes and the mode of gene transfer, a look at the gross structure would be desirable.

Gross Structure

When viewed under a light microscope, chromosomes appear in a highly condensed state during the metaphase of cell division. The metaphase chromosome retains a specific and characteristic shape so that its morphological details can be best described. Each species has a fixed number of chromosomes. (Table 6.1).

Table 6.1 Chromosome numbers found in some organisms

Organisms	*Chromosome number*
Prokaryotes (haploid)	
Bacteria	1
Eukaryotes (diploid)	
Drosophila melanogaster	8
Ascaris megalocephala	2
Cat	38
Mouse	40
Rat	42
Man	46
Chicken	78

The shape and size of chromosomes change in a cyclic fashion during cell growth and division. They are most relaxed in the non-dividing cells. A metaphasic chromosome has a primary constriction called *centromere*, which divides the chromosome into two arms (Fig. 6.4a). A chromosome is *metacentric* if the centromere is exactly in the middle, *acrocentric* when the two arms are unequal, and *telocentric* when the centromere is terminal to give a one-armed appearance (Fig. 6.4b).

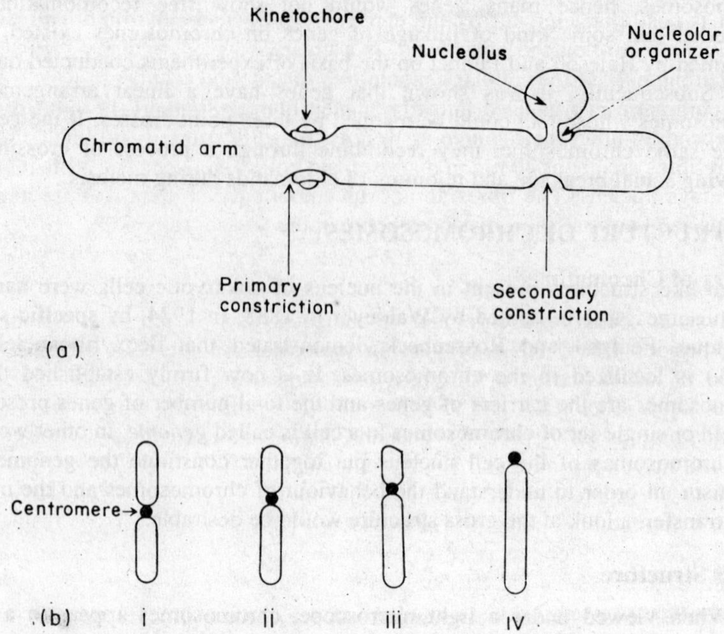

Fig. 6.4 Diagramatic representation of (a) the gross anatony of chromosome and (b) types of chromosomes on the basis of centromere locations, viz metacentric (i), sub-centric (ii), Telocentric (iii) and acrocentric.

The centromere and the kinetochore are two related regions, the former representing a primary constriction and the latter located at the centromere as a three-layered structure in most cases. The kinetochore is the site for attachment of spindle fibres where force is exerted on the chromosome to pull it towards opposite poles. The kinetochore is believed to be rich in repetitive DNA, which would permit its division when the chromatids are pulled apart. In many chromosomes a secondary constriction appears near the end, which is a non-staining region called secondary constriction. This secondary constriction is the nucleolar organizing region. The nucleolus is involved in protein synthesis.

Euchromatin and Heterochromatin

Each chromosome is made up of two halves or chromatids, which are connected to the centromere. Each chromatid is composed of chromatin material organized into parallel fibres. The term chromatin is generally applied to the DNA-containing substance, differentiated into *heterochromatin* and *euchromatin* regions on the basis of staining reactions. Heterochromatin is a deeply staining region and consists of two regions: constitutive heterochromatin having no structural genes and inactive in transcription; the other region is facultative heterochromatin that contains structural genes that are genetically inactive and cytologically condensed.

The constitutive heterochromatin is stable, condensed and late replicating in nature. It is best characterized by the chromatin present around the centromeric region. Euchromatin, on the other hand, is a greatly relaxed, actively replicating type containing bulk of the cellular genome.

The chromatin is organized into fibres, their diameter ranging between 3 and 25 nm depending upon their source as well as the stage of the life cycle. Human chromatin fibres are about 20 nm in diameter and collectively occupy 20% volume of the average nucleus. The fibres do not run smooth and straight, but are looped into bumps and knobs at short distances between the smooth runs.

Chemistry of Chromatin

Analysis of eukaryotic chromosome indicates that the chromatin material contains, besides DNA, RNA of varying molecular weights and proteins, the proportion of which varies in different organisms. Human chromatin contains about 15% DNA, 10% RNA and 75% proteins. The proteins belong to two major classes: *histones* or basic proteins, and *non-histone* proteins or acidic proteins. Histones are rich in arginine and lysine and constitute five types: H1, H2a, H2b, H3 and H4. These are low molecular weight molecules bound to the DNA by electrostatic interactions. The primary structure of all five histones is highly conserved through millions of years of evolution, which is suggestive of their role in structural and functional integrity of the chromatin.

Non-histone proteins occur in hundreds of varieties, about 2-4% of these are bound to homologous DNA. About half of non-histone proteins are structural and the remaining are functional proteins. some important functional proteins include DNA polymerase, RNA polymerase, serine protease, acetyl transferase and various other factors involved in replication, translation and regulation of gene activity.

Organization of chromatin Fibres

The chromatin fibre of a eukaryotic chromosome is formed by a linear duplex DNA molecule complexed with histones. In 1974, Kornberg showed that a chromatin fibre is formed by a series of repeating beaded units, now called *nucleosomes* Fig. 6.5. Thus the nucleosomes form subunits of chromatin. A nucleosome consists of a core particle of 10nm diameter with a *linker* fragment of about 2 nm. On an average, a nucleosome contains 200 base pairs, but may vary from 154 to 260. The number of base pairs in the DNA fragment associated with the core DNA is always 146 irrespective of the source. The core particle has a central histone octamer formed by a pair of each of H2a, H2b, H3 and H4 histones. The core DNA fragment of the DNA helix twists around the histone octose in about two loops and continues as linker DNA (Fig. 6.5) The total length of core DNA is about 67 nm, but becomes condensed into a length of 11 nm due to the loop formation.

C-value Paradox

The amount of DNA in the genes, is species specific and is called the *C-value*, expressed in picograms (pg) or base pairs or daltons. The Discrepency between the size of the genome and morphological complexities of the species is called the C-value paradox.

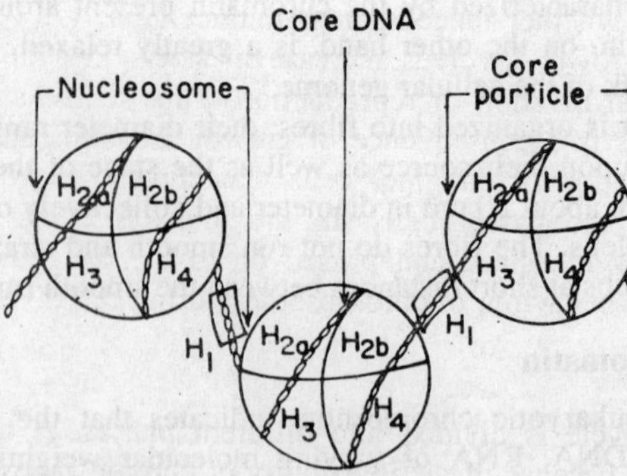

Fig 6.5 The structure of nucleosomes suggesting relationship between DNA and histone core particle consisting of histones H2a, H2b, H3 and H2 and also linker role of H1.

Arrangement of Genes

The chromosomes have a linear arrangement of genes, which represent specific lengths of DNA. Each gene (a specific length on one strand of the helix) is divided into coding and non-coding regions. The coding region carries a nucleotide sequence with hereditary information, flanked on either side by non-coding regions. The non-coding region upstream is organized into a *promotor* region with a set of seven nucleotides in the middle, called the *Pribnow Box* (Fig. 6.6). Between the promotor and the coding sequence is a short stretch of nucleotides called *initiator site* which signals transcription of the coding sequence. The non-coding region down stream has a set of nucleotides called the *terminator* which terminates transcription. The promotor allows binding of the enzyme for transcription, the initiator gives the start signal for transcription of the coding region, whereas the terminator acts as a stop signal for transcription.

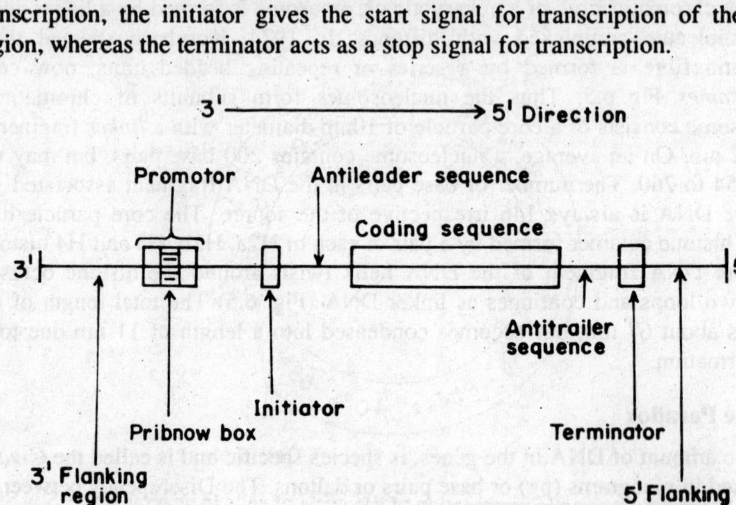

Fig. 6.6 Schematic representation of the functional organization of gene(s) in general.

6.4 CHROMOSOME DIVISION AND BEHAVIOUR

All organisms, whether unicellular or multicellular, grow and reproduce. At the cellular level growth involves increase in size and synthesis of cellular mass, followed by cell division or reproduction. Thus reproduction or increase in the number of cells is a consequence of growth. In multicellular organisms growth involves production of daughter cells that contain exactly the same hereditary information as the parent cells. In the division process copies of the DNA molecules are synthesized and passed on to the daughter cells. Our interest is focussed on the behaviour of chromosomes during cell division.

The Cell Cycle

The cell cycle is divided into different phases. A non-dividing cell has a resting nucleus because of the invisibility of chromosomes with usual stains. This stage is often termed *interphase* or more correctly the metabolic stage, because during this time the chromosomes are very active in transcribing RNA for protein synthesis and DNA replication prior to chromosome division. In the actively dividing cells the interphase nucleus is engaged in DNA synthesis prior to the next mitotic division.

All living cells repeatedly grow and divide in a cyclic fashion—a process called cell cycle. Four stages are recognized in the cell cycle (Fig. 6.7), namely the mitosis (M), DNA synthesis (S) phase, presynthetic (G_1) and postsynthetic (G_2) periods. The duration of the cell cycle and its constituent phases vary from cell to cell and are affected by certain metabolic features such as respiration and nuclear DNA content. A newly produced daughter cell enters a period of interphase involving synthesis of proteins, lipids, carbohydrates and other cell contents. This is the G_1 phase which varies in duration in different cells. It is very short in rapidly dividing cells, lasting only a few minutes, and longest in differentiated cells. In the cells which enter the divisional cycle, DNA molecules begin to duplicate. This

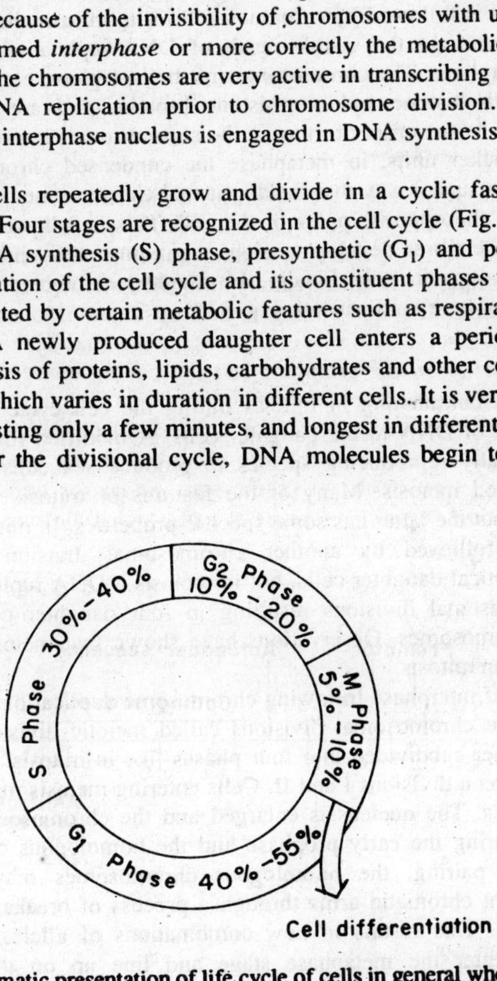

Fig. 6.7 Diagramatic presentation of life cycle of cells in general where S is Synthetic phase, G1 and G2 are Gap I and II, and D representing divisional phase along with generalized range (%) of each phase in the life-Cycle of Cells in general.

phase of DNA synthesis is called the S phase, which lasts for 6-8 hours in eukaryotes. The S phase is followed by the G_2 phase during which RNA synthesis occurs to produce proteins essential for division. The G_2 phase is generally of 2-5 hours duration, followed by the mitotic phase. The entire cell cycle may be completed within 8 to 12 hours, but in some plants it may last for as long as 17 to 32 hours.

Cell Division

The DNA doubles in amount prior to mitosis and meiosis. Mitosis involves extensive visible changes in the cytoplasm and chromosomal structure. Although mitosis is a continuous process, it is divided into four phases: prophase, metaphase, anaphase and telophase, which is the longest, the others being progressively shorter. Mitosis is followed by cytokinesis. Prior to mitosis the two chromatids of each chromosome are the visible result of DNA replication that has occurred during the interphase. The chromosomes of the interphase contain DNA subunits complexed with histone and non-histone proteins and are visible in a most extended form. At the onset of mitosis, the chromosomes in prophase condense in shorter and thicker units, in metaphase the condensed chromosomes attach to spindle fibres in such a way that the duplicated chromosomes show attachment to the fibres from the opposite poles of the cell. This is followed by the anaphase, when the spindle fibres develop tension, separating each chromatid and pulling each one to the opposite pole. Finally, the daughter chromosomes decondense and become enclosed in separate nuclei (Fig. 6.8).

Meiosis

The events culminating in mitosis during the cell cycle ensure transfer of identical copies of DNA to the daughter cells. A modified form of cell division occurs in sexually reproducing species to produce sex cells or gametes. This division is called meiosis. Many of the features of mitosis are shared by the meiotic cycle, but the latter has some specific properties. In mitosis a single DNA replication is followed by another chromosomal division to produce two genetically identical daughter cells, but in meiosis a DNA replication is followed by two chromosomal divisions resulting in four daughter cells with half the number of chromosomes. Observations have shown that meiosis generally takes much longer than mitosis.

After a brief interphase following chromosome duplication, meiosis proceeds in two complete chromosomal divisions called meiotic division I and II. Each division is further subdivided into four phases like in mitosis. There is no DNA replication between divisions I and II. Cells entering meiosis may have prolonged S and G_2 phases. The nucleus is enlarged and the chromosomes condense into thicker units during the early prophase and the homologous chromosomes start pairing. While pairing, the homologous chromosomes may exchange parts between adjacent chromatid arms through a process of breakage and reunion of reciprocal parts. This results in new combinations of alleles. Now the paired chromosomes enter the metaphase stage and line up on the spindle in the equatorial plane. This is followed by anaphase when the spindle fibres develop tension and pull the homologous chromosomes towards the poles.

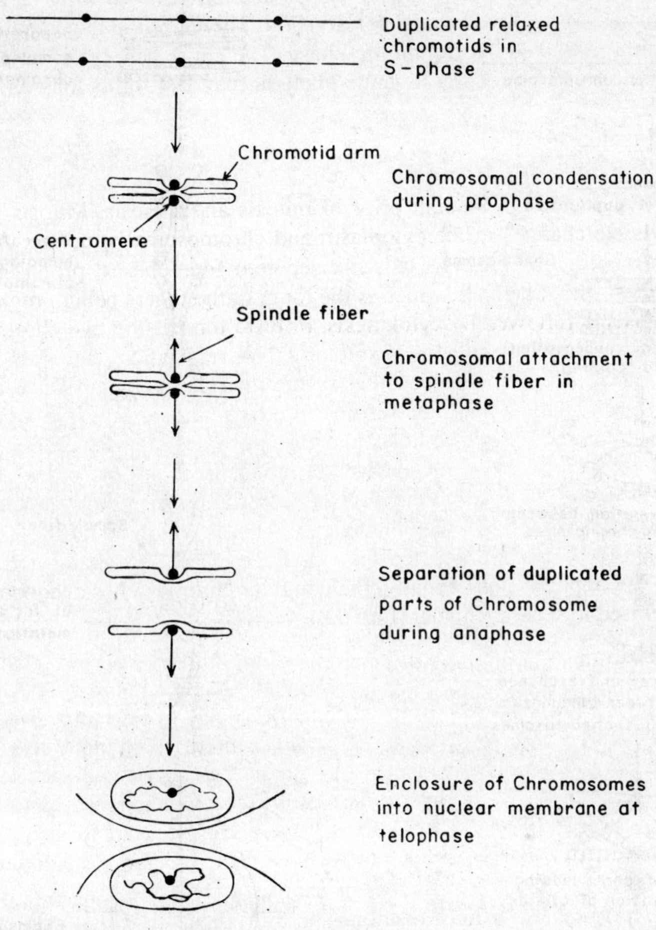

Fig. 6.8 Schematic presentation of the chromosomal behaviour during mitosis.

The orientation of the centromeres on each side of the plane makes this precise separation of homologous chromosomes possible. Each set of chromosomes contains one chromosome from the mother, and hence is haploid. Each chromosome contains two chromatids attached to a single centromere. In the second meiotic division each chromosome again lines up on the middle of the spindle attached by its centromere to the spindle fibres. There is no DNA synthesis, but the centromere of each chromosome duplicates and pulls apart carrying a chromatid towards the pole of the spindle. As the chromatids reach the ends of the spindle, each group becomes organized into separate nuclei. The formation of nuclei is followed by cytokinesis to form four daughter cells (Fig. 6.9).

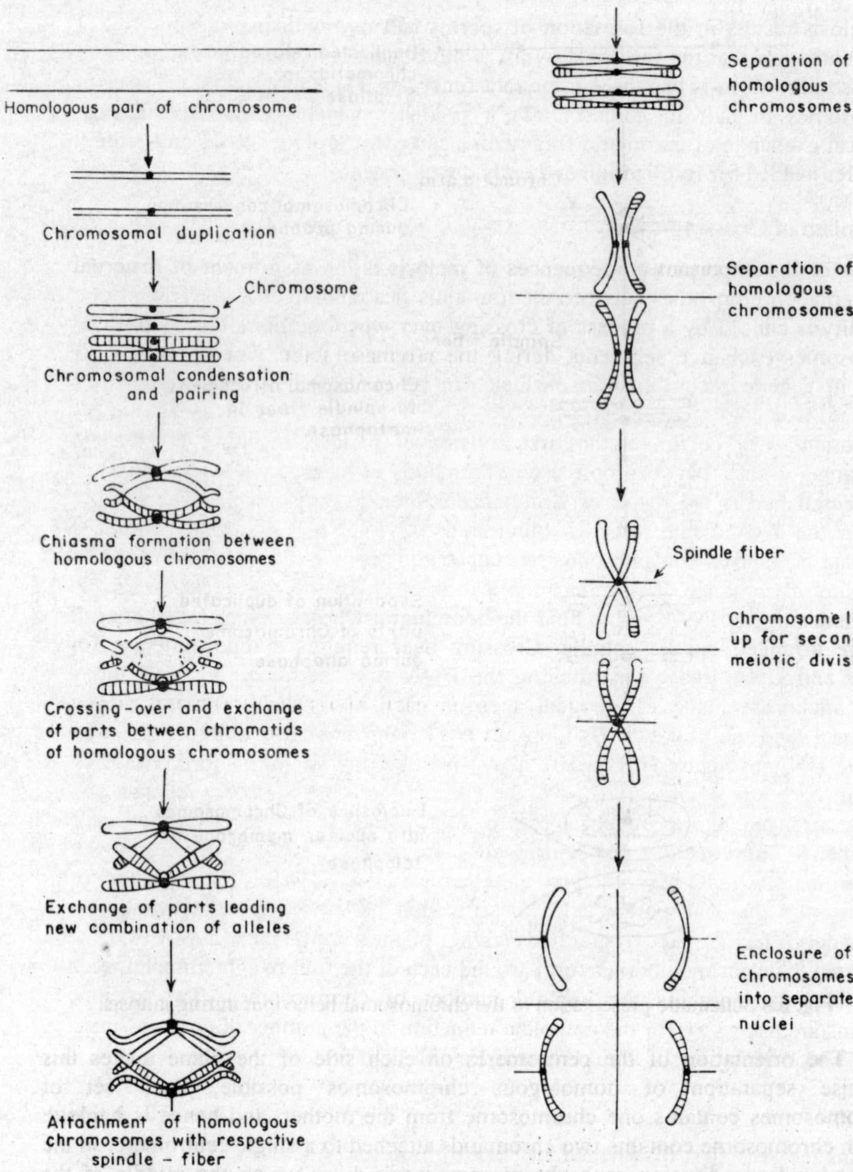

Fig. 6.9 Sequence of events during meiosis I and II. (a) relaxed homologous chromosomes, (b) homologous chromosomes form a pair, (c) pairing of homologous chromosomes, (d) Chiasma formation (e) crossing over and exchange of parts, (f) separation of homologous chromosomes, (g) to (i) daughter cells produced, (j) chromosomes line up for second meiotic division, and (k) chromatids in each cell showing haploid nature. Four daughter cell are produced.

Meiosis results in the formation of sperms and ova with haploid number of chromosomes so that the fertilized ovum would again acquire diploid number of chromosomes. Meiosis has three important functions: (1) it reduces the number of chromosomes to half in gametes, (2) it produces genetic variability through reciprocal exchange of chromatid fragments , and (3) it provides RNA and protein molecules needed for fertilization and early development.

Mechanism of Crossing-over

One of the important consequences of meiosis is the assortment of maternal and paternal chromosomes between the four cells as a random process. The genetic variability is caused by a process of crossing-over when the homologous pairs of chromosomes exchange segments during the prophase stage. The crossing-over results in genetic recombinations, which can be explained by Robin Holiday's model.

Crossing-over occurs in the first division of meiosis resulting in physical interchange of parts between non-sister chromatids of homologous chromosomes. This interchange is the result of simultaneous breaking and reciprocal rejoining between the broken ends of two chromatids. Cytological studies suggest that chiasmata occur between two nonsister chromatids of homologous chromosomes at the site of exchange. The actual implication of chiasmata during meiosis is to permit exchange of DNA and to hold the homologous chromosomes together until they are arranged on the spindle. Crossing over requires a topoisomerase for nicking and DNA ligase for stitching the DNA strands. According to Holiday model, nicks are made at identical sites on each non-sister chromatid during prophase I, and the broken ends loop out and rejoin in a cross arrangement with opposite DNA by ligase (Fig. 6.10). There may be one or more chiasma in each bivalent. After the exchange of parts the homologous chromosomes separate and become arranged in an equatorial plane, still attached to spindle fibres by their centromeres. Subsequently, the centromere pairs move to opposite poles dragging the chromatids resulting in two daughter cells with half the number of chromosomes than the parent cell. In the second meiotic division, the daughter cells undergo further subdivision and halving of the chromosome number is also completed. Nuclear membranes form around each of the four daughter nuclei, each of which now contains the haploid number of chromosomes. Two meiotic divisions are necessary for the complete reduction in the number of chromosomes, failing which the two cells would remain diploid.

Linkage

Linkage was first discovered in peas by Bateson and Punnet in 1906, who observed that two pairs of genes controlling flower colour and length of pollen grains were transmitted to the offspring in the same combination and rarely separated. This pattern of inheritance was explained by Morgan and his associates, who suggested that the genes present on the homologous pair of chromosomes are transmitted together. Such genes were named *linked* genes. Occasionally, however, new combinations may emerge in linked genes due to crossing-over or chiasmata formation between homologous chromosomes during prophase I of meiosis. The

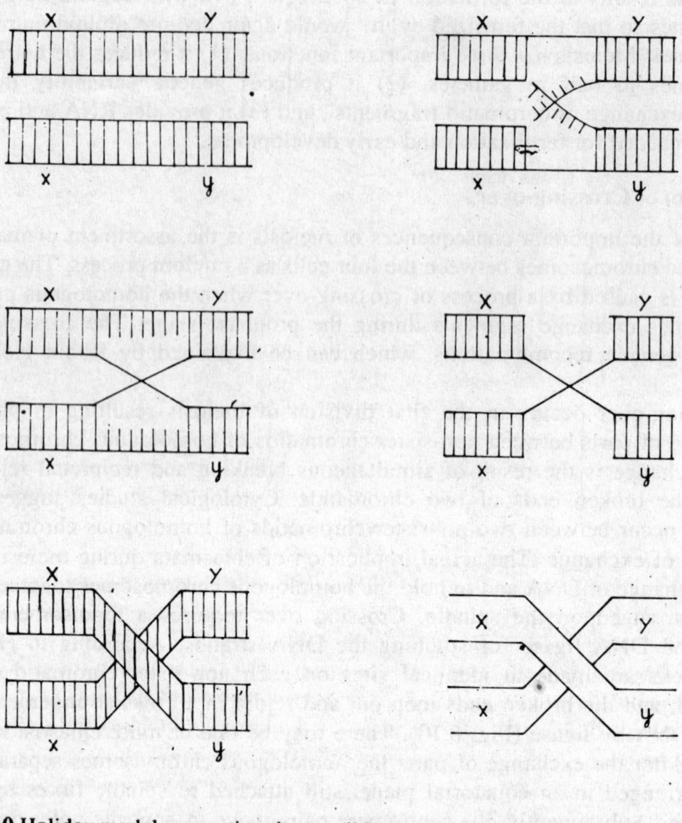

Fig. 6.10 Holiday model suggesting the possible steps in the molecular mechanism for recombination.

process of linkage and crossing-over may be explained with the help of Figure 6.11. Let the curved wing (s) and black body colour (b) be represented as recessive genes on the same chromosome in *Drosophila*, and their dominant alleles on the homologous chromosome as S and B. During first meiotic division, the two homologous chromosomes may not exchange parts, but sometimes an exchange may occur between the non-sister chromatids. In the absence of crossing-over, all the daughter cells will have the two genes in the parental arrangement, i.e. 50% gametes will have two dominant genes S and B, while the rest will have s and b combination. But in case of crossing-over recombinants will be produced in which 50% gametes will have S b and s B combination and the remaining will have parental combinations. The probability of crossing-over is called the crossing-over frequency and is expressed as a percentage.

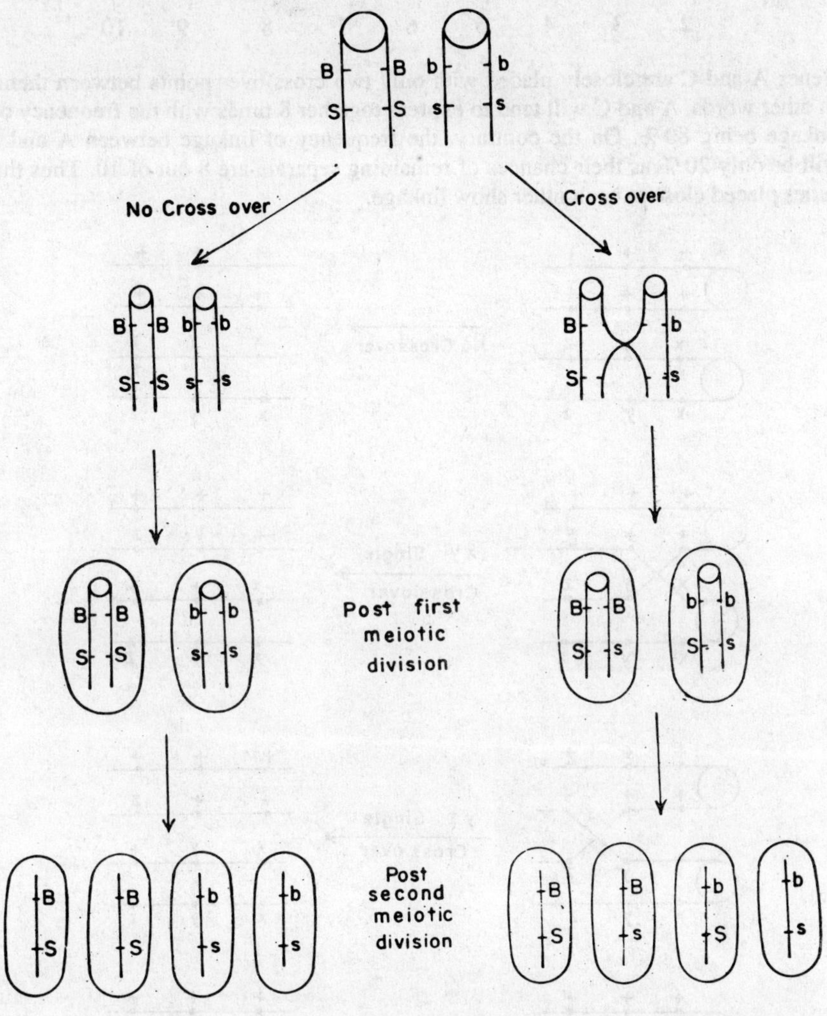

Fig. 6.11 Recombination of genes during meiosis resulting into genotype variability.

It has been observed that exchange between two chromatids always results in exchange of the entire gene segment and is never partial. The frequency of crossing-over between two genes is dependent upon the relative distance between two genes. If the genes are located on the same chromosome and placed at a greater distance, the frequency of crossing-over will be more, but if the distance between the two loci is very small the frequency of crossing-over is less and the two genes remain linked together.

Let us assume 11 genes A to K located on a chromosome in a linear fashion with 10 points or loci where crossing-over is possible:

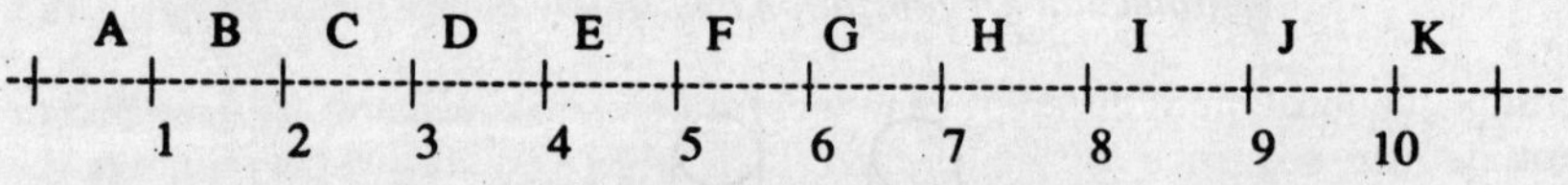

Genes A and C are closely placed with only two cross-over points between them. In other words, A and C will tend to express together 8 times with the frequency of linkage being 80%. On the contrary, the frequency of linkage between A and I will be only 20% as their chances of remaining separate are 8 out of 10. Thus the genes placed close to each other show linkage.

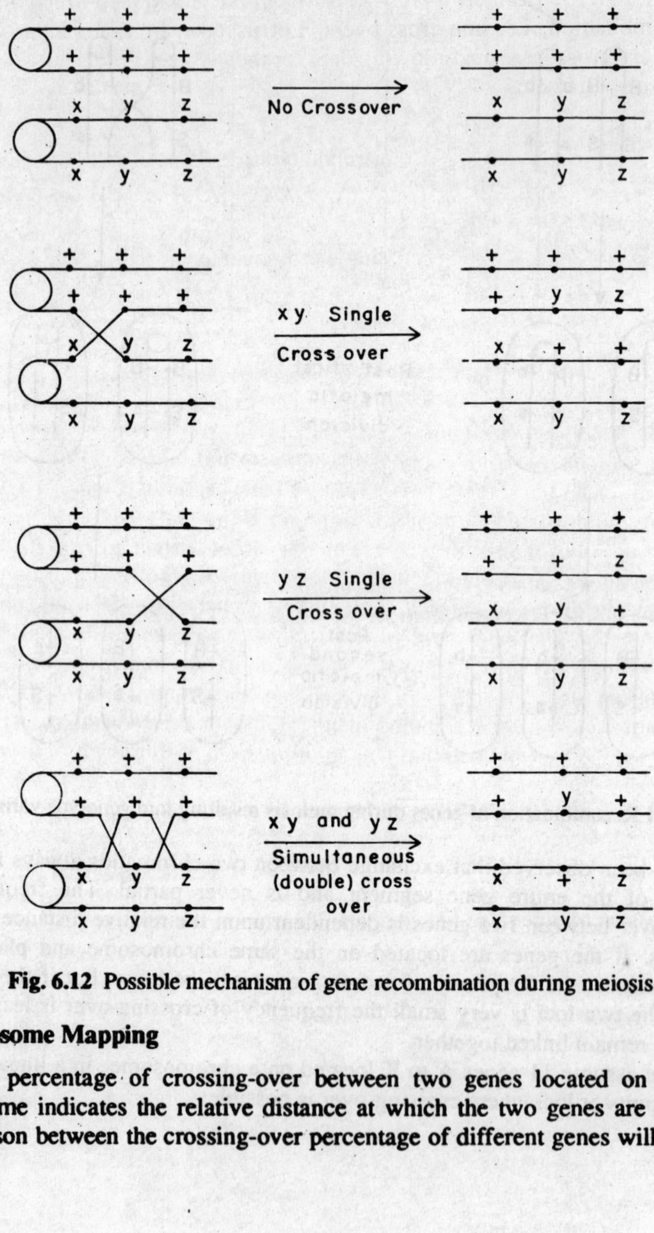

Fig. 6.12 Possible mechanism of gene recombination during meiosis.

Chromosome Mapping

The percentage of crossing-over between two genes located on the same chromsome indicates the relative distance at which the two genes are placed. A comparison between the crossing-over percentage of different genes will also help determine the location of different gene loci for correct sequencing. The method to determine the sequence of genes and their relative distances is known as chromosome mapping. Let three genes x, y, and z be located on a chromosome with two cross-over points between them (Fig. 6.12).

Their corresponding dominant alleles, called wild type (+) are present on the homologous chromosome in the same loci. Assuming that crossing-over occurs during gamete formation, there will be three possibilities: (a) single cross-over between x and y, (b) single cross-over between y and z, and (c) double cross-over between x y and y z. Besides, there will be about 50 % gametes which will show parental combination, i.e. non-cross-overs. Let us take an example in which a three-point test cross gave the following data for the gametes produced:

Genotype of gametes		
1. + + +	290	parental types
2. x y z	310	
3. + + z	95	single crossovers, region I
4. x y +	105	
5. x + +	80	single crossovers, region II
6. + y z	90	
7. + y +	16	double crossovers
8. x + z	14	
	1000	

From the above data we are in a position to calculate the crossover percentage and gene mapping. The total number of crossovers between x and y may be calculated:

+ y z	90
x + +	80
+ y +	16
x + z	14
	200 = 20 %

The crossover percentage between y and z can be calculated:

+ + +	95
x y +	105
+ y +	16
x + z	14
	230 = 23 %

The crossover percentage between x y and y z is 20% and 23% respectively. For chromosome mapping, each crossover percentage is taken as equal to one map unit, hence x and y are placed 20 units apart, whereas y and z are placed 23 units apart. In the double crossover only the middle gene is exchanged, therefore y is present in between x and z. The chromosome map may be constructed thus:

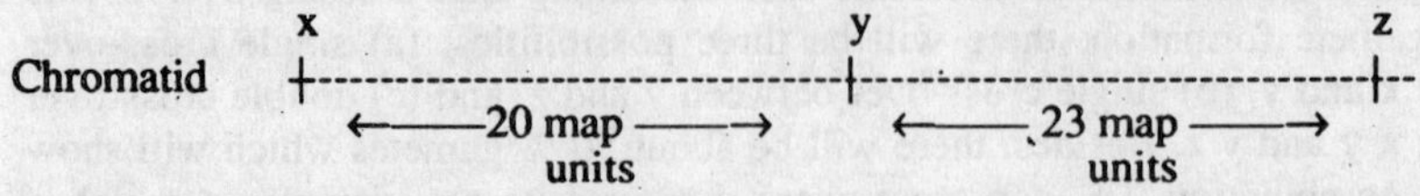

Consequence of Meiosis

In sexually reproducing organisms, fusion of two specialized cells, the sperm and the ovum, takes place to produce a zygote. If nuclear divisions were always mitotic, then fusion of male and female sex cells or *gametes* would result in doubling of the chromosome number in the offspring. Since the number of chromosomes and the amount of DNA is constant for a given species, it is necessary that the number of chromosomes is halved before fusion takes place. This would ensure constancy of chromosome number from generation to generation. Therefore meiotic division precedes the gamete formation.

Meiosis takes place in the testis and the ovary where sex cells or gametes are formed. These gonads contain germ cells with a full diploid number of chromosomes, which undergo two successive divisions to produce four haploid daughter cells, but the chromosomes are duplicated only once in the S-phase. Each daughter cell contains one half the number of chromosomes in the original cell, so that when a zygote is formed by the fusion of two dissimilar gametes, the original diploid number of chromosomes is restored. Thus meiosis is responsible for maintaining the constant chromosome number in a species generation after generation.

Another consequence of meiosis is the assurance of genetic variability among species that may give rise to new species. Sometimes in hybrids, the chromosomes derived from the two parents do not pair normally during meiosis, leading to unequal distribution of chromosomes into gametes. On occasions, defective meiosis may give rise to doubling of chromosomes, resulting in abnormal numbers. Polyploidy or defective meiosis gives rise to abnormalities. Genetic variability offers greater chances that some heritable characters will be more advantageous in a changing environment. Genetic variability may be adjudged from the fact that no two human individuals are identical unless produced from a single fertilized egg. The same is also true for other sexually reproducing species. This is because all members of a species in a population share a common gene pool, but each gene may exist in one or more alternative forms. Recombination of alleles during meiosis produces virtually infinite number of genotypes.

DNA : THE GENETIC MATERIAL

We shall now explore the molecular nature of the genetic material and its functional components, because only in the molecular structure of the gene lies the explanation of its actual working. Genes which are components of the DNA determine the characteristics of organisms and it is through them that these characteristics are handed down to the offspring. It therefore follows that genes must contain the information that guides an organism to develop along certain lines. It is also imperative that the genetic material, of which genes are the constituents, must be able to replicate or reproduce without losing any of its information content. The question arises as to how this genetic information is replicated and how it is ultimately transcribed and translated into protein molecules.

In searching for the chemical and molecular nature of the gene, we must look for a molecule that is endowed with the properties of *information storage, replicability,* and *mutability*. For many years proteins were considered the possible candidates because of the structural diversity which is essential for carrying the hereditary messages. However, the emergence of *one gene one enzyme* hypothesis from the work of G.W. Beadle and E.L. Tatum in 1941, and also the work carried out by O.T. Avery, C.M. Macleod and M. McCarty in 1944, produced evidences that DNA rather than protein is the carrier of hereditary information.

7.1 DNA CARRIES GENETIC INFORMATION

Let us look briefly at the work done by O.T. Avery and his colleagues. The bacteria that cause pneumonia, *Pneumococcus*, exist in two different strains, capsulated and non-capsulated. The former or the virulent *pneumococci* are characterized by the cells being surrounded by a thick polysaccharide capsule. In contrast, in the avirulent or non-pathogenic strains of pneumococci, the capsule is completely missing. Avery and his colleagues prepared a medium containing disrupted encapsulated bacteria. Into this medium non-capsulated strains were introduced and allowed to grow. The remarkable observation was that the non-capsulated strains developed capsules, as did their descendents, proving that a chemical constituent in the extract of capsulated bacteria is capable of bringing about a permanent and heritable change in the non-capsulated avirulent strain. This effect was investigated more closely by Avery and coworkers when they isolated DNA in a highly purified form from the virulent strains and added it to the avirulent cells *in vitro*. However, DNA from the avirulent strain was unable to transform avirulent cells. Moreover, the former DNA induced-transformation was permanent and heritable characteristic over many generations. Avery and his colleagues, therefore, argued that DNA is the active agent in carrying the genetic information (Fig. 7.1).

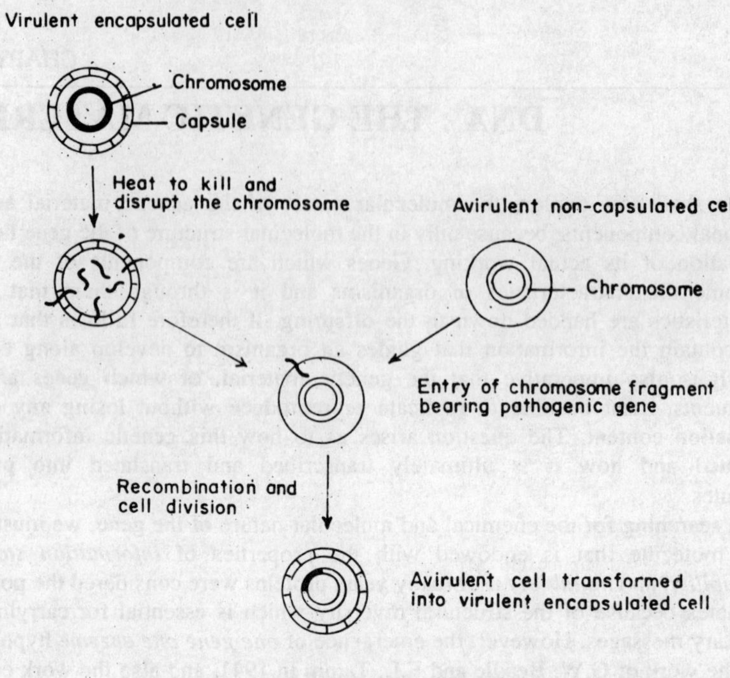

Fig. 7.1 Avery's experiment-transformation of genetic characteristics of a *Pneumococcus* bacterium.

Hershey-Chase Experiment

So much evidence is now available that DNA as the hereditary material has become a foundation principle of biology. One of the most original experiments of this principle was an experiment performed by A.D. Hershey and M. Chase in 1952. They took two kinds of organisms in their experiment: bacteria and bacterial viruses, also called bacteriophages. Bacteriophages or simply the phages are minute infectious agents that invade the host cells and use the host cell mechinery to make copies of themselves. Hershey and Chase took phage particles which were isotopically labelled either in the protein coat or in the DNA by incubating the host bacteria in a medium containing appropriate labelled precursors (Fig. 7.2). They cultured bacteria on growth media containing either ^{32}P or ^{35}S so that DNA would be labelled with ^{32}P since DNA contains phosphorus. Bacteria grown on the ^{35}S medium would have their protein labelled, but there would be no label on the DNA.

Bacteria grown on ^{32}P were then infected with phages, which reproduced inside them using their synthetic machinery. It was observed that the new progeny of phages incorporated radioactive phosphorus into their DNA. Similarly bacteria grown in ^{35}S medium were infected with phages, which reproduced inside the host and incorporated labelled sulphur in the protein coat. In the subsequent step,

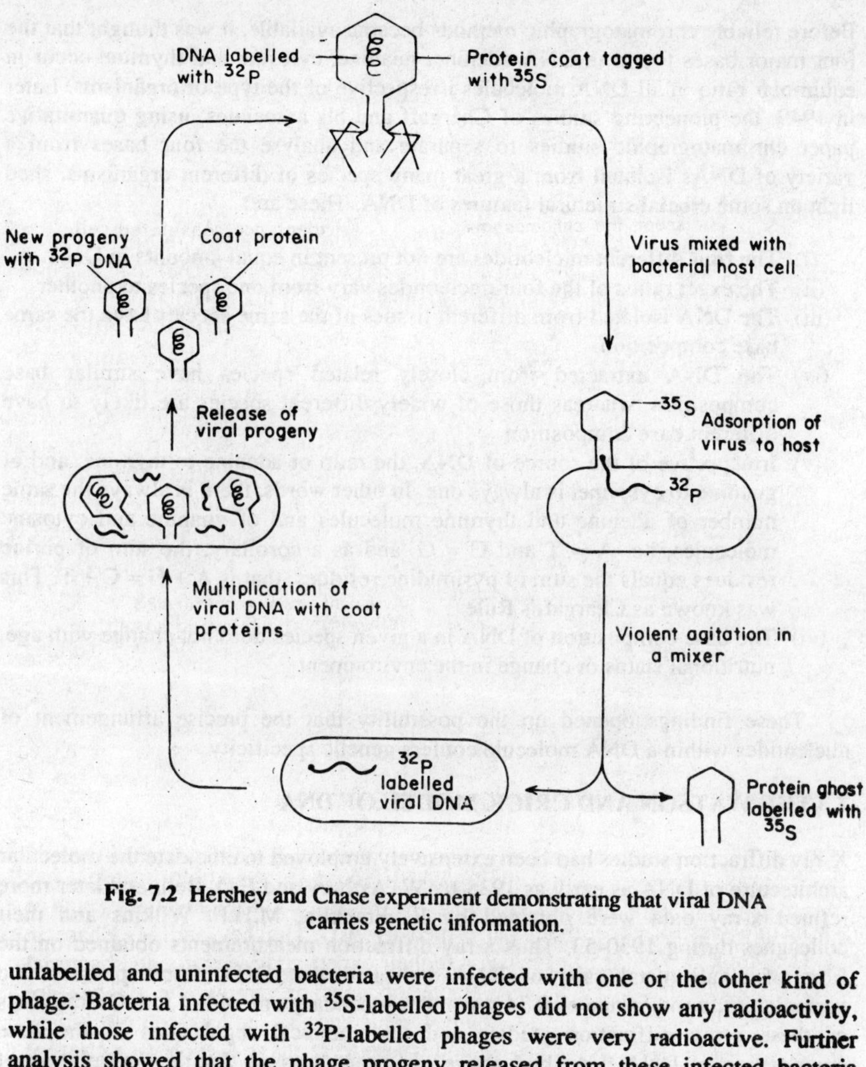

Fig. 7.2 Hershey and Chase experiment demonstrating that viral DNA carries genetic information.

unlabelled and uninfected bacteria were infected with one or the other kind of phage. Bacteria infected with ^{35}S-labelled phages did not show any radioactivity, while those infected with ^{32}P-labelled phages were very radioactive. Further analysis showed that the phage progeny released from these infected bacteria contained only ^{32}P label. With such labelled phage particles they showed conclusively that the viral DNA rapidly enters the host cell, whereas the viral protein does not.

Although the chemical components of nucleic acids were known at the time when Avery and coworkers did their experiments on pneumococcus, their significance was not appreciated and some biologists were reluctant to accept that nucleic acids were of genetic importance.

An important piece of biochemical evidence supporting the concept that DNA is the carrier of hereditary messages came from the work of E. Chargaff and his colleagues between the period 1949-1953, which explained that the base composition of DNA is related to the species.

7.2 BASE EQUIVALENCE IN DNA

Before reliable chromatographic methods became available, it was thought that the four major bases found in DNA-adenine, guanine, cytosine and thymine-occur in equimolar ratio in all DNA molecules irrespective of the type of organisms. Later in 1949, the pioneering studies of Chargaff and his associates, using quantitative paper chromatographic studies to separate and analyse the four bases from a variety of DNAs isolated from a great many species of different organisms, shed light on some crucial structural features of DNA. These are:

(i) The four different nucleotides are not present in equal amounts in DNA.
(ii) The exact ratios of the four nucleotides vary from one species to another.
(iii) The DNA isolated from different tissues of the same species have the same base composition.
(iv) The DNA extracted from closely related species have similar base composition, whereas those of widely different species are likely to have different base composition.
(v) Irrespective of the source of DNA, the ratio of adenine to thymine, and of guanine to cytosine, is always one. In other words, there is always the same number of adenine and thymine molecules and of guanine and cytosine molecules, i.e. A = T and C = G, and as a corollary, the sum of purine residues equals the sum of pyrimidine residues, that is A + G = C + T. This was known as Chargaff's Rule.
(vi) The base composition of DNA in a given species does not change with age, nutritional status or change in the environment.

These findings opened up the possibility that the precise arrangement of nucleotides within a DNA molecule confers genetic specificity.

7.3 THE WATSON AND CRICK MODEL OF DNA

X-ray diffraction studies had been extensively employed to elucidate the molecular architecture of DNA as early as 1938 by W. Astbury and F.O. Bell, and later more refined x-ray data were obtained by R. Franklin, M.H.F. Wilkins and their colleagues during 1950-53. Thus x-ray diffraction measurements obtained on the fibre of alkali metal salts of DNA indicated that DNA fibres possess two periodicities, a major one of 0.34 nm and a secondary one of 3.4 nm. On the basis of these x-ray diffraction studies and the chemical regularities in the base composition of DNA described above, the stage was set for the formulation of three-dimensional conformation of the DNA molecule. In 1953, J.D. Watson and F.H.C. Crick postulated a precise three-dimensional, double-helical model for DNA, the accuracy of which was later attested by more refined x-ray data of M.H.F. Wilkins and associates. The model accounted for two important things: the chemical and physical properties of DNA, and the precise mechanism by which the genetic information can be replicated accurately.

The Double Helix

The Watson and Crick model of DNA proposed that the two right-handed polynucleotide chains are coiled in a helical fashion around the same axis, forming a double helix (Fig. 7.3). The two chains or strands are antiparallel, i.e. their 3', 5'-

internucleotide phosphodiester bridges run in opposite directions (Fig. 7.4). The coiling of two strands is such that they cannot be separated from each other except by unwinding the coils; such a coiling is termed *plectonemic*.

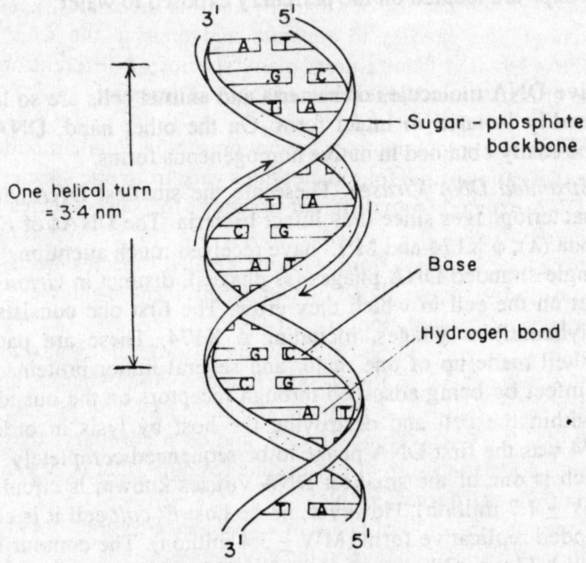

Fig. 7.3 Watson and Crick model of DNA helix.

In the double helix, the two DNA strands are held together by hydrogen bonds (weak, non-covalent chemical bonds) between pairs of bases on the opposite strands. However, the collection of these weak bonds along the entire length of the molecule makes the double helical structure of DNA very stable. The base pairing is very specific. The purine adenine only base-pairs with the pyrimidine thymine, while the purine guanine base-pairs with the pyrimidine cytosine (Fig. 3.16). It may be noted that between adenine and thymine there are two hydrogen bonds, whereas between cytosine and guanine there are three. The double helix is stabilized not only by the hydrogen bonds between complementary base pairs, but also by electronic interactions between the stacked bases, as well as by hydrophobic interactions. The purine and pyrimidine bases of each strand are stacked on the inside of the double helix with their planes parallel to each other and perpendicular to the long axis of the double helix.

As enunciated by Chargaff's rule, the number of adenine residues must be equal to the number of thymine residues and the number of guanine and cytosine residues must likewise be equal. The sequence of bases of the two strands of a given helix has a complementary relationship with the sequence on one strand exactly defining that of its partner strand. To account for 0.34 nm periodicity observed by X-ray diffraction, Watson and Crick postulated that the bases are stacked at a centre to centre distance of 0.34 nm from each other. The number of base pairs between two turns of the helix is exactly ten for the *relaxed* DNA. This is known as the linkage number which can easily account for the secondary repeat distance of 3.4 nm. The helix has a diameter of about 2.0 nm and has one shallow and one deep groove. The relatively hydrophobic bases are within the double helix, shielded from water, and the polar sugar residues and negatively charged phosphate groups are located on the periphery exposed to water.

Viral DNA

The native DNA molecules of bacteria and animal cells are so large that they cannot be readily isolated in intact form. On the other hand, DNA of bacterial viruses can be easily obtained in native homogeneous forms.

Single Stranded DNA Viruses. These are the smallest DNA viruses and are also called bacteriophages since they infect bacteria. The DNAs of *E. coli* phages, such as lambda (λ), ϕ X174 and M 13 have received much attention. There are two classes of single stranded DNA phages (ss phages), distinct in *virion* structure and in their effect on the cell in which they grow. The first one consists of spherical (actually polyhedral) ss phages, including ϕ X174. These are packaged in an icosahedral shell made up of one major and several minor proteins. Like typical phages they infect by being adsorbed through receptors on the outside of the cell, replicating within the cell and destroying the host by lysis in order to escape. Phage ϕ X174 was the first DNA phage to be sequenced completely. The DNA of ϕ X174, which is one of the smallest DNA viruses known, is circular and single stranded (MW $\simeq$ 1.7 million). However, in the host *E. coli* cell it is converted into a double stranded replicative form (MW $\simeq$ 3.4 million). The contour length of this ss form is only 1.77 μm. Other examples of ss DNA phages are G 4 and S 13.

The second category consists of linear, filamentous ss DNA phages including phage λ and M 13. The former is a linear duplex DNA molecule with a molecular weight of 32 million. Its contour length is about 17.2 μm. The other examples of this class, M 13, fd, and fl are thin particles, as long as 1 μm, and their DNA normally forms an extended loop within the cell. They also infect by adsorbing to an external receptor. Although their DNA is replicated and transcribed like that of spherical phages, filamentous ss DNA phages do not kill the cell or even make a lysis protein. Instead, phage particles are extruded continuously as the cell continues to grow. Simultaneously, the synthesis of phage proteins and DNA also continues at a moderate rate that allows cellular metabolism to continue. (Fig. 7.4)

Some linear viral DNAs, such as that of phage λ, have cohesive (sticky) ends. The 5' ends of these duplex DNAs project as single strands beyond the 3' end. These ends have twelve nucleotides which are complementary and can readily pair to form circles (Fig. 7.5). Moreover, some linear viral DNAs show terminal repetition, i.e. the presence of a base sequence at the 3' end repeats the sequence at the 5' end of the strand. Another characteristic of viral DNAs is that some of their bases are modified, e.g. presence of hydroxymethyl cytosine instead of cytosine, and apart from that the hydroxyl group of hydroxy-methyl cytosine is often glycosylated. These modifications protect the viral DNA from degradation by host-cell endonucleases.

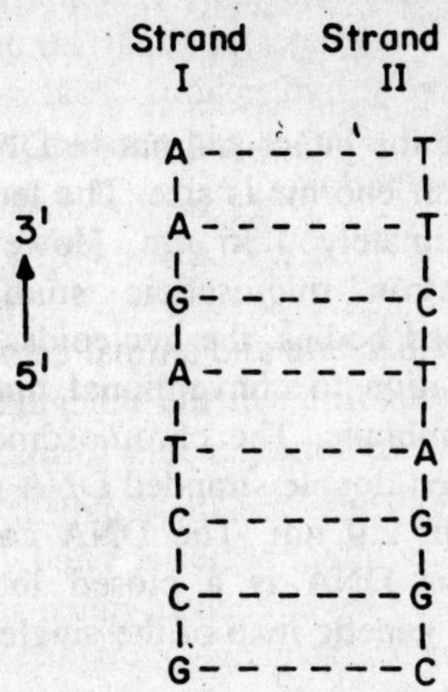

Fig. 7.4 Complementary antiparallel strands of DNA

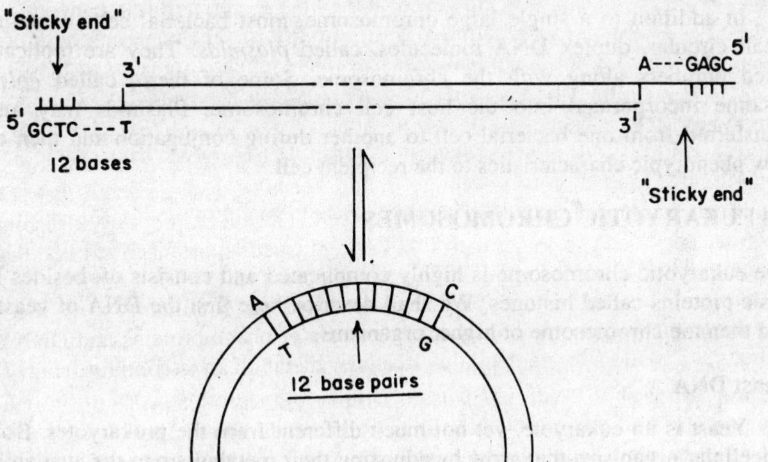

Fig. 7.5 DNA of phage lambda. The circular DNA has been transformed in a linear fashion showing sticky ends.

Adenovirus DNA is also linear and double stranded and has specialized proteins covalently attached to its 5' ends. Pox virus has closed hairpin loops at its ends.

Bacterial DNA

It is often difficult to isolate the intact and native DNA molecules of *E. coli* and other bacteria because of their enormous size. The length of a fully extended *E. coli* chromosome is approximately 1.36 nm. However, it never exists in partially extended form. Electron microscopic studies have revealed its compaction into irregularly shaped bodies, the nucleoids. Despite their similarity in name, nucleoids bear no relation to conventional nuclei since they are not surrounded by any form of membrane. The chromosome of *E. coli* appears to consist of a single, enormous sized double stranded DNA molecule of about 2.8 x 10^9 MW and thickness of about 2.0 nm. The DNA contains about 4 million nucleotide pairs. This enormous DNA is a closed loop, called a circle, in accordance with the fact that the genetic map of the single *E. coli* has been found to be circular.

For many decades it was believed that bacterial DNA, unlike eukaryotes, occurs naked and does not have a compact chromatin-like arrangement. Recently, however, two small histone-like basic proteins have been found in *E. coli*. These proteins are semi-tightly bound to DNA, leading to its compaction. The bacterial chromosome which is rather tightly and compactly looped in a nuclear zone is normally held together by some RNA. The DNA molecule is thrown into 50 or more highly twisted or supercoiled loops.

In addition to a single large chromosome, most bacterial cells contain 1-20 small circular, duplex DNA molecules, called *plasmids*. They are replicated in fixed numbers along with the chromosome. Some of them, called *episomes*, become incorporated into the host cell chromosome. Plasmids may also be transferred from one bacterial cell to another during conjugation and then confer new phenotypic characteristics to the recipient cell.

7.4 EUKARYOTIC CHROMOSOMES

The eukaryotic chromosome is highly complicated and consists of, besides DNA, basic proteins called histones. We shall describe here first the DNA of yeast cells and then the chromosome of higher organisms.

Yeast DNA

Yeast is an eukaryote, yet not much different from the prokaryotes. Both are unicellular organisms that grow by adjusting their metabolism to the availability of nutrients in a changing environment. However, the difference can be accounted for by the increased complexity of the eukaryotic cell. DNA molecules of baker's yeast, *Saccharomyces cerevisiae*, has 16 chromosomes, ranging in size between the T2 phage and *E. coli* DNA. Like other yeasts, *S. cerevisiae* can exist in either the haploid or the diploid state. Under conditions of nutritionaal deficiency, the diploid cells undergo meiosis to yield four haploid progeny cells, which become encapsulated as spores within a thick-walled sac, called the *ascus*. When the ascus ruptures, the spores are released and subsequently germinate to commence a new population of haploids (Fig. 7.6). While under satisfactory nutritional conditions, the diploid cells grow and divide, giving rise to diploid progeny.

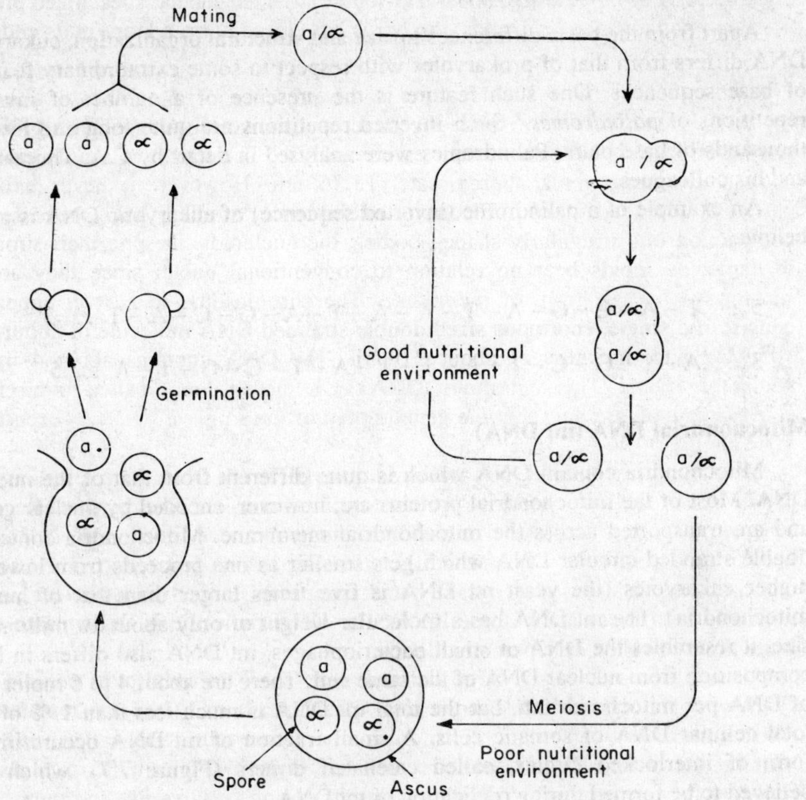

Fig. 7.6 Life cycle of *Saccharomyces cerevisae*.

Chromosomes of Higher Organisms

The chromosomes of higher eukaryotes are highly complex, huge in size and therefore highly folded. The largest chromosome of *Drosophila melanogaster* is about 40 times larger than *E. coli* chromosome. Human chromosomes are even much bigger, about 700 to 800 μm long and in a super-coiled fashion.

Unusually large giant chromosomes of the salivary glands of fruit fly have given important clues to the molecular organization of the chromatin fibres. They contain extended segments of DNA-protein fibres separated by a series of bands (about 500 in number) in which the fibres are intensely folded. Each band is flanked by interbands of varying widths. These multistranded giant chromosomes are called *polytene chromosomes*, which are visible during interphase of the cell-cycle. It has been suggested that the puffs in the folded regions are loops of bare and unwound DNA exposed for transcription, i.e. the DNA decondenses into a much more open state forming a distinctive *puff*. The larger and more diffuse a puff appears, the higher is the rate of transcription.

Palindromes in Eukaryotic DNA

Apart from the basic differences in size and structural organization, eukaryotic DNA differs from that of prokaryotes with respect to some extraordinary features of base sequences. One such feature is the presence of a number of inverted repetitions of *palindromes*.* Such inverted repetitions are quite long and involve thousands of base pairs. Palindromes were analysed in detail by C.A. Thomas, Jr. and his colleagues.

An example of a palindrome (inverted sequence) of eukaryotic DNA is given below:

5'. . . .T – A – C – G – A – T – A – A – T – A – G – C – A – T. . . .3'

3'. . . .A – T – G – C – T – A – T – T – A – T – C – G – T – A . . . .5'

Mitochondrial DNA (mt DNA)

Mitochondria contain DNA which is quite different from that of the nuclear DNA. Most of the mitochondrial proteins are, however, encoded by nuclear genes and are transported across the mitochondrial membrane. Mitochondria contain a double stranded circular DNA which gets smaller as one proceeds from lower to higher eukaryotes (the yeast mt DNA is five times larger than that of human mitochondria). The mt DNA has a molecular weight of only about 10 million. In size, it resembles the DNA of small bacteriophages. mt DNA also differs in base composition from nuclear DNA of the same cell. There are about 4 to 6 molecules of DNA per mitochondrium, but the total mt DNA is much less than 1 % of the total cellular DNA of somatic cells. A small fraction of mt DNA occurs in the form of interlocked circles, called catenated dimers (Figure 7.7), which are believed to be formed during replication of mt DNA.

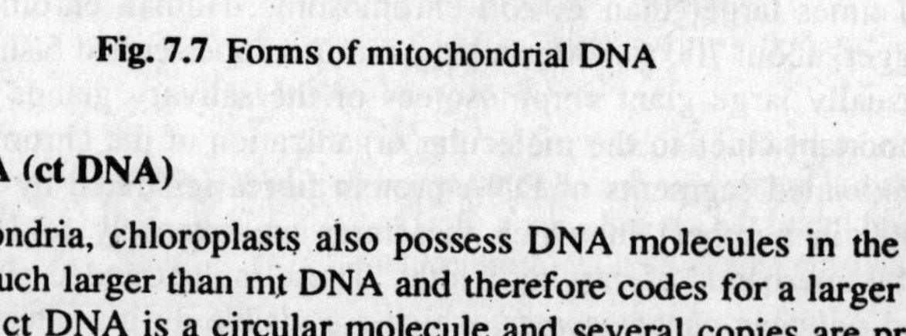

Fig. 7.7 Forms of mitochondrial DNA

Chloroplast DNA (ct DNA)

Like mitochondria, chloroplasts also possess DNA molecules in the stroma. The ct DNA is much larger than mt DNA and therefore codes for a larger number of proteins. Each ct DNA is a circular molecule and several copies are present in the chloroplast. Chloroplasts have DNA-dependent RNA polymerase, which synthesizes mRNA molecules. The ct DNA alone is not capable of synthesizing all proteins, but is partly controlled by nuclear DNA.

---

* A palindrome is a word or phrase which reads the same forward or backward.

Supercoiling of DNA

The linear DNA molecule, because of its free ends, can freely rotate to allow changes in the number of times the two strands of the double helix twist around each other. Once the two ends become convalently linked to form circular DNA molecules, the absolute number of times the two chains twist around each other, the linkage number, cannot change. Normally, changes in the average number of base pairs per turn of the double helix will be accomodated by the formation of appropriate number of supercoils in the opposite direction. Supercoiling, therefore, refers to further twisting of double helical DNA molecules. Untwisting of the double helix usually leads to supercoiling in the negative direction, while overtwisting leads to positive (right handed) supercoiling (Fig. 7.8). The supercoiled state is inherently less stable than the uncoiled DNA and nicking of a single strand instantly converts a supercoiled molecule into the simple, relaxed, circular state.

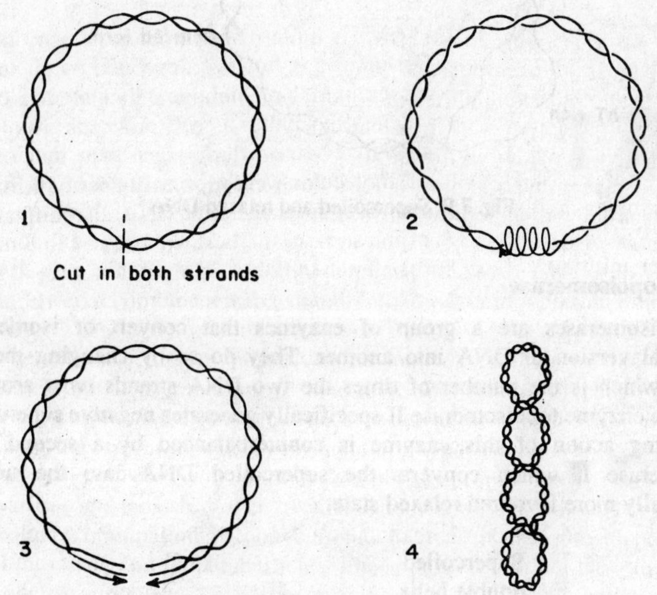

Fig. 7.8 Supercoiling of DNA

1. Circular relaxed double stranded DNA
2. Rotation in opposite direction from that of the twist of the helix.
3. Rejoining of the cut ends.
4. Negatively supercoiled DNA molecule.

As positive supercoiling is introduced, the DNA becomes increasingly compact and thus the relaxed DNA can be easily distinguished from the supercoiled one by its low sedimentation rate and by the longer time it takes to move through an agarose gel. Due to extra stress within the negatively supercoiled DNA, there is a tendency of localized denaturation in which the complementary strands come apart. Regions undergoing denaturation are rich in relatively weak AT base pairs since they are held together by two hydrogen bonds, while segments rich in GC base pairs are usually the last to denature (Fig. 7.9).

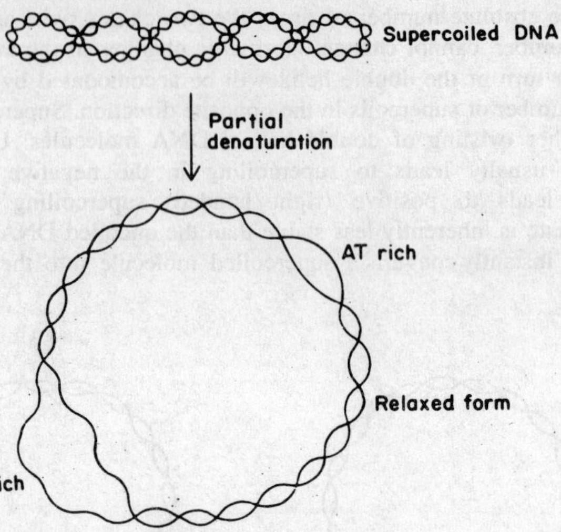

Fig. 7.9 Supercoiled and relaxed DNA

Role of Topoisomerase

Topoisomerases are a group of enzymes that convert or isomerize one topological version of DNA into another. They do so by changing the linkage number, which is the number of times the two DNA strands twist around each other. The enzyme topoisomerase II specifically generates negative supercoils. The supercoiling action of this enzyme is counterbalanced by a second enzyme, topoisomerase I, which converts the supercoiled DNA into the unstrained, energetically more favoured relaxed state:

$$\text{Supercoiled double helix} \; \underset{\text{Topoisomerase I}}{\overset{\text{Topoisomerase II}}{\rightleftharpoons}} \; \text{Relaxed double helix}$$

The relative amounts of the two topoisomerases are finely balanced and both catalyze the breakage and rejoining of phosphodiester bonds of DNA molecule.

Denaturation of DNA

Solutions of the native double helical DNA molecules, when subjected to extremes of pH or heated to near 100°C, undergo significant changes in their physical properties. In the process, covalent bonds remain intact while the weak hydrogen bonds break, separating the complementary strands. This physical change is known as *denaturation* (Fig. 7.10).

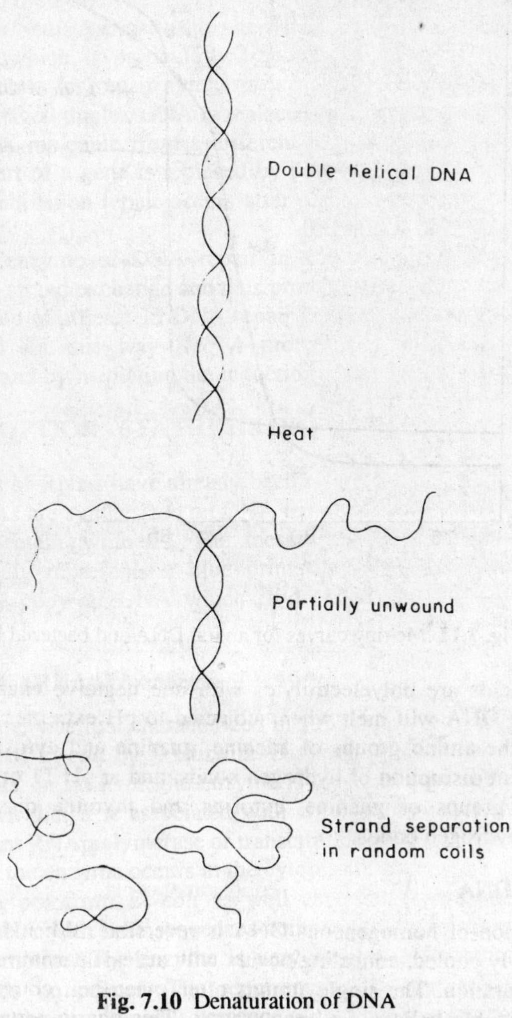

Fig. 7.10 Denaturation of DNA

Treatment of DNA with urea also effects denaturation. During denaturation the viscosity of DNA solution decreases, light absorption at 260 nm increases and the native double helix undergoes transition into a randomly looped form which is a single stranded or denatured DNA. In fact, thermal denaturation of DNA is often termed as melting, as over a range of a few degrees rise of temperature the double helix melts forming single stranded random coils. DNA from different cell types has different melting points. The mid-point temperature (Tm) of this process is the melting temperature of the helix of a specific DNA polymer (Fig. 7.11) Tm increases in a linear fashion with the content of GC base pairs since higher temperatures are needed to separate GC-rich strands. Hence, the higher the content of GC-base pairs, the more stable is the molecule.

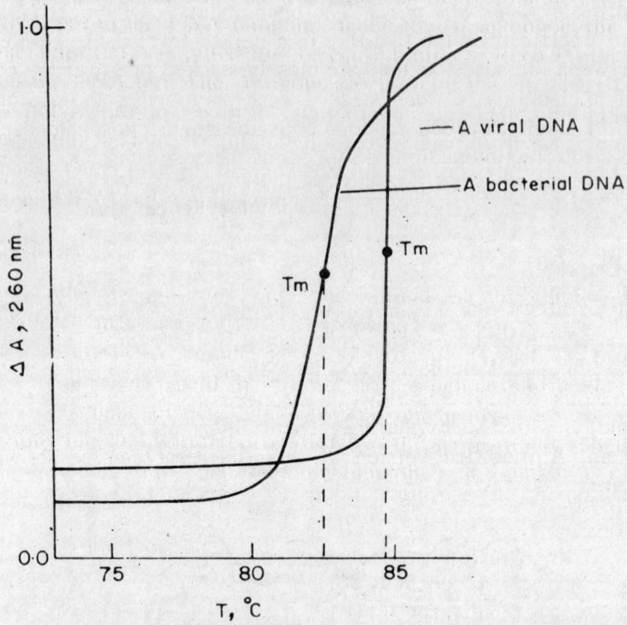

Fig. 7.11 Melting curves for a viral DNA and bacterial DNA.

Nucleic acids are polyelectrolytes with one negative charge per nucleotide residue, hence DNA will melt when subjected to pH extremes (pH range: 4-11). At low pH, the amino groups of adenine, guanine and cytosine are protonated with consequent disruption of hydrogen bonds, and at pH 11 or above, protons of the hydroxyl groups of guanine, cytosine and thymine dissociate leading to disruption of hydrogen bonds.

Annealing of DNA

Denaturation of homogeneous DNA is reversible. When heated solutions of DNA are slowly cooled, annealing occurs with at least partial return of the coil to helical configuration. The single strands often meet their complementary strands and re-form double helices, i.e. renaturation. This has important implications in molecular biology. The ability to renature DNA molecules has been employed to form hybrid DNA molecules in hybridization experiments.

Hypochromic Effect

The native double stranded DNA absorbs much less ultraviolet light at 260 nm as compared to the combined absorbance of its constituent mononucleotides. Thus, denatured DNA will show higher absorbance at 260 nm, nearly 35-40% more. The increase in the amount of absorbed UV-radiations also depends upon the AT base pairs and the greater the number of AT base pairs, the higher is the absorbance. This is called the *hypochromic effect*. The phenomenon of hypochrornism in the native DNA is explained by the fact that in a helical form there is an electronic interaction between the bases, as they are closely stacked one over the other. The interaction of π electrons between stacked bases diminishes the absorbance by each residue. In single stranded random coils the sugar-phosphate backbone is disrupted and electronic interactions disappear as the bases unstack, increasing the absorbance. This property is extremely useful in calculating the degree of helicity of DNA.

The Central Dogma

The Watson and Crick hypothesis explained that DNA functions as the template for RNA molecules, which subsequently move to the cytoplasm and along with ribosomes determine the arrangement of amino acids in proteins. In 1956 Francis Crick termed this flow of information as the *central dogma* of molecular biology, which states that the genetic information flows from DNA to RNA to protein. A modified version of this statement can be shown in the following manner:

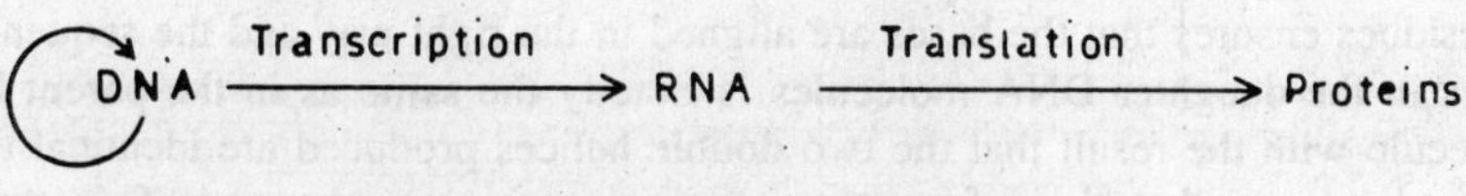

The arrows indicate the direction of flow of information.

The central dogma, therefore, defined three major processes for preservation and transmission of genetic information:

(1) Replication—A process which is indicated by the arrow encircling DNA, signifying that DNA is the template for self-replication.

(2) Transcription—The arrow between DNA and RNA indicates that all cellular RNAs are made on DNA templates.

RNA chains sometimes act as templates for DNA strands of complementary sequence as is the case in *reverse transcription*. But such reversals of the normal flow of information are rare events compared to the enormous number of RNA molecules made on DNA templates.

(3) Translation—The process by which all proteins are determined by or translated on RNA templates on the ribosomes.

Thus the central dogma of molecular biology as propounded more than three decades ago still remains valid.

DNA REPLICATION AND TRANSCRIPTION

Since the discovery of the double helical DNA molecule, an important feature that excited geneticists was the complementary relationship between the base sequences on the intertwined polynucleotide chains. Moreover, Watson and Crick, on the basis of the double helical model of DNA, also predicted a suitable mechanism of its duplication while preserving its information content which is very important from the genetic point of view. They proposed that each of the complementary DNA strands serves as a template for the formation of a new strand and this can only be achieved by the separation of the parent chains and subsequent synthesis of new DNA strands on each chain, finally generating two double helical DNA molecules. The Watson and Crick model demonstrates a most probable method of replication in which the two complementary strands of the DNA first separate from each other like the two halves of a zip-fastener and new complementary strands of DNA are assembled from the four available deoxyribonucleotide triphosphates on each of the two separated parent strands.

8.1. MECHANISM OF REPLICATION

In fact, it was the self-complementary nature of the DNA that finally led most biologists to accept Avery's conclusion that DNA, not protein, is the carrier of genetic information. The complementary relationship between A and T, and C and G residues ensures that the bases are aligned in the right way and the sequence of bases in the daughter DNA molecules is exactly the same as in the parent DNA molecule with the result that the two double helices produced are identical to the original one so that the information content remains unchanged. This type of replication has been called a semi-conservative replication (Fig.8.1 a), a neat and economical way of explaining replication. A second possibility is that the final duplication product consists of a double-helix of the original two strands and a second double-helix consisting of newly synthesized chains, a process called conservative type of replication (Fig 8.1b), that is, instead of unzipping, the original double-helix remains intact and in some way stimulates the synthesis of a second double-helix identical to the first. A third possibility, dispersive replication, could also be considered if the nucleotides of the parent DNA are randomly scattered among the components of the daugther DNA material so that the new DNA consists of a mixture of old and new nucleotides scattered along the chains (Fig. 8.1.c).

Meselson and Stahl's Experiment

The above three possibilities were put to the test by Meselson and Stahl in 1958 in an elegant experiment which conclusively proved that in the intact living *E. coli* cells DNA is replicated in a semi-conservative manner as postulated by Watson and Crick. *E. coli* cells were grown for several generations in a medium in which the sole source of nitrogen was NH_4Cl, whose normal nitrogen (^{14}N) was replaced with the heavy isotope ^{15}N. This was continued until it was ascertained that heavy nitrogen had been incorporated into all the DNA. After several generations of growth, heavy nitrogen was replaced by the normal nitrogen.

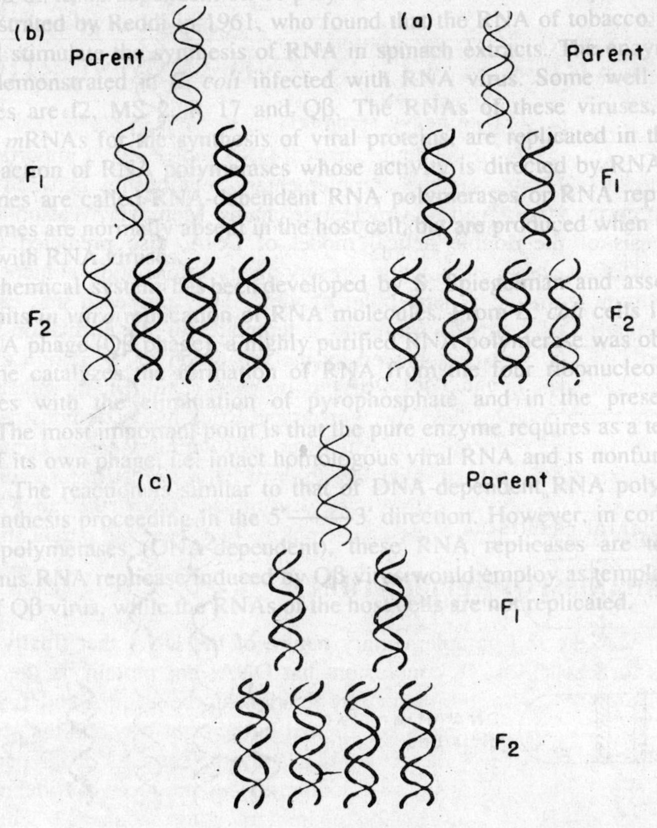

Fig. 8.1 Possible mechanisms of DNA replication. (a) Semiconservative replication, (b) Conservative replication, (c) Dispersive replication.

Samples of bacteria were then taken at short intervals and their DNA was carefully extracted and analysed for relative ^{14}N and ^{15}N content by density gradient centrifugation, a technique which depends on the fact that the compounds containing ^{15}N are slightly heavier than those containing ^{14}N. After exactly one generation in the ^{14}N medium which resulted in doubling of the number of cells, the DNA isolated showed but a single band in cesium chloride (CsCl) density gradient, midway in density between the normal ^{14}N DNA and the heavy ^{15}N DNA of cells grown exclusively on ^{15}N medium (Fig. 8.2).

These results definitely eliminated the other two possibilities and conclusively established the semi-conservative mode of replication of DNA. In the description that follows it will be seen that replication of even the simplest DNA molecule is a complicated, multistep process involving the role of many enzymes and proteins. The precise types and number of enzymes and other proteins required differ somewhat in the replication of viral, bacterial and eukaryotic DNA. We shall however, examine the details of the replication in prokaryotic DNA.

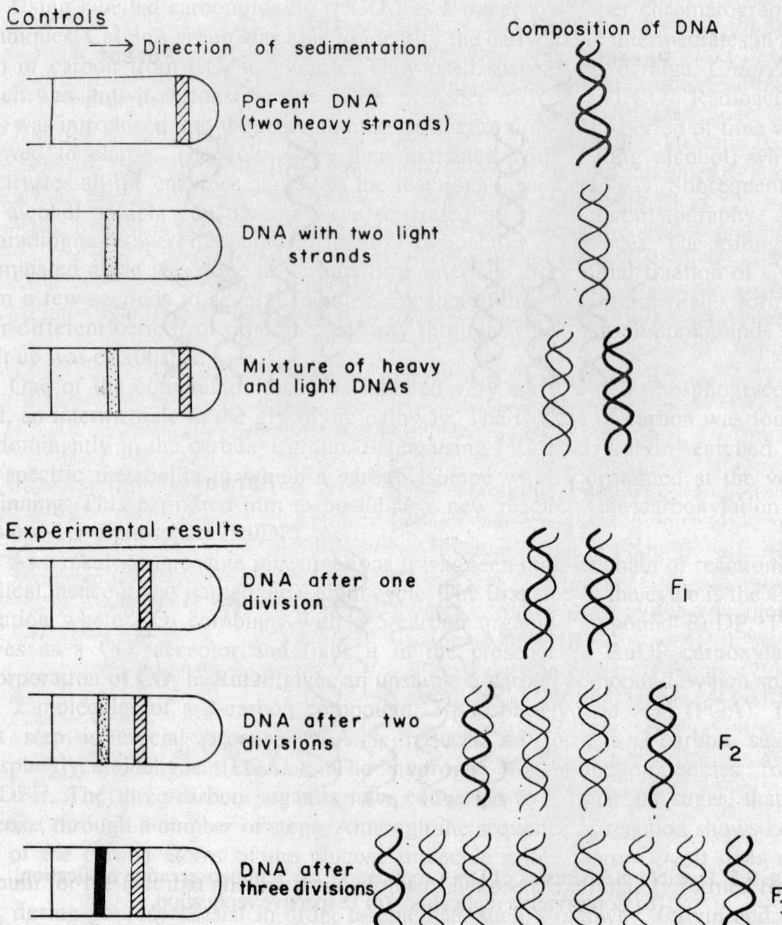

Fig. 8.2 Results of Meselson-Stahl experiment to demonstrate semiconservative replication of DNA

8.2 ENZYMES INVOLVED IN DNA REPLICATION

Efforts have been made to understand the nature of enzymes involved in DNA replication. Actually the way in which these enzymes are used to make complete DNA chains is rather complicated.

Three Types of DNA Polymerases

The enzyme which was first investigated by A. Kornberg and his colleagues in 1956 was named DNA polymerase (now called DNA polymerase I). Now at least three different DNA polymerases are known from *E. coli* which have almost similar catalytic properties. Today it appears that DNA polymerase III is the main enzyme concerned in the replication process and is the major polymerizing agent. Although DNA polymerase I also participates, it is largely used in the repair of DNA, i.e. for filling up gaps between the small okazaki fragment precursors that are formed during synthesis. So far, no specific task has yet been attributed to DNA polymerase II.

E. coli DNA polymerase I (DNA Pol I) has a molecular weight of 109,000 and it catalyzes the addition of mononucleotide units to the free 3′-OH end of a DNA chain. It has a single polypeptide chain of about 1000 amino acid residues. Apart from the polymerizing action, DNA Pol I also catalyzes two other reactions. One of them is 3′——→5′ exonuclease activity, namely hydrolysis of single nucleotide residues from the 3′-terminal of a DNA chain, and the other is 5′——→3′ exonuclease activity catalyzing hydrolysis of mononucleotide residues from the 5′-terminal.

DNA pol II has a molecular weight of about 120,000 and possesses only 3′——→5′ exonuclease activity. The enzyme is unable to replicate long single strands with a short complementary primer. Biological functions of DNA pol II are not yet known.

DNA pol III is by far the most active of the three polymerases. It is 15 times more active than DNA pol I and 300 times more active than DNA pol II. It has a molecular weight of about 140,000 and has about the same template primer dependence as DNA pol II. DNA pol III normally occurs in a more complex but biologically active form which is known as DNA pol III* and is believed to be a dimer of DNA pol III. DNA pol III* requires an auxilliary protein DNA copolymerase III*. Evidence has been provided by A. Kornberg and his colleagues that DNA pol III* combines with DNA copol III* to yield a DNA pol III*-copol III* complex (Fig. 8.3). This complex, a DNA template, an RNA primer and ATP are required to initiate replication. After initiation, Copol III* dissociates itself with the release of ADP and Pi.

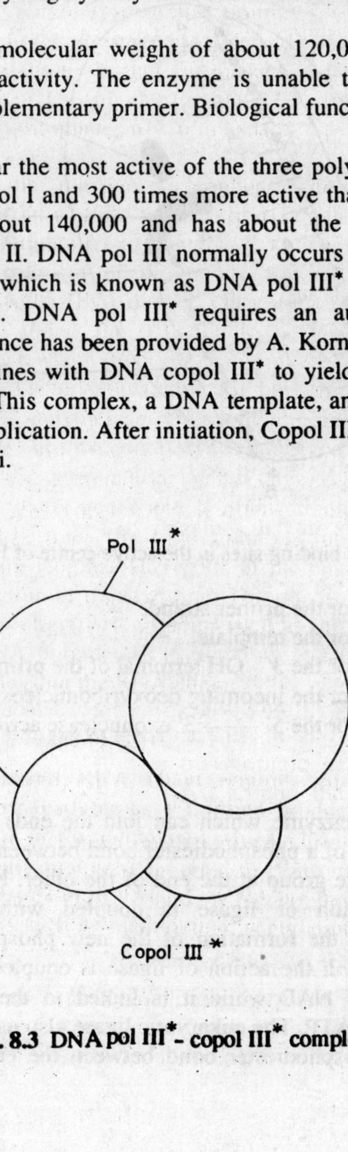

Fig. 8.3 DNA pol III* - copol III* complex

An important characteristic of DNA polymerase is that it requires the presence of pre-existing DNA and in the absence of the latter no polymer is formed. Usually the preformed DNA either serves as a primer or as a template. When it functions as a primer it provides a growing point to which additional nucleotide residues are added by the polymerase. On the other hand, when the pre-existing DNA serves as a template, the enzyme DNA polymerase makes a parallel strand of DNA, complementary in base composition and sequence.

DNA polymerase has five different binding loci on its active site (Fig. 8.4). These loci have been classified according to the specific tasks they perform:

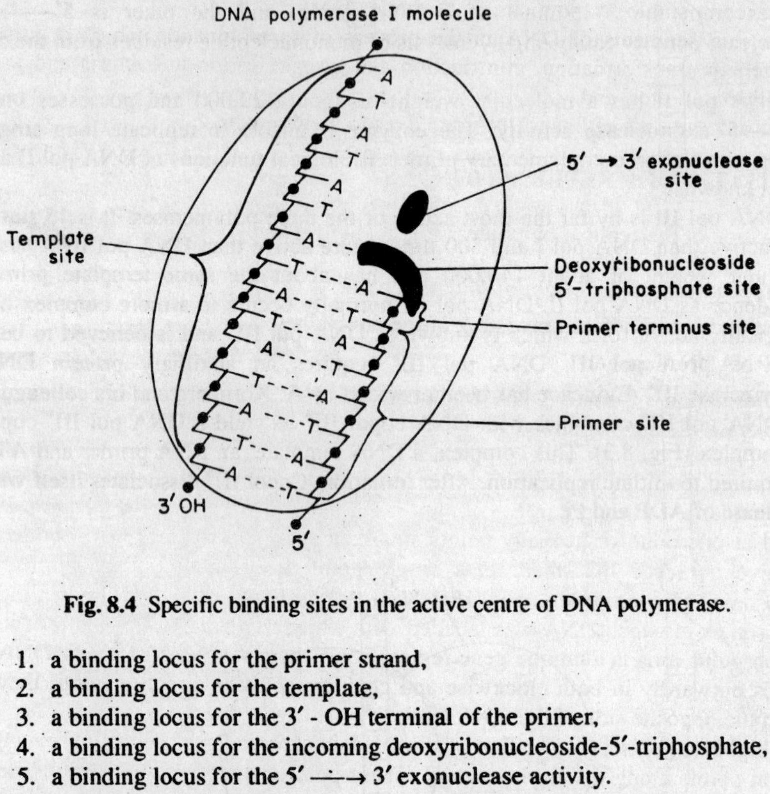

Fig. 8.4 Specific binding sites in the active centre of DNA polymerase.

1. a binding locus for the primer strand,
2. a binding locus for the template,
3. a binding locus for the 3′ - OH terminal of the primer,
4. a binding locus for the incoming deoxyribonucleoside-5′-triphosphate, and
5. a binding locus for the 5′ ⟶ 3′ exonuclease activity.

DNA Ligase

DNA ligase is an enzyme which can join the ends of two DNA chains by catalyzing the synthesis of a phosphodiester bond between 3′-OH at the end of one chain and a 5′-phosphate group at the end of the other. Normally, joining of two molecules by the action of ligase is coupled with the breakdown of a pyrophosphate bond as the formation of the new phosphodiester bond requires energy. In case of *E. coli* the action of ligase is coupled to the cleavage of the pyrophosphate bond of NAD^+while it is linked to the cleavage of the α, β-pyrophosphate bond of ATP. The eukaryotic ligase also uses ATP. DNA ligase can form only a single phosphodiester bond between the ends of the chains and it requires that the chains to be joined be associated in a double helix with a complementary DNA strand, and the residues which are joined should be paired to adjacent bases in the other strand.

E. coli DNA ligase consists of a single polypeptide chain of about 77,000 molecular weight, Each *E. coli* cell has about 200 to 400 copies of the DNA ligase. Fulfillment of the energy requirement during the reaction mechanism is unusual in the sense that NAD^+ which is a coenzyme for the oxidation-reduction reactions functions as a source of adenylyl group whose main role in this reaction is in the energy transfer.

Several functions that have been assigned to DNA ligase are:

1. to cooperate with DNA polymerases in the replication of DNA ,
2. to repair single-strand nicks in the duplex DNA,
3. to join segments of DNA in the process of recombination which occurs during genetic transformation, transduction and lysogenization in bacteria and in meiosis in eukaryotes, and
4. to link the ends of linear DNA duplexes to yield circles.

8.3 INITIATION OF REPLICATION

The initiation of the replication process can be divided into the following steps:

(a) recognition of the origin of initiation point,

(b) strand separation or untwisting of the double helix,

(c) RNA priming, and

(d) Formation of small DNA fragments on the RNA primers—the leading versus the lagging strand.

Meselson and Stahl's experiment demonstrated semi-conservative replication of the circular chromosome in *E. coli*, but it did not explain whether replication is initiated at one point or at many points simultaneously. Cairns in 1963 provided conclusive evidence that there is a single point of initiation in the circular chromosome. His evidence was based on visual observation of replicating chromosomes in which DNA was labelled with tritiated thymidine (^{3}H-TdR). The initiation point (ori) is a unique gene region from where synthesis of a new DNA proceeds outwardly in both clockwise and counterclockwise directions until they meet on the opposite side of the DNA circle.

In higher organisms replication of the DNA in chromosomes begins at several initiation points along the duplex strand. These points on a given DNA molecule are precisely defined. One or more specific initiator proteins are required to recognize these initiation points. However, for the initiation to begin the duplex strand must first be separated into single strands for the specific polymerases to function. Therefore, besides requiring the breakage of hydrogen bonds holding base pairs together, the replication process necessarily involves untwisting of the intertwined parental chains. In linear DNA molecules, replication on a specific initiation point is found to commence internally and generates bidirectionally growing Y-shaped molecules until replication is completed (Fig. 8.5).

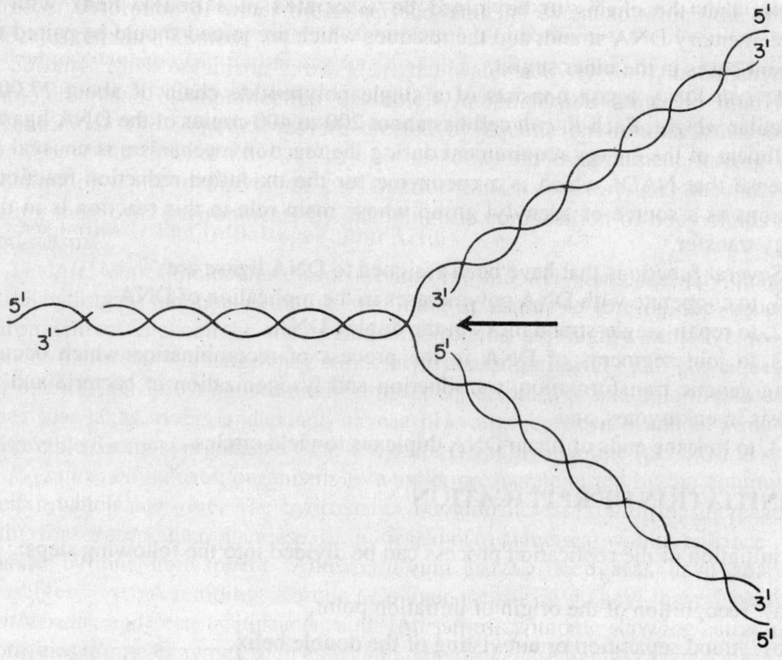

Fig. 8.5 Direction of progeny chain growth and partial unraveling of parental helix at the replication fork, *Ori.*

DNA replication is discontinuous and always occurs in the 5′ ——→ 3′ direction on both strands of a duplex molecule. The specific DNA pol III can add nucleotide precursors to DNA and extend the chain in the 5′ ——→ 3′ direction (Fig. 8.6). Preceding the action of the replicating enzymes, presumably at several points, unique proteins of low molecular weights in both prokaryotic and eukaryotic organisms, specifically bind to one of the two strands. These are unwinding proteins which bind to those regions of the duplex which are rich in A-T base pairs. These regions are more susceptible to *melting* or separation of the two strands.

The next step is the generation of RNA primers i.e., fragments of RNA. These are generated by a DNA-directed RNA polymerase which requires ribonucleoside-5′-triphosphates as precursors of the ribonucleotide residues. These priming strands are complementary to the DNA strands and are 50-100 nucleotides long. (Fig. 8.7)

Once the RNA primer has been made, the DNA polymerase begins to add deoxyribonucleotide units generated from deoxyribonucleotide-5′-triphosphates to the 3′-terminal of the RNA primers. These RNA-DNA fragments, often called *Okazaki* fragments, grow in the conventional 5′——→3′ direction. (Fig. 8.7). There is, however, a difference in the type and rates of growth between the two daughter strands formed on the two parent DNA templates. The strand which is formed on the 3′——→5′ template strand is called the *leading strand*. This strand grows continuously and moves slightly ahead of the other strand. As the leading strand is made in the 5′——→3′ direction, the other strand known as the *lagging strand* is also replicated in the direction opposite to the direction of movement of the replicating fork. That is, the 3′——→5′ strand grows discontinuously, being assembled from a group of DNA fragments made individually in the 5′——→3′ direction (Fig. 8:8). Both the leading and the lagging strands are made in short segments of 1000-2000 nucleotides.

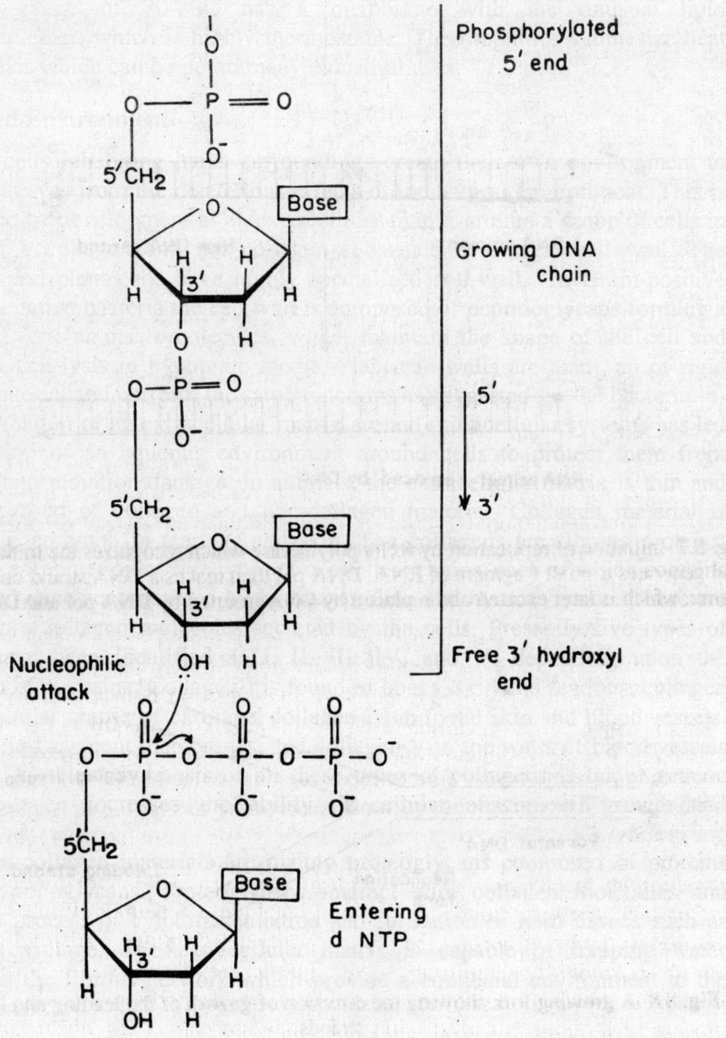

Fig. 8.6 Chain extension by DNA polymerase.

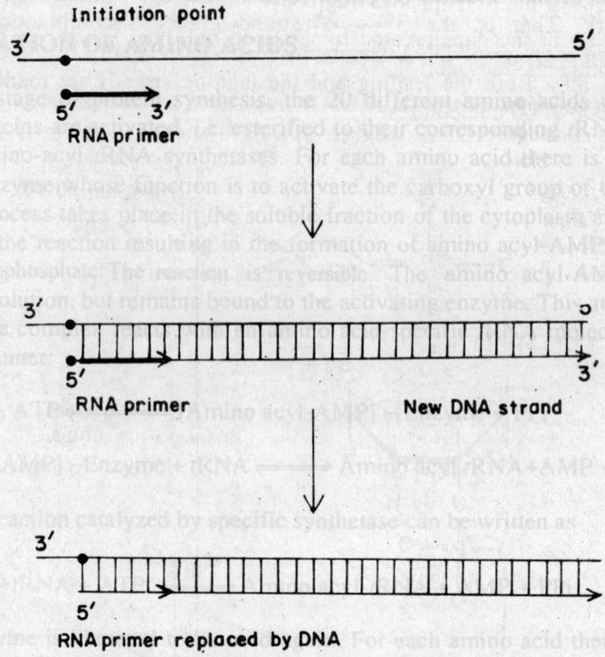

Fig. 8.7 Initiation of replication by RNA polymerase which recognizes the initiation point and generates a small fragment of RNA. DNA pol then makes a DNA strand on the RNA primer, which is later excised and replaced by DNA inserted by DNA pol and DNA ligase.

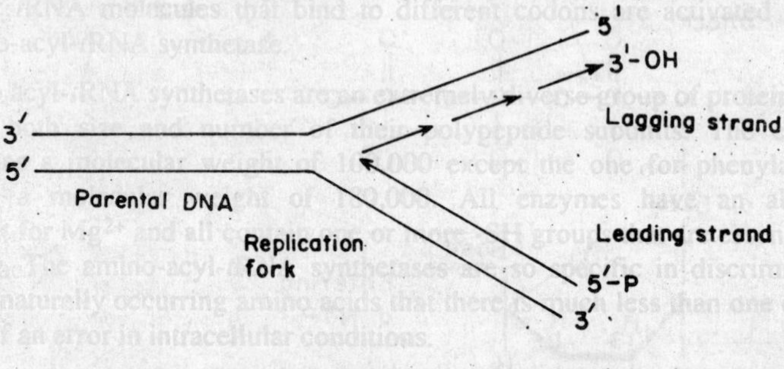

Fig. 8.8 A growing fork showing the direction of growth of the leading and lagging strands.

Elongation and Termination

Following initiation, a number of short segments of DNA are made on RNA primers in the presence of the holoenzyme DNA polymerase and ATP. During the process, ATP is hydrolyzed to ADP and Pi. As soon as the elongation process i.e., growth of chains, is completed three important events occur.

(a) excisions of the RNA primer fragments,

(b) filling up of the gaps created by excision of RNA primers by deoxyribonucleotide residues, and

(c) joining together of the short DNA fragments by phosphodiester bonds to form a continuous DNA daughter strand.

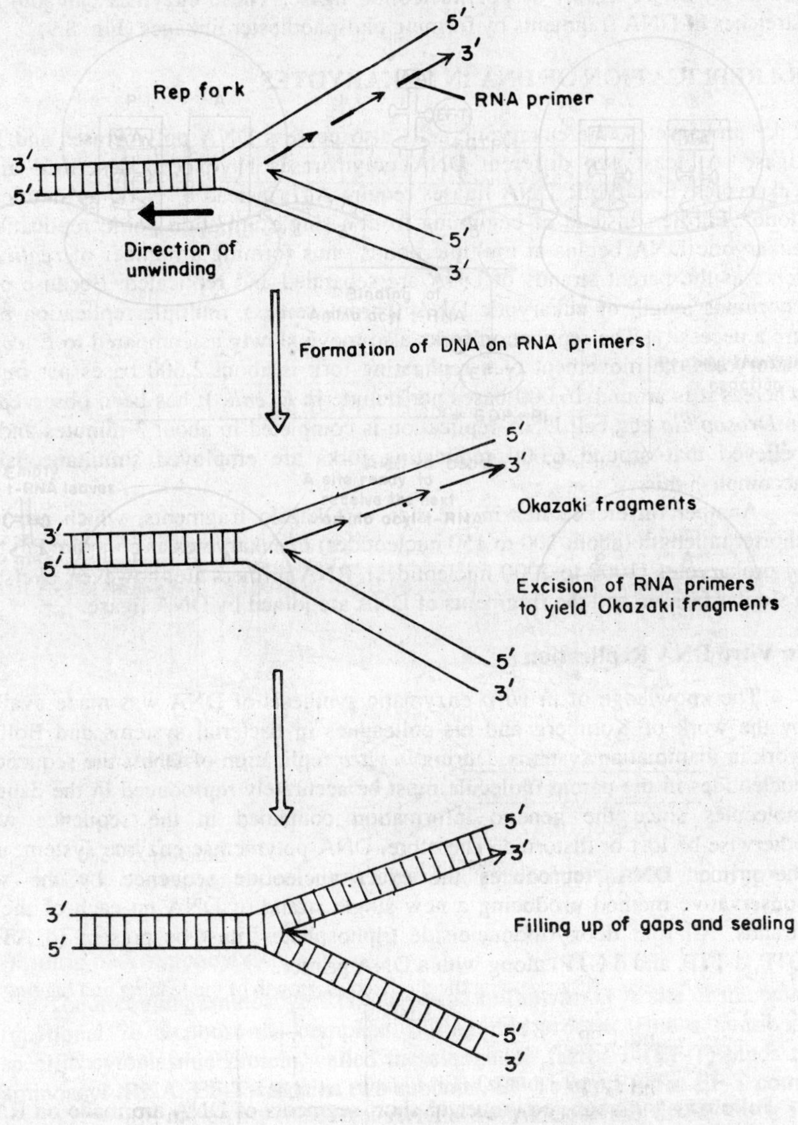

Fig. 8.9 Formation of DNA on RNA primers by DNA Pol III. The gaps between Okazaki fragments are filled by DNA Pol I and the nicks sealed by DNA ligase.

Excision of the RNA primer takes place in the presence of specific enzyme proteins. It is possible that the RNA primer is excised residue by residue from its 5′-end, by 5′——→3′ exonuclease activity of DNA pol I whose primary task is to fill in the gaps left between the RNA-primed DNA fragments (Okazaki fragments). In doing so it adds to the 3′-end of the fragment and simultaneously cuts away the RNA primer that lies ahead. Later, the adjacent 5′ and 3′ termini are joined by DNA ligase, or polynucleotide ligase. These enzymes can join short stretches of DNA fragments by forming phosphodiester linkages (Fig. 8.9).

8.4 REPLICATION OF DNA IN EUKARYOTES

Like prokaryotes, the eukaryotic cells also possess DNA polymerases and DNA ligase. At least two different DNA polymerases have been described among eukaryotes. Eukaryotic DNA ligases require ATP instead of NAD as an adenylyl donor. Further, instead of beginning from a single initiation point, replication of eukaryotic DNA begins at multiple points, thus forming a number of *replicating forks* as the parent strands of DNA are separated and replicated. Because of the enormous length of eukaryotic DNA (chromosomes), multiple replication points are a necessity. The replication forks also move slowly as compared to *E. coli*. In eukaryotes the movement of a replicating fork is about 2,600 bases per minute, whereas it is around 16,000 bases per minute in *E. coli*. It has been observed that in *Drosophila* egg cell DNA replication is completed in about 3 minutes and it is believed that around 6,000 replicating forks are employed simultaneously to accomplish this.

Another difference lies in the size of Okazaki fragments, which are much shorter in length (about 100 to 150 nucleotides) in eukaryotes as compared to those of prokaryotes (1000 to 2000 nucleotides). RNA primers are, however, excised in a similar fashion and the fragments of DNA are joined by DNA ligase.

In Vitro DNA Replication

The knowledge of *in vitro* enzymatic synthesis of DNA was made available by the work of Kornberg and his colleagues in bacterial systems and Bollum's work in mammalian systems. During *in vitro* replication of DNA, the sequence of nucleotides in the parent molecule must be accurately reproduced in the daughter molecules since the genetic information contained in the sequence would otherwise be lost or distorted. Therefore, DNA polymerase enzyme system, using the primer DNA, reproduces the exact nucleotide sequence by the semi-conservative method producing a new single strand of DNA on each of the two strands. All four deoxyribonucleiside triphosphates must be present (d ATP, d GTP, d TTP, and d CTP) along with a DNA primer.

$$\begin{matrix} \text{nd TTP} \\ \text{nd GTP} \\ \text{nd ATP} \\ \text{nd CTP} \end{matrix} + \text{DNA} \longrightarrow \text{DNA} \begin{bmatrix} | \\ \text{d TP} \\ | \\ \text{d GP} \\ | \\ \text{d AP} \\ | \\ \text{d CP} \\ | \end{bmatrix}_n + \text{4n PPi}$$

The DNA primer could be from any source, but it must have a high molecular weight. Synthesis may be accomplished to the extent of 20 times or more of the weight of the primer added until one of the substrates is exhausted. If one of the substrates is omitted the extent of the reaction is diminished by a factor of 100 or so. In this so called limited reaction, the addition of a few nucleotides only occurs at the end of the primer chain, but this is also governed by the base pairing rules.

8.5 REVERSE TRANSCRIPTASE

Until recently it has been a basic tenet of the *central dogma* of molecular biology that genetic information can pass from DNA to RNA and then to a protein, or from DNA to a daughter DNA molecule. In other words, it was thought that a new DNA molecule could receive its genetic information only from a previously existing DNA molecule. However, a new situation has been encountered in the replication of certain RNA viruses by D. Baltimore and H. M. Temin. A number of RNA viruses are known to cause cancer in susceptible animals. They include the Rous Sarcoma virus (RSV), which produces cancer in chickens, avian myeloblastosis virus (AMV) which produces leukemia in birds, and Rauscher mouse leukemia virus producing cancers in rodents. These viruses are distinctive in causing permanent, in vitro heritable transformation of certain normal cells into malignant cells. In early 1960s Temin postulated that the RNA of these viruses contains cancer genes which are transcribed into the form of DNA in order to become incorporated as a permanent, heritable part of the host cell genome, possible by the action of a DNA polymerase directed by the tumour virus RNA. It was observed that multiplication in animal cells of such RNA viruses can be prevented by inhibitors of DNA synthesis during the early stages of infection. This observation suggested that transfer of information is possible from the infecting RNA to a DNA copy. This DNA could then be used as a template for the synthesis of viral RNA by a regular transcription process in the cell. However, normal uninfected animal cells were not expected to have the unique enzyme required for the transfer of information from RNA to DNA. Both groups of investigators, therefore, looked for such an enzyme in the virus particle itself—the *virion,* which would carry the enzyme into the host cell. Following the discovery of S. Mizutani and H. M. Temin, and D. Baltimore in 1970, the idea was generally accepted that information can flow back from RNA to DNA. They successfully showed that retroviruses, earlier known as RNA tumour viruses, themselves contain RNA dependent DNA polymerases, also called *reverse transcriptase.* The infecting RNA of retroviruses serves as a template for the synthesis of single-stranded DNA complement that in turn functions as a template for a complementary DNA chain. The resulting double stranded DNA, generally after insertion in a host cell chromosome, then acts as a template for the synthesis of an RNA chain identical to the one which initiated the viral life cycle (Fig. 8.10). The flow of genetic information in the reverse direction (i.e. RNA to DNA) does not pose any conceptual problems, since RNA-DNA hybrid double-helices also exist.

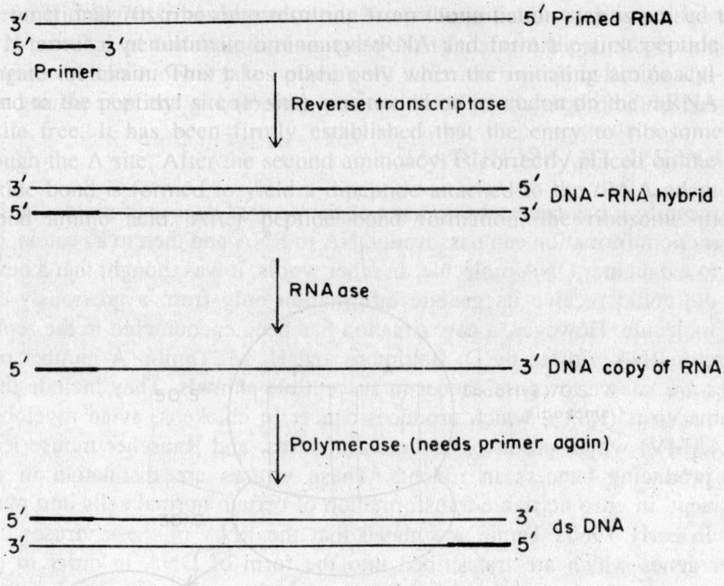

Fig. 8.10 Synthesis of double stranded DNA (ds DNA) from an RNA molecule by reverse transcriptase.

Reverse transcriptase has been isolated and purified from several different RNA-tumour viruses. They have molecular weights ranging from 70,000 to 160,000. They closely resemble DNA dependent DNA polymerase which makes DNA in the 5′——→3′ direction, utilizing deoxyribonucleoside 5′-triphosphates as precursors and require both template and a primer strand with a free 3′-hydroxyl terminus.

The recognition of reverse transcriptase has broadened the basic concepts of molecular biology.

8.5 REPAIR OF DNA

DNA is a relatively fragile molecule and can be easily damaged. Organisms can survive only if their DNA is replicated faithfully and protected from chemical and physical damage that would change its information content. Bacterial genes can survive many millions of replication cycles without any damage. All living cells possess machinery by which damage to DNA can be repaired. The integrity of DNA is protected in a large measure by enzymes capable of accurately repairing certain types of damage or defects in one strand of the DNA so long as the other strand containing complementary information is intact. Despite the safeguards, mistakes of several types occur, such as base substitutions, small and large additions and deletions and gross rearrangements of DNA sequences.

DNA can undergo several types of damages *in vivo*:

(1) X-irradiation and ultraviolet radiations inflict considerable damage on cell DNA. X-irradiations cause nicks or breaks in DNA and ultraviolet

radiations frequently produce dimerization of two adjacent thymine residues to produce a *thymine dimer*.

(2) Highly reactive chemicals, including alkylating agents like nitrosoamine and nitrosoguanidine may modify one or more bases so that pairing mistakes are made.

(3) Purine bases may be lost as a result of local pH change.

(4) Single-strand and double-strand breaks can result from bending or shear forces.

Cells have a large range of enzymes devoted to repairing damage to DNA that would otherwise be lethal. There are three principal types of repair mechanisms.

Photoreactivation: This enzymatic mechanism is catalyzed by photolyase, an enzyme that reverses the photochemical fusion of adjacent pyrimidine bases into pyrimidine dimers caused by ultraviolet light. The enzyme binds to the defective site on the DNA. Illumination of the cell with visible blue light results in the absorption of light energy by the enzyme. The absorbed energy in some manner promotes cleavage of covalent bonds at the defective site in one of the strands. The most common photo-reactivation process is the cleavage of the abnormal cyclobutane ring of a thymine dimer to yield two free thymine residues (Fig. 8.11). By this process the damage caused by ultraviolet light is reversed by the visible light.

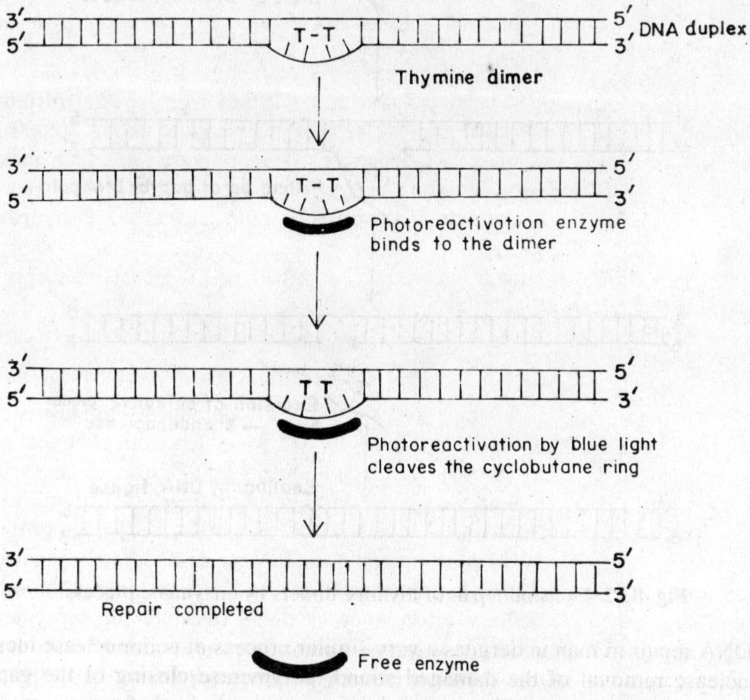

Fig. 8.11 Repair of thymine dimer by photoreactivation.

Enzymatic Processes: These repair mechanisms are independent of light and are known as *dark repair* processes. Although there are a few enzymes that simply reverse the damage, a more universal enzymatic process is *excision repair* in which a damaged segment is removed and replaced through new DNA synthesis for which the undamaged strand acts as a template. For example, if the defect is a thymine dimer, it is detected by 5′——→3′ endonuclease activity of DNA pol I which has a patrolling function and makes an incision at the 5′-side of the defect. The defective segment is then replaced with the correct base pairs by the polymerase action of DNA pol I, which then cuts off or excises the defective segment. The new strand is then joined at its 3′ end to the 5′ end of the remaining intact strand through the action of DNA ligase (Fig. 8.12). This process is also known as *cut patch, cut and seal* process.

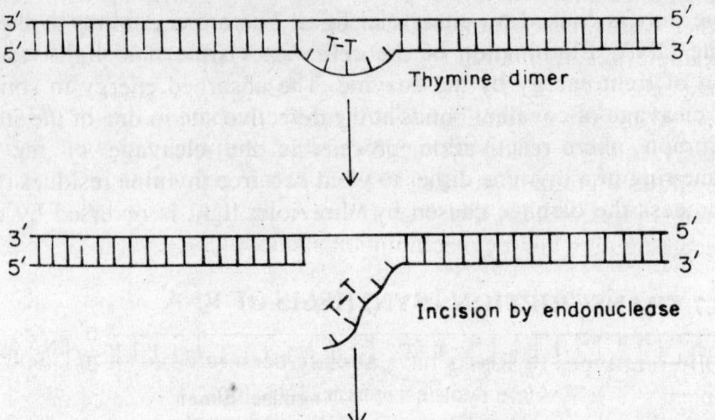

Fig. 8.12 Excision repair of thymine dimers by enzymatic process.

DNA repair in man undergoes a very similar process of endonuclease incision, exonuclease removal of the damaged strand, polymerase closing of the gap and fusion by ligase action. These conclusions were derived from a study of *Xeroderma pigmentosum* , a rare human genetic disease in which the skin is abnormally sensitive to the ultraviolet component of sunlight, resulting in severe skin reaction. Such patients have a tendency to develop skin cancer. It was discovered by J. E. Clever that this disease is due to a genetic deficiency of an enzyme, probably an endonuclease, normally participating in excision repair of UV-induced thymine dimers. When skin fibroblasts (tissue culture cells) of such patients are irradiated with ultraviolet light, thymine dimers are rapidly formed and are not repaired. However, if these fibroblasts were first irradiated with X-rays and then with UV-light to form the dimers, DNA repair occurred rapidly. Evidently X-rays caused chain breaks and then endogenous exonuclease, polymerase and ligase activity repaired the DNA.

Recombinational Repair: An alternate repair method is repair through recombination, which is particularly important if excision repair cannot occur because a template for repair synthesis is not available. In this process a large segment of damaged duplex DNA is replaced by the corresponding intact segment of another DNA molecule from a different but homologous duplex of DNA. In other words, part of a gene is replaced by part of another molecule of DNA. For example, recombination repair occurs after damage to a viral chromosome in the host cell.

A central enzyme of DNA repair in *E. coli* is the *Rec A* protein which catalyzes DNA strand exchange and thus promotes recombinational repair and also regulates a group of at least 15 DNA repair genes called the SOS genes leading to their induction. In this way *Rec A* protein promotes DNA repair both by recombination and by mediating the induction of about 15 SOS genes.

8.7 TRANSCRIPTION - SYNTHESIS OF RNA

Different types of RNAs have already been referred to in chapter 3. While all the species of RNAs are synthesized on DNA templates, each one of them has specific functions in protein synthesis. The messenger RNA is particularly important because it truely represents a blue print of DNA. Thus mRNA is an exact complementary copy of DNA which is decoded and translated into a protein molecule.

DNA-dependent RNA Polymerase

This enzyme was first characterized in 1959 from rat liver by S.B. Weiss, and in the extracts of *E. coli* by J. Hurwitz. Distribution of the enzyme appears to be quite ubiquitous. In rat liver, the enzyme is located in the cell nuclei. In plants it has been shown that it is associated with chromosomal DNA. In eukaryotes the DNA-dependent RNA polymerase or transcriptase is found in the nucleus; whereas in prokaryotes the enzyme occurs in the cytoplasm.

The RNA pol from *E. coli* as well as other organisms catalyzes the net synthesis of RNA with base sequences complementary to the strand of DNA that serves as a template. The enzyme is very similar in its action to DNA pol. It, however, needs no primer. The requirements for the DNA dependent RNA polymerase are the simultaneous presence of the four ribonucleoside triphosphates: ATP, CTP, GTP, and UTP, a divalent cation Mg^{2+} or Mn^{2+}, and native double-stranded DNA. RNA pol catalyzes the reaction

$$\text{NTP} + (\text{NMP})_n \xrightleftharpoons{Mg^{2+} / Mn^{2+}} (\text{NMP})_{n+1} + \text{PPi}$$

The polymerase adds mononucleotide units to the 3′-hydroxyl end of the RNA chain and thus builds RNA in 5′——→3′ direction, antiparallel to the DNA strand used as the template.

The enzyme from *E. coli* has been obtained in homogeneous form by R.R. Burgers and his colleagues. It has a molecular weight of 490,000 daltons. Extensive purification has revealed that the enzyme contains five separable units of four types designated as α, β, β′, and σ. Although the functions of all the subunits are not completely understood, it is clear that the β′ subunit is required for binding the RNA pol to the DNA template, while the β subunit is the catalytic site. The sigma factor (σ) is required for correct initiation of RNA synthesis at the specific site on the DNA template. Without a σ protein, the enzyme (α_2 ββ′) is called the core polymerase and with the sigma factor it is called the holoenzyme (α_2 ββ′ σ), indicating that the σ subunit does not function in catalysis but is a regulatory subunit. Other proteins designated as omega (ω), rho (ρ) and kappa (κ) also cooperate with RNA pol in transcription.

Template Specificity

Under intracellular conditions only one strand of the duplex DNA is transcribed into RNA messages by DNA dependent RNA polymerase. This must be so since if both the DNA strands served as templates, two RNA products of complementary sequences would be transcribed which would code for two different proteins. Genetic evidence tells us that one gene codes for only one protein, hence only one DNA strand also known as the sense strand, functions as a template while the other strand that does not function as a template is called the *antisense strand* (Fig. 8.13).

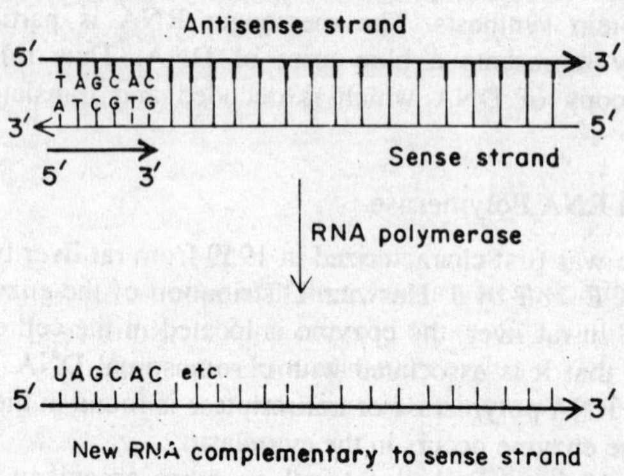

Fig. 8.13 Sense and antisense strands of the DNA duplex and the growing RNA chain as related to the DNA template (sense strand).

Association of RNA Pol with DNA Template

In contrast to DNA pol that requires a primer interaction with the enzyme, RNA pol needs only a duplex DNA and no primer. The transcription begins at specific promoter sites in the DNA template and terminates at the end of a defined gene sequence. Promoter sites have specific sequences of bases that are recognized by the holoenzyme RNA pol at suitable initiation sites which appear to be pyrimidine-rich structures of about 10 residues. Such initiation sites are widely separated in the DNA template. An essential component of binding at the promoter site is the sigma (σ) factor which is probably involved in keeping the holoenzyme in the proper conformation to bind to the initiation site on the DNA template and in opening the DNA (duplex DNA) at the promoter site. Thus binding of the polymerase to the template is followed by strand separation and a conformational change in the holoenzyme.

Initiation of Transcription

The first or the initiating nucleoside 5′-triphosphate which is always ATP or GTP is bound to the polymerase at the promoter site to form the *initiation complex*. The complex in the presence of nucleoside 5′-triphosphates now initiates transcription at the initiation site that is adjacent to the promoter site. The first internucleotide linkage or phosphodiester linkage is formed after binding of the second nucleoside 5′-triphosphate, which is usually a pyrimidine nucleoside, UTP or CTP.

Elongation: Once the first two nucleotide residues have been joined, chain elongation proceeds rapidly with transcription taking place in the 5′——→3′ direction (Fig. 8.14), antiparallel to the 3′——→5′ strand of the template DNA. During elongation a segment of the newly formed RNA forms a transient complementary hybrid RNA-DNA duplex. Once transcription begins and about 10 nucleotide residues are added, the σ factor dissociates from the holoenzyme to yield the core polymerase which proceeds to complete the process of transcription. The free σ factor then becomes available for initiating a new RNA chain with another molecule of core polymerase. The elongation of the RNA chain continues along the DNA template until the core polymerase gets a signal for termination.

Termination: Specific termination signals are necessary because the DNA under transcription has many genes and is often circular. Without a termination signal, the RNA would be transcribed indefinitely. Moreover, *t*RNAs and *r*RNAs have quite specific chain lengths. Therefore, at the end of a gene a sequence of bases signal the completion of transcription. The precise nature of the termination signal is not known, but oligomeric proteins such as rho (ρ) and kappa (κ), which are auxiliary to the RNA polymerase presumably attach to the enzyme blocking further transcription, releasing the RNA chain.

Post-transcriptional Changes in RNA

The newly synthesized RNA chain requires enzymatic processing after transcription before it is available as a finished product. In prokaryotes, certain modifier enzymes occur in the cytoplasm which cleave the precursor RNAs to shorten their lengths until they are converted to ribosomal, messenger or transfer RNA species. Specific bases in *t*RNA and *m*RNA are then modified enzymatically by a variety of reactions such as methylation, acetylation, deamination, reduction etc.

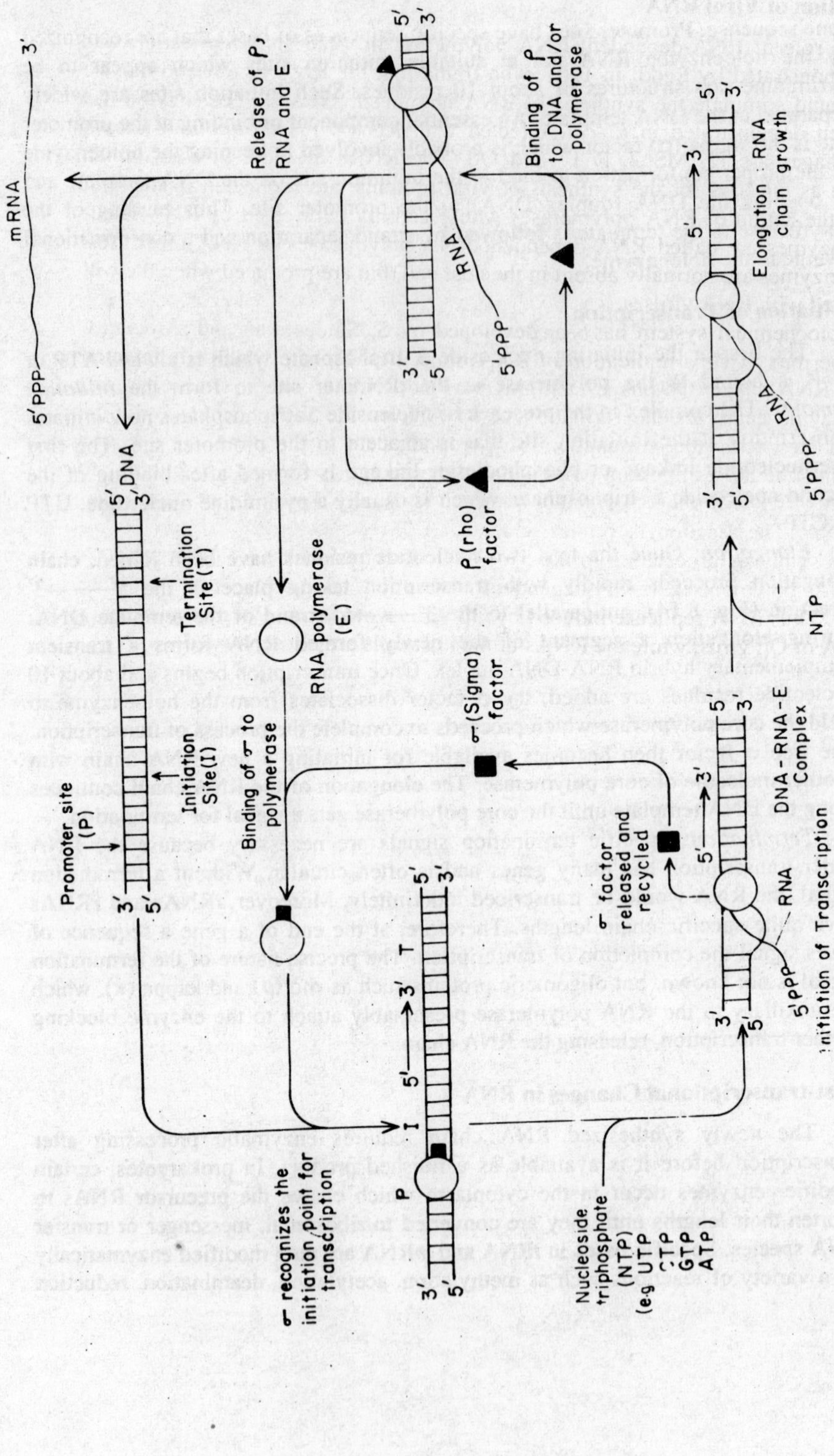

Fig. 8.14 Synthesis of RNA by *E. coli* RNA polymerase.

Replication of Viral RNA

The role of RNA-dependent RNA polymerase in viral RNA replication was first demonstrated by Reddi in 1961, who found that the RNA of tobacco mosaic virus would stimulate the synthesis of RNA in spinach extracts. The enzyme has also been demonstrated in *E. coli* infected with RNA virus. Some well known RNA phages are f2, MS 2, R 17 and Qβ. The RNAs of these viruses, which function as *m*RNAs for the synthesis of viral proteins, are replicated in the host cell by the action of RNA polymerases whose activity is directed by RNA itself. Such enzymes are called RNA-dependent RNA polymerases or RNA replicases. These enzymes are normally absent in the host cell, but are produced when the cell is infected with RNA viruses.

A biochemical system has been developed by S. Spiegelman and associates, which permits *in vitro* replication of RNA molecules. From *E. coli* cells infected with an RNA phage (Qβ phage), a highly purified RNA polymerase was obtained. This enzyme catalyzes the formation of RNA from the four ribonucleoside-5′-triphosphates with the elimination of pyrophosphate and in the presence of Mg^{2+}ions. The most important point is that the pure enzyme requires as a template the RNA of its own phage, i.e. intact homologous viral RNA and is nonfunctional with DNA. The reaction is similar to that of DNA-dependent RNA polymerase with the synthesis proceeding in the 5′——→3′ direction. However, in contrast to the DNA polymerases (DNA-dependent), these RNA replicases are template specific. Thus RNA replicase induced by Qβ virus would employ as template only the RNA of Qβ virus, while the RNAs of the host cells are not replicated.

CHAPTER 9

PROTEIN SYNTHESIS

Nucleic acids are the informational molecules where the genetic information is encoded in the four-letter alphabet and gets translated into proteins through a precise and complicated mechanism. Amino acids are the alphabets of proteins which are assembled into meaningful sequences dictated by the nucleotide sequences on the nucleic acids. The synthesis of peptide bonds is the principal chemical event in protein synthesis which is coupled to a sophisticated machinery for the precise translation of specific nucleotide sequences. Of primary interest is the mechanism by which information from DNA is used in a specific manner for the synthesis of proteins. We have already discussed synthesis of DNA and RNA in Chapter 8 and will now attempt to summarise the orderly process by which proteins are formed.

As early as 1950, Paul Zamecnik first succeeded in achieving cell-free protein synthesis in the extracts of rat liver. Zamecnik and his colleagues were also able to show a requirement of ATP, GTP and the soluble components, and they also identified ribonucleoprotein particles, the ribosomes, as the actual site of protein synthesis. We would also see the involvement of single-stranded RNA molecules in this process of translation. One type of RNA i.e., *m*RNA serves as a template and the other two types namely *t*RNA and *r*RNA play vital roles in protein synthesis.

9.1 ESSENTIAL FACTORS AND STAGES IN PROTEIN SYNTHESIS

Zamecnik and his colleagues found that the soluble fraction of liver furnishes essential factors in protein synthesis. One such heat-labile soluble factor in liver cytoplasm was found to be an enzyme catalyzing an ATP-dependent activation reaction by which free amino acids are esterified to a heat-stable component also present in the soluble fraction, known as *t*RNA. Thus, with the discovery of *t*RNA and amino-acyl *t*RNA synthetases, the synthesis of amino-acyl *t*RNA was investigated as independent of the overall process of protein synthesis. The *t*RNAs are a group of relatively small molecules known as *adapter molecules*, to which amino acids are attached. Experiments with chemically defined amino acid residues, for example, isotopically labelled amino acids, of an amino-acyl *t*RNA provided proof for the hypothesis first propounded by Crick and Hoagland, that *t*RNA is a molecular *adapter* into which the amino acid is plugged so that it may interact in the nucleotide triplet language of the genetic code. The *t*RNA adapters for different amino acids have different structures, each usually adapted for fitting a different nucleotide sequence on to the template. There are about 60 different types of *t*RNA molecules.

Subsequent discoveries of the initiator f-Met *t*RNA and the initiating codon AUG revealed that the synthesis of polypeptide chains takes place in four major stages and that specific enzymes and cofactors are required for different events in each stage. The whole machinery of protein synthesis, thus, consists of more than a hundred components which operate in a coordinated fashion in the cytoplasm of both the prokaryotic and eukaryotic organisms.

To start with, we would deal with the mechanism by which the polypeptide chain is built and by which ribosomes make possible the transfer of genetic information into proper amino acid sequences. The first stage, also called the activation step, takes place entirely in the soluble cytoplasm and in this step amino acids are enzymatically esterified to their corresponding *t*RNAs at the expense of ATP. Activation of amino acids would simultaneously require *t*RNAs, amino-acyl *t*RNA synthetases, ATP and Mg^{2+}.

In the second stage, which is also called the initiation of the polypeptide chain, the messenger RNA which carries the genetic information specifying amino acid sequences and the initiating amino-acyl *t*RNA are bound to the smaller subunit of the ribosome. The process requires three specific proteins or initiating factors, IF-1, IF-2, IF-3, and GTP and Mg^{2+}. Finally in this step the large ribosomal unit attaches to form a functional ribosome. Therefore, the necessary components required for the initiation process are f-Met *t*RNA in prokaryotes, Met-*t*RNA in eukaryotes, *m*RNA, GTP, Mg^{2+}, the three initiation factors and 30S and 50S ribosomal subunits.

In the third stage, also called the elongation step, the polypeptide chain is lengthened by stepwise addition of new amino-acyl residues transferred enzymatically from amino-acyl *t*RNA and each amino acid is bound to the ribosome in response to a specific codon or base triplet in the *m*RNA. The protein which promotes the binding of amino-acyl *t*RNA to the ribosome in the presence of GTP was termed as the elongation factor, EF-T, for this transfer function. Further, it was discovered that EF-T could be separated into two proteins, EF-Ts and EF-Tu. The protein responsible for hydrolyzing GTP in the presence of ribosomes was named EF-G for GTPase activity. The energy is provided by GTP for the ribosomes to move along the *m*RNA to bring the next codon in position for alignment of the next amino-acyl *t*RNA. Therefore, for the elongation process the necessary components are functional 70S ribosomes, amino-acyl tRNA specified by codons, Mg^{2+}, EF-T, EF-G and GTP.

In the last stage of protein synthesis, that is termination, the polypeptide chain is completed when appropriate signals in the *m*RNA are reached and the product is then released from the ribosome, a process that requires specific releasing factors. The necessary components for the chain termination process are functional 70S ribosomes, termination codon on *m*RNA and polypeptide release factors (R_1, R_2 and R_3).

Polarity of Translation

All biopolymers have a polarity in the sense that one end of the molecule is different from the other. A completed polypeptide chain has at one end an amino acid bearing a free amino group (the N-terminus) and at the other end an amino acid bearing a free carboxyl group (the C-terminus). Protein synthesis by ribosomes always involves stepwise addition of single amino acids in relation to the polarity of the messenger template. The polarity of the amino acid assemblies was first analyzed by H. Dintzis from the distribution of radioactivity in the polypeptide chain. Radioactive amino acid was found preferentially at the C-terminus of the protein, meaning that the C-terminal residues are added in the last, and therefore synthesis must start at the N-terminal. However, a clear gradient of increasing radioactivity is observed as one moves from the amino to the carboxyl end. The message for translation is read with $5' \rightarrow 3'$ polarity, hence it follows that in bacteria nascent *m*RNA molecules can be translated while they are still being transcribed at their 3′-terminus on DNA by RNA polymerase.

9.2 ACTIVATION OF AMINO ACIDS

In the first stage of protein synthesis, the 20 different amino acids commonly found in proteins are activated, i.e. esterified to their corresponding *t*RNAs by the action of amino-acyl *t*RNA synthetases. For each amino acid there is a specific activating enzyme whose function is to activate the carboxyl group of that amino acid. This process takes place in the soluble fraction of the cytoplasm and ATP is required for the reaction resulting in the formation of amino acyl-AMP, releasing inorganic pyrophosphate. The reaction is reversible. The amino acyl-AMP, is not liberated in solution, but remains bound to the activating enzyme. This amino acyl-AMP-enzyme complex reacts with an amino acid-specific *t*RNA molecule in the following manner:

$$\text{Amino acid} + \text{ATP} \rightleftharpoons [\text{Amino acyl-AMP}] - \text{Enzyme} + \text{PPi}$$

$$[\text{Amino acyl-AMP}] - \text{Enzyme} + t\text{RNA} \rightleftharpoons \text{Amino acyl } t\text{RNA} + \text{AMP} + \text{Enzyme}$$

The overall reaction catalyzed by specific synthetase can be written as

$$\text{Amino acid} + t\text{RNA} + \text{ATP} \xrightleftharpoons{Mg^{2+}} \text{Amino acyl-}t\text{RNA} + \text{AMP} + \text{PPi}$$

The enzyme is liberated to be used again. For each amino acid there appears to be at least one specific *t*RNA molecule. In fact, there are several types of *t*RNAs for the same amino acid. Such *t*RNAs are called *isoacceptors*. Their existence is connected with the fact that the genetic code often uses more than one nucleotide sequence or codon to specify a given amino acid, a characteristic called degeneracy (Table 9.2). However, there need not be a separate activating enzyme for each of the several *t*RNAs corresponding to a given amino acid. In fact, isoacceptor *t*RNA molecules that bind to different codons are activated by the same amino-acyl-*t*RNA synthetase.

Amino acyl-*t*RNA synthetases are an extremely diverse group of proteins with respect to both size and number of their polypeptide subunits. The enzyme normally has a molecular weight of 100,000 except the one for phenylalanine which has a molecular weight of 180,000. All enzymes have an absolute requirement for Mg^{2+} and all contain one or more -SH groups that are essential for the activity. The amino-acyl-*t*RNA synthetases are so specific in discriminating among the naturally occurring amino acids that there is much less than one chance in 10,000 of an error in intracellular conditions.

As mentioned earlier the activation reaction occurs in two steps on the catalytic site of the enzyme. The first step is the formation of an enzyme bound intermediate-amino acyl adenylic acid or amino acyl AMP.

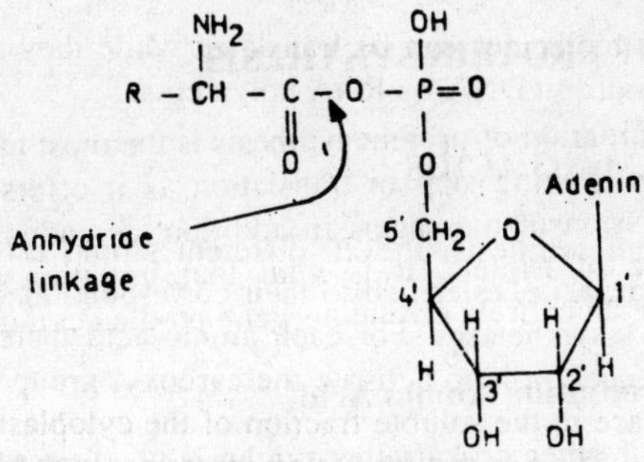

Amino acyl adenylic acid

In this step the carboxyl group of the amino acid is bound to the 5′ phosphate group of AMP through an anhydride linkage with the displacement of pyrophosphate. In the second step, the amino acyl group is transferred to the free 2′ or 3′-hydroxyl group of *t*RNA, i.e. 2′ or 3′-hydroxyl group of the ribose in the terminal adenosine group of the *t*RNA molecule (Fig. 9.1).

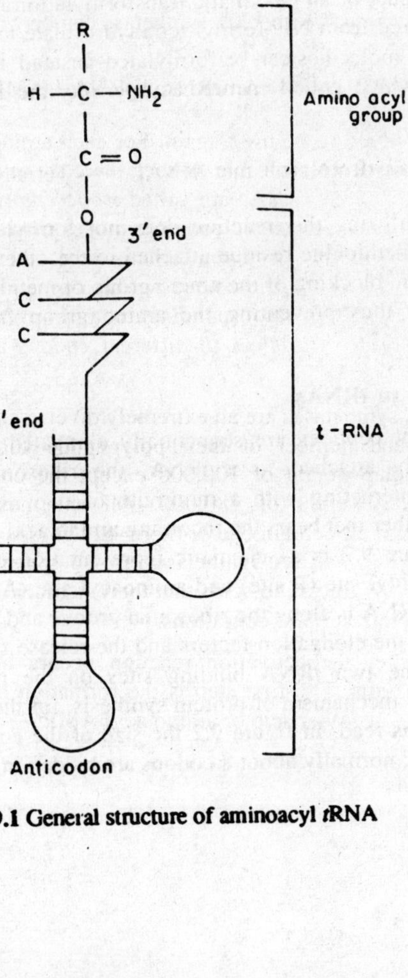

Fig. 9.1 General structure of aminoacyl *t*RNA

9.3 INITIATION OF PROTEIN SYNTHESIS

The mechanism of initiation of protein synthesis is the most interesting event since it is usually the rate-limiting step in translation as it offers a potential site for regulating the expression of messages. In addition, as codons are translated as a continuous sequence of triplets, it is vital that initiation occurs at a correct nucleotide, otherwise an entirely erroneous gene product could result.

Methionine—The Initiating Amino Acid

In *E. coli* and all other prokaryotes synthesis of all proteins begins with the amino acid methionine, coded by the initiating codon AUG. In almost all bacteria the starting amino acid is N-formyl methionine (f-Met). This is a formylated methionine that has a formyl group attached to its terminal amino group. There are evidences to show that almost all proteins are initiated with methionine: formyl methionine in bacteria and subcellular organelles, and methionine itself in the cytoplasm of eukaryotes.

In *E. coli* the initiating methionine residue does not enter as methionyl-*t*RNA but rather as N-formyl-methionyl-*t*RNA (symbolized as f-Met-tRNA). This f-Met-*t*RNA is the product of an enzymatic transformylation reaction in which a formyl group is transferred from N^{10}-formyl tetrahydrofolate to methionyl-*t*RNA. Not all methionyl-*t*RNA molecules can be formylated. Instead, there are two types of met-*t*RNA, one of which called met-*t*RNA$_F$ permits the formylation reaction. The reaction is

$$N^{10}\text{-formyl tertrahydrofolate} + \text{met-}t\text{RNA}_F \longrightarrow \text{Tetrahydrofolate} + \text{f-met-}t\text{RNA}_F$$

The enzyme catalyzing this reaction does not formylate free methionine or as stated earlier, a methionine residue attached to the other species of met-*t*RNA. In the above reaction, blocking of the amino group of methionine is carried out by the N-formyl group, thus preventing the amino group from entering the peptide linkage.

Ribosomes Bind to *t*RNAs

Since nascent peptides are sequentially elongated at the C-termini and are always covalently attached to a *t*RNA, the ribosomes must be capable of simultaneously interacting with a minimum of two *t*RNAs: one that bears the peptide and the other that bears the incoming amino acid that will serve to elongate the peptide. Figure 9.2 is a schematic representation of the ribosome structure showing the peptidyl site (P site) and aminoacyl site (A site) on the 50S subunit. Binding of the *m*RNA is along the ribosomal groove and other binding sites for the initiating factors, the elongation factors and the release factors are also present on the ribosome. The two *t*RNA binding sites on the ribosome are the key to understanding the mechanism of protein synthesis, for they are the points at which the genetic code is read. In figure 9.2 the size of the codons is not shown in the proper proportion; normally about 8 codons are hidden in the ribosomal groove.

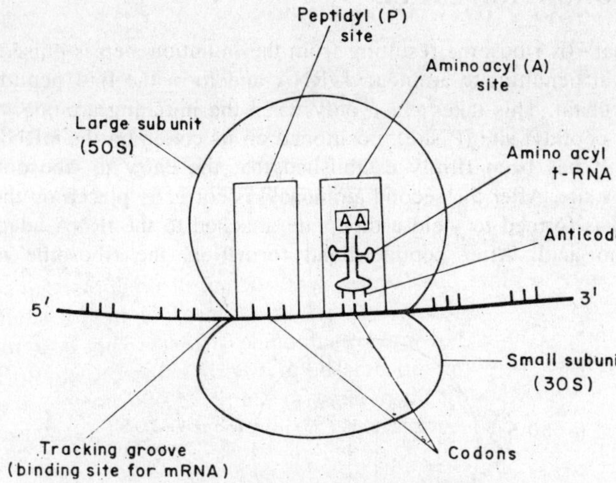

Fig. 9.2 An *E. coli* ribosome showing major binding sites.

Formation of the Initiation Complex

Initiation of protein synthesis in bacteria starts with the formation of a complex between the smaller 30S ribosomal unit, f-Met-*t*RNA$_F$ and an *m*RNA molecule. During protein synthesis there is continuous dissociation of 70S ribosomes into their 50S and 30S subunits and continuous reassociation of the units to form intact ribosomes again. First Kaempfer and his colleagues and then Nomura and his associates showed that the 70S ribosome must first dissociate into 50S and 30S subunits before polypeptide synthesis can begin. The messenger RNA and the initiating aminoacyl *t*RNA cannot bind directly to 70S ribosomes but are first bound to the 30S subunits, which then combine with the 50S subunit. Subsequent research has, however, shown that this is a rather complex multistep process requiring the participation of three specific initiation factors (IF-1, IF-2, and IF-3) having molecular weights of about 9,000, 65,000 and 21,000 respectively. These factors are not permanently bound to the 30S subunit, but are loosely bound to ribosomes and undergo cycles of attachment and release. The sequence of events in the formation of a complete functional 70S ribosome in prokaryotes is shown in Figure 9.3. The 30S ribosomal subunit along with the initiation factors first forms an initiation complex with the initiator *t*RNA and the *m*RNA at an appropriate initiator codon. In the next step the 50S subunit joins this complex and the initiation factors dissociate to form the complete functional 70S ribosome. In this process GTP is hydrolyzed to GDP and Pi. Presumably this elaborate initiation process is required to ensure that the initiating aminoacyl *t*RNA is bound at the peptidyl site and positioned at the initiation codon AUG so that the ribosomes start translation at the correct point on the *m*RNA template. The initiation factors are used over and over again to start new chains. As already established, translation of the codons of *m*RNA takes place in the 5′——→3′ direction, thus the P site and the A site on the ribosomes recognize the polarity of the *m*RNA chain.

9.4 THE ELONGATION CYCLE

The functional 70s ribosome resulting from the initiation step is poised to accept the N-terminal penultimate aminoacyl-*t*RNA and form the first peptide bond to elongate the chain. This takes place only when the initiating aminoacyl-*t*RNA is bound to the peptidyl site (P-site), positioned on its codon on the *m*RNA with the A site free. It has been firmly established that the entry to ribosomes occurs through the A site. After the second aminoacyl is correctly placed on the A site, a peptide bond is formed to yield a dipeptide attached to the *t*RNA adaptor of the second amino acid. After peptide bond formation, the ribosome moves the

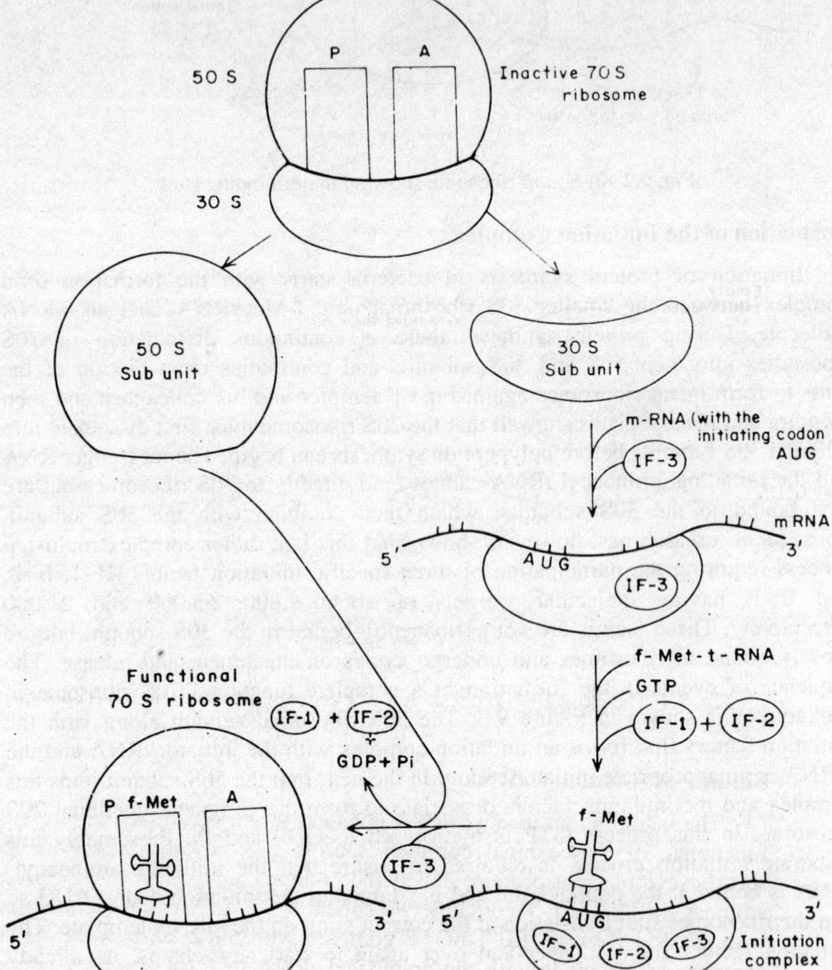

Fig. 9.3 Formation of the initiation complex and the functional 70 S ribosome.

message along so that the *t*RNA with the dipeptide is shifted to the P site and the next adjacent codon triplet is positioned at the vacant A site; a new aminoacyl-*t*RNA is bound to it and the second peptide bond is formed. This process of addition of the amino acid is repeated over and over again, adding one amino acid at a time to form a complete chain. In the elongation cycle a sequence of three reactions is repeated with each adjacent codon (Fig.9.4).

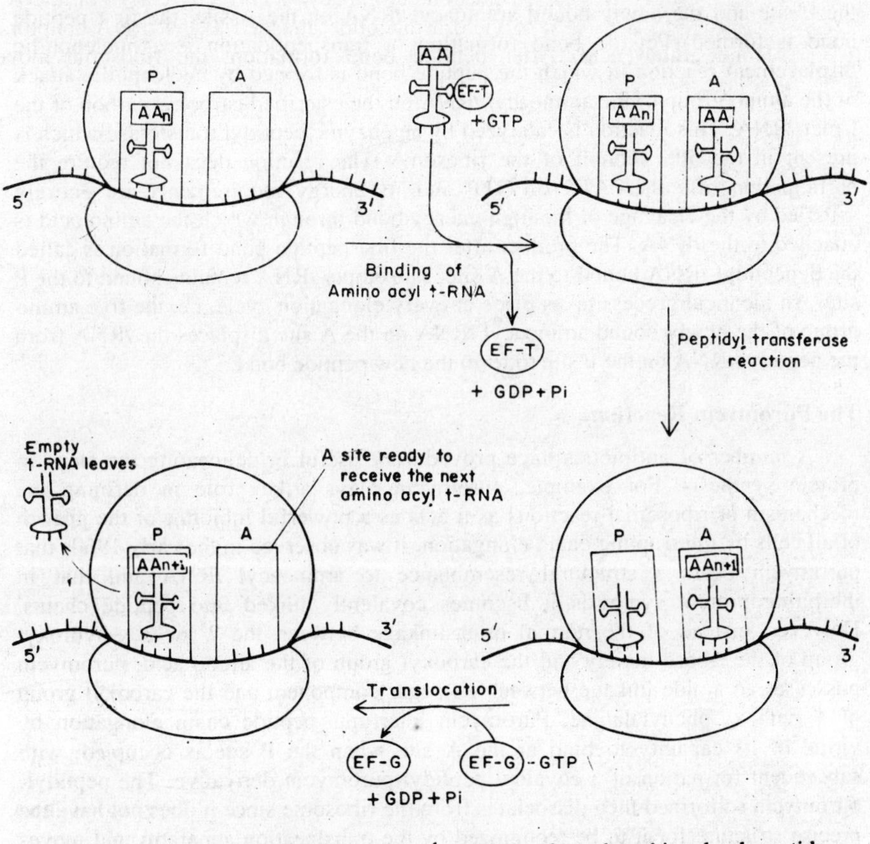

Fig. 9.4 The elongation process. AA n^{+1} denotes a formyl methionyl polypeptide.

Binding of Aminoacyl *t*RNA

The incoming aminoacyl *t*RNA is bound to the vacant A site of the complete functional 70 S ribosomal complex. During this process, first a nonribosomal specific cytoplasmic protein, called the elongation factor T (EF-T) binds to the aminoacyl *t*RNA. EF-T contains two subunits, EF-Ts and EF-Tu. EF-T combines with GTP to form an EF-Tu-GTP complex with the release of EF-Ts. It is this complex which binds with the aminoacyl *t*RNA to form a ternary complex (i.e. EF-Tu-GTP-aminoacyl *t*RNA complex). This complex binds to the ribosome in such a way that the aminoacyl *t*RNA is positioned on the A site. Simultaneously GTP is hydrolyzed to GDP and Pi. GDP leaves the ribosome in the form of an EF-

Tu-GDP complex. This mechanism, at the expense of GTP hydrolysis, enhances the accuracy of amino acid placement at the correct position.

Peptide Bond Formation

The formation of the peptide bond is the only subreaction of protein synthesis. This reaction, given appropriately bound substrates, is catalyzed by a component contained entirely within the large ribosomal subunit. With the f-met-*t*RNA on the P site and the newly bound aminoacyl *t*RNA on the A site, the first peptide bond is formed. Peptide bond formation or transpeptidation is a nucleophilic displacement reaction in which the peptide bond is formed by nucleophilic attack of the amino group of the aminoacyl *t*RNA on the esterified carboxyl carbon of the f-met-*t*RNA. This reaction is catalyzed by an enzyme peptidyl transferase which is present in the 50S subunit of the ribosome. The reaction does not require the participation of either ATP or GTP and its energy requirements are entirely satisfied by the cleavage of the high-energy bond through which the amino acid is attached to the *t*RNA. The product after the first peptide bond formation is called the depeptidyl-*t*RNA bound to the A site. The empty *t*RNA remains bound to the P site. An identical process takes place at every elongation cycle, i.e. the free amino group of the newly bound aminoacyl *t*RNA on the A site displaces the *t*RNA from the peptidyl *t*RNA on the P site to form the new peptide bond.

The Puromycin Reaction

A number of antibiotics have proved very useful in delineating the steps in protein synthesis. For example, puromycin plays a key role in defining the mechanism of ribosomal reactions as it acts as a powerful inhibitor of the growth of all cells by interrupting chain elongation. It was observed in the early 1960s that puromycin bears a structural resemblance to aminoacyl *t*RNA and that in inhibiting protein synthesis it becomes covalently linked into peptide chains. However, instead of the normal ester linkage between the 2′ or 3′—hydroxyl group of the ribose moiety and the carboxyl group of the amino acid, puromycin possesses an amide linkage between its sugar component and the carboxyl group of 4-methoxy-phenylalanine. Puromycin interrupts peptide chain elongation by virtue of its capacity to bind at the A site when the P site is occupied, with subsequent formation of a covalent peptidyl-puromycin derivative. The peptidyl-puromycin so formed then dissociates from the ribosome since it does not have the precise structure for it to be recognized by the translocation apparatus and moves to the P site.

Translocation Process

The stepwise sequential reading of the genetic code requires the ribosome to move along the message on *m*RNA. After the peptide bond has been made, the peptidyl *t*RNA which has just been elongated is still bound to the A site, positioned at the codon of the newly added amino acid. In this step of the *m*RNA elongation cycle, the ribosome moves to the next codon of the *m*RNA and simultaneously shifts the peptidyl-*t*RNA from the A to the P site. This translocation is one of the most mechanically complex reactions of protein synthesis. A specific protein, the elongation factor G (EF-G), also known as translocase, is required which catalyzes the hydrolysis of GTP. In this reaction,

GTP first binds to EF-G and the resulting complex binds to the ribosome. GTP is hydrolyzed to GDP and Pi and the energy of hydrolysis is used for the conformational changes that move the ribosomes to the next codon on the *m*RNA and thus shift the peptidyl *t*RNA from the A site to the P site. Normally, at each translocation step the *m*RNA template is advanced precisely three nucleotides, i.e. A site becomes vacant with a new codon ready to receive a new aminoacyl *t*RNA and start a new elongation cycle. After the translocation step the factor EF-G dissociates from the ribosome.

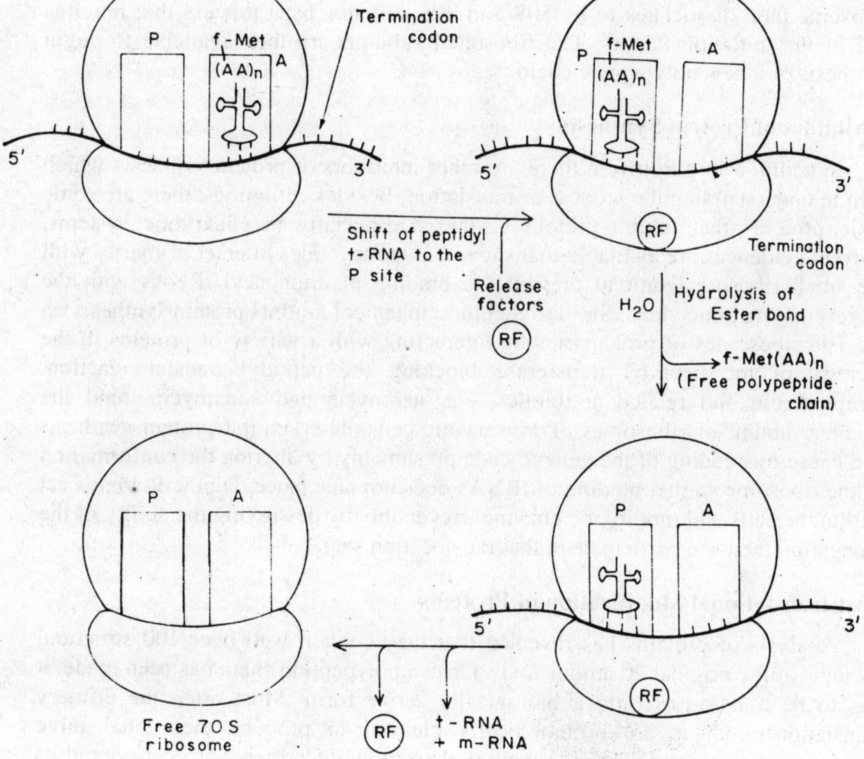

Fig. 9.5 Termination of protein synthesis-steps in release of the completed polypeptide chain.

9.5 TERMINATION OF PROTEIN SYNTHESIS

Two conditions are necessary for termination of protein synthesis. One is the presence of a codon that specifically signals that polypeptide elongation should stop. The other is the presence of a release factor (RF) which reads the chain terminating signal. Termination of polypeptide chains is signalled by one of the three special terminating codons in the *m*RNA. Behind all this complexity is the fact that after the polypeptide chain has reached its full length, its carboxyl end is still bound to its *t*RNA adapter. Termination must therefore involve the splitting of the terminal *t*RNA. Release of the peptidyl *t*RNA from the ribosome is promoted by three specific release factors, RF-`1, RF-2, and RF-3, isolated by Caskey and Capechi from *E. coli*. RF-1 responds specifically to the triplets UAA and UAG, while RF-2 responds to UAA and UGA, and RF-3 does not have release activity of its own but it stimulates the reaction catalyzed by RF-1 or RF-2. The release factors bind to terminating codons to cause a shift of the polypeptidyl *t*RNA from the A site to the P site (Fig. 9.5). Whether GTP hydrolysis is required for termination in *E. coli* is not yet firmly established, although the release factor RF-3 which appears to stimulate the reaction does bind GTP and GDP. The ester bond between the polypeptide chain and the last *t*RNA is then hydrolyzed. Once the polypeptide chain is released, the last *t*RNA and the *m*RNA leave. The free 70S ribosome then dissociates to its 50S and 30S subunits by a process that requires one of the initiation factors. The ribosomal subunits are then available to begin synthesis of a new polypeptide chain.

Inhibitors of Protein Synthesis

In addition to puromycin there are other inhibitors of protein synthesis which help in understanding the process of translation. Besides antibiotics, there are some toxic proteins that inhibit protein synthesis, especially in eukaryotic systems. Current evidences are available that show that tetracyclines interact primarily with the small ribosomal unit to prevent the binding of aminoacyl *t*RNAs with the corresponding anticodons. Similarly, chloramphenicol inhibits protein synthesis on the 70S ribosomes of prokaryotes by interacting with a variety of proteins in the vicinity of the peptidyl transferase blocking the peptidyl transfer reaction. Streptomycin and related antibiotics, e.g. neomycin and kanamycin, bind the smaller subunit of ribosomes of prokaryotic cells. They inhibit protein synthesis and cause misreading of the genetic code presumably by altering the conformation of the ribosome so that binding of *t*RNAs does not take place. Diptheria toxins act within the cells and modify the enzyme irreversibly by destroying the ability of the elongation factor to participate in the translocation step.

Post-translational Modification of Proteins

Analysis of proteins has revealed that they contain well over 100 structural variants of the original 20 amino acids. Once a polypeptide chain has been made, it has to be transformed into a biologically active form. Most often the primary translation products are proteolytically cleaved to produce their final three dimensional structures. These structural aberrations of proteins are considered as post-translational modifications.

The formyl group of f-met residue does not appear in the finished proteins of prokaryotes. It is removed by the action of an enzyme, *deformylase*. The initiating methionine residue of some polypeptides may also be removed by a methionine aminopeptidase. A number of other simple modifications are also observed on different amino acids particularly as they occur at the termini of proteins. Such examples are acetylated and amidated amino acids. Some proteins are acetylated on the amino terminal, on the o-amino group, such as the coat protein of the tobacco mosaic virus. In addition, a number of interesting single amino acid transfer reactions have been observed at the amino and carboxyl termini of proteins. These one-step reactions occur by non-ribosomal means and result in amino acids that are joined at the ends of proteins by peptide bonds not specified by the message. The functional significance of these modifications is obscure, but they result in either altered protein functions or are essential to the protein function. Interchain disulphide cross-linkages are also formed enzymatically after the polypeptide chain has been built. A wide variety of carbohydrates become attached to proteins in many different ways, especially in cell membranes. During the post-translation period, methylation, phosphorylation and hydroxylation of proteins by specific enzymes also occurs. This may involve attachment of coenzymes or prosthetic groups that are essential to enzyme function. Their covalent attachment to the polypeptide is by means of the amino acid side chains.

9.6 THE GENETIC CODE

In the preceding discussion we have seen the role of ribosomes and *t*RNAs in aligning amino acids in the correct sequence to synthesize a specific polypeptide. Mention has also been made of the codons and anticodons. Let us now define the term *code* more precisely and also consider how the four-letter language of nucleic acids is translated into the twenty-letter language of proteins. The genetic dictionary of codes for the amino acids is now well-known and fully confirmed by genetic and biochemical evidences. While calculating the size of the coding units, it has been seen that a three-letter or *triple code* is most feasible for coding amino acids. This means that a triplet code can theoretically account for a total of 4^3 or 64 different combinations. Therefore, a codon coding for a given amino acid consists of a group of three adjacent nucleotide residues on *m*RNA which directs the precise synthesis of amino acid sequence of a protein.

During the 1960s, a number of experiments were performed in the laboratories of Nirenberg and Khorana to examine the composition of various triplet codes. Finally, these biochemists were able to assign specific amino acids to 61 out of the 64 possible codons. The remaining three codes are now designated as terminating codons (Table 9.1).

Generalizations of the Codon Dictionary

A few generalizations about the genetic code may be considered here.

1. The code is universal and is applicable to all organisms.
2. The genetic code is commaless, without any punctuation. There is no signal to indicate the end of one codon and the beginning of the next.
3. The code is non-overlapping, i.e. the adjacent codons do not overlap.
4. The code is degenerate, implying that there is more than one code for most of the amino acids. For example, UUA, UUG, CUU, CUC, CUA and CUG are the codes for leucine.

There is no code that specifies more than one amino acid, hence degeneracy is not imperfect. But degeneracy is not uniform either. The code for arginine, leucine and serine has six-fold degeneracy, i.e. there are six codes for each one of them, whereas some amino acids may have two-fold degeneracy. Of all the amino acids only tryptophan and methionine have but a single code (Table 9.2).

Table 9.1 The Codon Dictionary

	U	C	A	G	
U	UUU Phe UUC Phe UUA Leu UUG Leu	UCU Ser UCC Ser UCA Ser UCG Ser	UAU Tyr UAC Tyr UAA Ter UAG Ter	UGU Cys UGC Cys UGA Ter UGG Trp	U
C	CUU Leu CUC Leu CUA Leu CUG Leu	CCU Pro CCC Pro CCA Pro CCG Pro	CAU His CAC His CAA Gln CAG Gln	CGU Arg CGC Arg CGA Arg CGG Arg	C
A	AUU Ile AUC Ile AUA Ile AUG Met	ACU Thr ACC Thr ACA Thr ACG Thr	AAU Asn AAC Asn AAA Lys AAG Lys	AGU Ser AGC Ser AGA Arg AGG Arg	A
G	GUU Val GUC Val GUA Val GUG Val	GCU Ala GCC Ala GCA Ala GCG Ala	GAU Asp GAC Asp GAA Glu GAG Glu	GGU Gly GGC Gly GGA Gly GGG Gly	G

5. In a given codon, the first two bases are specific but the third base is less specific and is flexible. For most of the cases, arginine, leucine and serine being the exceptions, the degeneracy of the code involves only the third base in the codon. For example, the codons for phenylalanine are *UU*U or *UU*C , and alanine is coded by *GC*U, *GC*C, *GC*A and *GC*G. In some cases the third base can simply be either a purine or a pyrimidine, for example, histidine is coded by CA*U* and CA*C* in which the third base is a pyrimidine. Similarly, the codons for glutamine are CA*A* and CA*G*, where the third base is a purine. The third base is therefore called a *wobble*. Nearly all codons can be represented as XY^{A}_{G} or XY^{U}_{C}.
6. Of the 64 triplet codes, three codes, namely UAG, UAA and UGA do not code for any amino acids and are used as signals for chain termination.
7. The codon AUG is a code for methionine. It serves as an initiator code as well as for insertion of methionine in a polypeptide.

Table 9.2 Degeneracy of codes

Amino acid	*No. of codons*	*Amino acids*	*No. of codons*
Ala	4	Leu	6
Arg	6	Lys	2
Asn	2	Met	1
Asp	2	Phe	2
Cys	2	Pro	4
Gln	2	Ser	6
Glu	2	Thr	4
Gly	4	Trp	1
His	2	Tyr	2
Ile	3	Val	4

CHAPTER 10

GENETIC RECOMBINATIONS AND GENE TRANSFER

DNA is the genetic material and specific regions on this molecule that direct the synthesis of polypeptide chains are called *genes*. Genes are composed of many codons. When a gene is considered as a unit of function, it is known as a *cistron*. Thus a cistron is the smallest unit of genetic material responsible for the synthesis of a specific polypeptide.

Experimental work with micro-organisms, such as *E. coli* and the mould *Neurospora crassa* has shown that the genetic make-up of the organisms could be altered by manipulation. The altered DNA expresses itself in the phenotype of the organisms. In this chapter we shall study the mechanisms by which new genetic recombinations could be achieved.

Genetic recombination involves some kind of movement of genetic markers between two distinct chromosomes derived from two different sources. The recombination events may occur during the crossing-over event of meiotic chromosomes, bacteriophage infection, bacterial conjugation and bacterial transformation. In all these cases genetic recombination occurs by different mechanisms, though some similarities do exist among them. Lederberg and Tatum who were pioneers in demonstrating genetic recombination in bacteria, concluded that genetic recombination in bacteria was not the result of fusion of two cells, but rather involved the transfer of a piece of the genetic material, i.e. DNA from one individual to another. This gave rise to an altogether new genetic constitution of the recipient cell or organism with new characters.

Genetic recombination is characterised by a high degree of precision ensuring neither gain nor loss of the genetic material. For some time it was thought that recombination might occur during replication. Recent evidence however, indicates that chromosomes were physically broken during recombination and that pieces from two chromosomes were exchanged. This has been proved both in virus and eukaryotic systems. It has been suggested that recombination is generally accompanied by DNA synthesis in both prokaryotic and eukaryotic cells.

10.1 TRANSFER OF CHARACTERS

In eukaryotes, single, complete sets of genes are brought together during fertilization and combined in the zygote. In the diploid zygote, after mitotic divisions, a recombination between the two sets of genes and a reduction (by meiosis) to the single set of genes takes place. The gametes so formed are haploid. These new combinations of genetic material are different from the parasexual processes, including the recombination of characters in prokaryotes. Bacteria are almost always haploid. They possess only one set of DNA molecule (chromosome) and during recombination, usually, only a part of the genetic material of a donor cell is transferred to a recipient cell, resulting in the formation of a partial zygote (merozoite). The chromosomes of the recipient and the partial chromosome of the donor pair and exchange segments. Subsequent division of the chromosome and the cell results in daughter cells that contain only the recombined chromosome.

Gene-mapping of the relative positions of different genes in bacterial chromosomes is made possible by the fact that under special circumstances a segment of the chromosome of one type of cell containing one or more genes may become detached and transferred to the chromosome of the second type of cell into which it becomes covalently incorporated. The identity of the recombined genes can be deduced from changes in the phenotype of the recipient cell and its progeny. Such recombination of genes is a common event in eukaryotes which usually reproduce by the sexual method, but it is relatively infrequent in prokaryotes.

The genes of *E. coli* and its viruses are now understood better than those of eukaryotes and other forms of life. They provide the simplest examples of what genes are and how the genetic code operates. However, the exploitation of the genome of *E. coli* and its properties reveal that the information, needed to properly regulate the rate at which specific genes work, is also encoded in the nucleotide sequence of DNA.

The use of mutants has been vital for precise genetic knowledge. Thus, by studying a specific aberration in a particular mutant of *E. coli*, the functioning of a normal cell is revealed. In contrast, single-cell eukaryotic microorganisms, such as the green alga *Chlamydomonas*, yeast *Saccharomyces cervisiae* and the mould *Neurospora crassa* were the first subjects of genetic analysis.

10.2 GENETIC RECOMBINATION

A simple experiment was devised to detect recombination in *E. coli* in which only the recipient cells were able to multiply but the parent strains, with specific growth requirements, could multiply in a minimal media. In these experiments two strains of bacteria, each possessing different growth requirements, were mixed together (Fig. 10.1). Neither strain was able to grow in the absence of its specific growth requirement; the amino acids threonine and leucine were required by one strain, and the vitamin biotin and the amino acid methionine by the other. After the two strains were mixed together, a small number of cells were able to grow without addition of any growth factors to the medium. This result showed that the growing cells had somehow acquired wild-type copies of each of their mutant genes, strongly suggesting that *E. coli* has a sexual phase that can bring together the chromosomes of two different cells. Crossing over could then take place between two chromosomes carrying mutations in different genes and in some cases, to a chromosome with wild-type alleles for all necessary genes.

According to the current state of research, at least three different mechanisms are recognized by which DNA is transferred to a bacterial cell and can be recombined with the recipient bacterial chromosome or with a plasmid in vivo:

(1) a general, homologous recombination,
(2) a site-specific recombination, and
(3) a non-homologous recombination.

General Homologous Recombination

Homologous recombination comprises the mechanism by which the DNA that has been transferred into the recipient cell recombines with the resident DNA by reciprocal exchange of DNA segments (Fig. 10.2). The recombining DNA partners must have more or less the same base sequence, i.e. they should exhibit maximum homology with the exception of any mutational differences. On reflection it is obvious what mechanism precisely aligns DNA molecules for crossing over or homologous recombination; complementary base pairing between strands unwound two different chromosomes and put them in exact register. Crossing over thus creates genetic diversity by producing different combinations of particular genetic alleles.

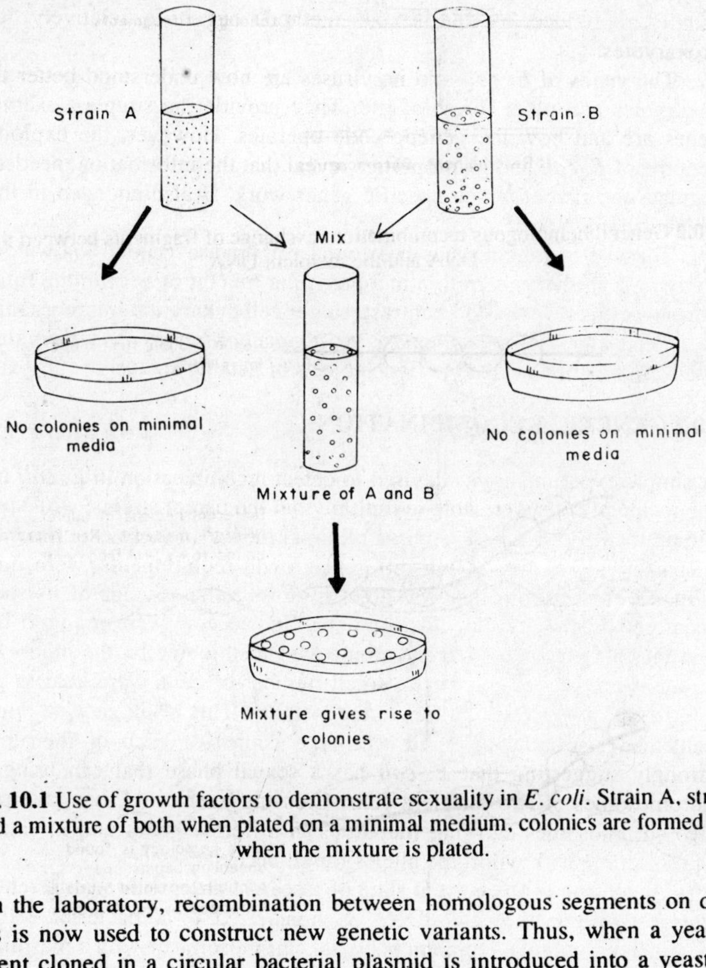

Fig. 10.1 Use of growth factors to demonstrate sexuality in *E. coli*. Strain A, strain B and a mixture of both when plated on a minimal medium, colonies are formed only when the mixture is plated.

In the laboratory, recombination between homologous segments on different DNAs is now used to construct new genetic variants. Thus, when a yeast DNA fragment cloned in a circular bacterial plasmid is introduced into a yeast cell, it finds and crosses over with its homologous sequence, thereby inserting the plasmid into a yeast chromosome. At least in bacteria and yeast, any two homologous segments will recombine in the cell, provided one of them has a break or a region of single-stranded DNA.

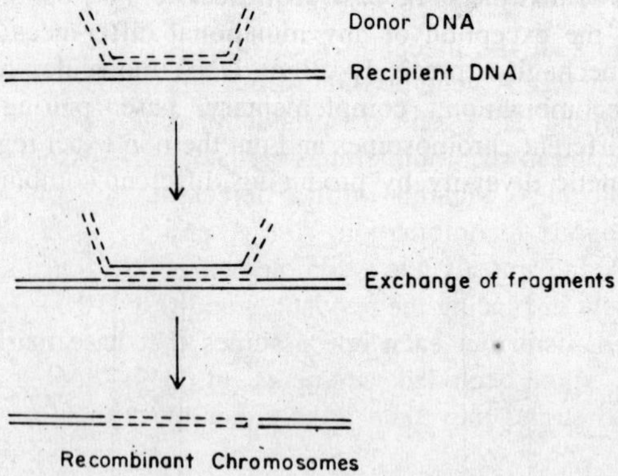

Fig. 10.2 General homologous recombination: exchange of fragments between the donor DNA and the recipient DNA

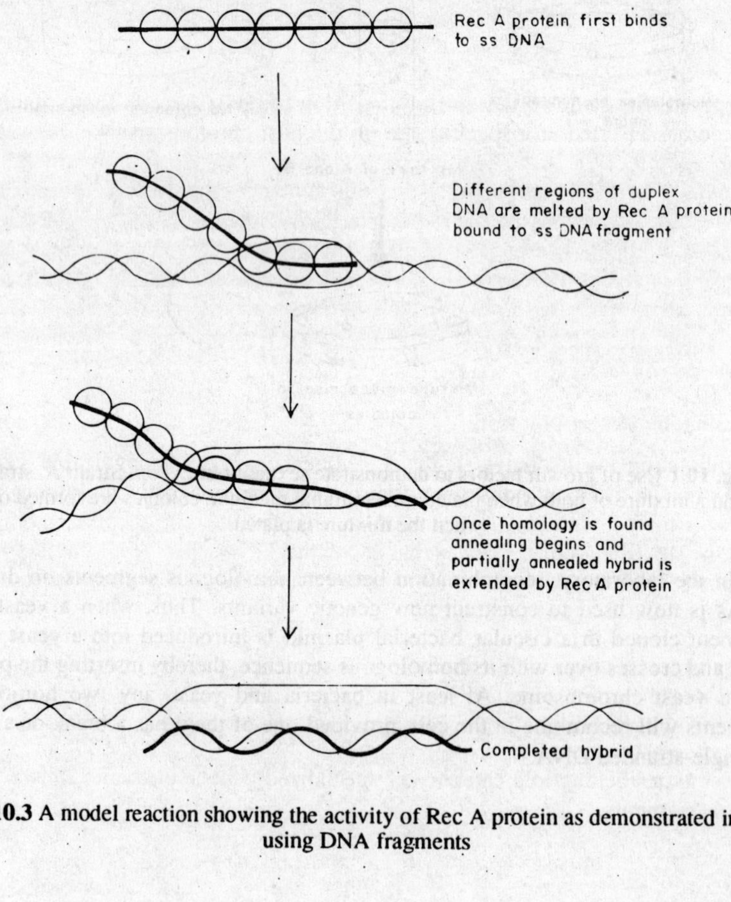

Fig. 10.3 A model reaction showing the activity of Rec A protein as demonstrated in vitro using DNA fragments

Role of Rec A Proteins in Homologous Recombination

Our understanding of the mechanism of homologous recombination comes from the studies of the Rec A gene product of *E. coli*. Homologous recombination is under the control of the *rec A* gene, the product of which, an enzyme, carries out the fundamental reaction in recombination, i.e. making the base paired hybrid that joins two DNA molecules. Mutants with a defect in this gene (*rec*⁻) lose the capacity for homologous recombination. *Rec A* protein specifically recognizes single-stranded DNA and anneals it to a complementary sequence in a homologous duplex, simultaneously displacing the resident strand (Fig. 10.3). There are several models for the mechanism, but each one assumes that base pairing takes place between unwound, single-stranded segments of both double-stranded DNA partners. The second strand may then arise either by replication or through the repair mechanisms.

Site-specific Recombination

Site-specific recombination is independent of homologous recombination and can occur in *rec*⁻ mutants. It consists of integration of a small, double-stranded DNA segment at a specific site on a large double–stranded DNA. The small partner loses its autonomy by this integration. A common example of site-specific recombination is the integration of the bacteriophage lambda (λ) into the chromosome of *E. coli*.

Experiments have shown that during transition of the lambda to the prophage state, it becomes inserted at a specific site on the host chromosome, i.e. between *gal* and *bio* operons (Fig.10.4). The insertion is initiated by the strands which are in close contact. Earlier it was thought that this contact is due to some homology between the base sequences of the segments. However, the homology of these segments is only slight. The process actually involves a phage coded-protein, the integrase. This catalyzes a break at a specific attachment site (*att B*) in λ DNA and another break at another site (*att* λ) of the host chromosome between the *bio* and *gal* operons, and the subsequent crossed reunion of double strands of the phage and the host genomes.

Non-homologous Recombination

Recombination events between DNA segments without recognizable gene homology are classified as non-homologous recombination. This represents an integrated form of recombination similar to that in the site-specific process involving the addition rather than exchange of the DNA. The non-homologous recombination is independent of the *rec A* gene. The following types of DNA are able to take part in such recombinations:

(a) insertion sequences or IS elements.
(b) transposons or Tn elements.
(c) bacteriophage Mu or Mutator elements.

All the three types belong to a category of specialized genetic elements known as transposable elements.

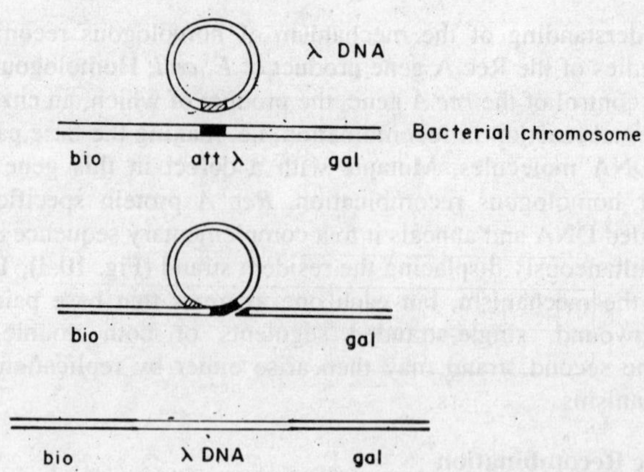

Fig. 10.4 Site-specific recombination: the integration of phage lambda into the chromosome of host bacterial cell.

Transposable Genetic Elements: It used to be thought that genes have a relatively fixed chromosomal location which changes only as a consequence of chromosomal rearrangement resulting from illegitimate crossing over between incompletely homologous short sections of DNA. In the late 1940s, experiments with maize (corn) showed that certain genetic elements regularly jump to new locations. In such jumping or transposition they often inactivate the genes with which they become attached. The existence of such mobile elements was thus first revealed by their capacity to disturb the functioning of other previously characterized genes. Through the use of newly developed recombinant DNA technology and DNA sequencing technique, the nature of *E. coli* transposing elements quickly became established at the molecular level. These elements were found to be DNA sequences that code for enzymes which bring about the insertion of an identical copy of themselves into a new DNA site. Transposition events involve both recombination and replication processes which frequently generate two daughter copies of the original transposable element. One copy remains at the parental site while the other appears at the target site.

The insertion of transposable elements invariably disrupts the integrity of their target genes. Since transposable elements carry signals for the initiation of RNA synthesis, they sometimes activate previously dormant genes. To summarize, these elements can be defined as small, mobile DNA sequences that move around chromosomes with near total disregard for homology and insertion of these elements may produce deletions, inversions, chromosomal fusions and even more complicated rearrangements. Often, they carry genes that allow them to become selective, such as specifically antibiotic resistant genes.

Types of Transposable Elements: Two types of transposable elements have been identified in bacteria (Fig. 10.5):

(1) simple transposons, also called as insertion sequences (IS),
(2) complex transposons, known simply as transposons.

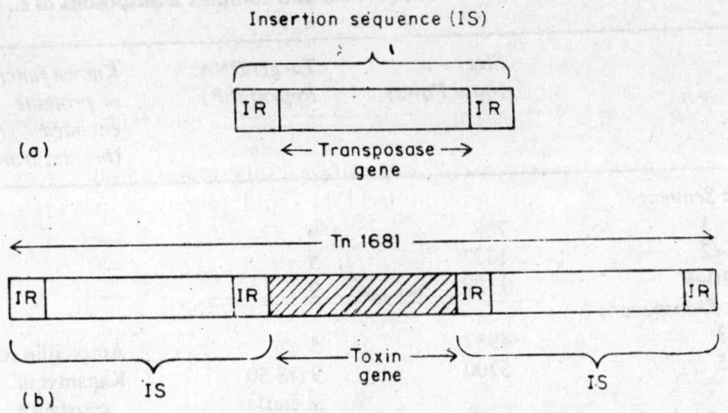

Fig. 10.5 (a) A simple transposon (IS) carries only the genetic information necessary for transposition which is often enclosed by short regions of inverted repeat (IR) sequences. (b) The complex transposon is composed of two insertion sequences and contains genetic information in addition to that needed for transposition. In Tn 1681 transposon carries the gene for a heat-labile toxin.

It was discovered about two decades ago that certain spontaneous mutants of *E. coli* result from insertion of extraneous DNA (alien DNA). Such mutations can occur in structural and regulatory genes anywhere on the chromosome. The extraneous DNA consists of insertion sequences (IS elements) which occur in chromosomes of bacteria and plasmids. These insertion sequences do not carry any genetic information except that necessary for transposition i.e., the transposase gene, which is the coding region for an enzyme transposase that recognizes the ends. The transposase gene region is often enclosed by specific terminal DNA sequences which are short regions of inverted repeat (IR) sequences. The IS elements have 800–1400 base pairs devoid of any code for recognizable phenotype characters. Little is known about their function, but their mutagenic action is due to their insertion in the DNA and consequent errors in transcription. It can be assumed that IS elements play a significant role in reorientation and joining of the genetic material.

Complex transposons are transposable elements in which terminal sequences enclose additional genetic material unconnected with transposition. In some complex transposons such as Tn 1681, two complete IS elements enclose one to several functional bacteria genes, often carrying remnants of IS elements at the ends. These are most easily recognized when they carry identifiable gene markers, such as those which specify antibiotic resistance production of toxin.

Transposons can be integrated into the genome at a number of places, though not randomly. They can integrate from a plasmid to a bacterial chromosome, to another plasmid or to a temperate phage. As the transposons contain genes that determine recognizable characters, they are more easily recognized as compared to IS elements.

There is a good evidence of the presence of transposons in all organisms. Some transposable elements of *E. coli* are given in Table 10.1

Table 10.1 Some Insertion sequences and complex transposons of *E. coli*.

	Size (base Pairs)	*Target DNA Repeat (bP)*	*Known functions or proteins encoded (besides transposase)*
Insertion Sequences			
IS – 1	768	9	—
IS – 2	1327	5	—
IS 10–R	1329	9	—
Complex Transposons			
Tn 3	4957	5	Ampicillin resistance
Tn 5	5700	9 (IS 50 at end)	Kanamycin resistance
Tn 10	9300	9 (IS 10 at end)	Tetracycline resistance
Tn 681	2100	9 (IS 1 at end)	Heat stable enterotoxin
Tn 2571	23000	9 (IS1 at end)	Resistance to chloramphenicol, fusidic acid, streptomycin, sulphonamides, and mercury.

Bacteriophage *Mu*

The bacteriophage Mu has the same unusual integration behaviour as transposons and IS elements. It is a temperate bacteriophage having typical phage properties and could be regarded as a giant transposon. At first, phage *Mu* (Mutator) was thought to behave in a unique fashion since it does not multiply in a lytic way. It always commences its life-cycle by inserting itself into the *E. coli* chromosome at random locations. The resulting *prophage* often mutates the genes into which they become inserted (Fig. 10.6). Thus phage *Mu* was originally known for its mutagenic properties. Later in the life-cycle, upon receipt of an appropriate signal, the *Mu* prophage somehow generates many new copies of itself, which are then reinserted randomly in the *E. coli* chromosome. DNA-cutting events subsequent release new copies which are then encapsulated forming a large progeny of *Mu* particles. In effect, phage *Mu* is a transposon that is supposed to have not only the genes that regulate its transposition, but also the genes encoding structural proteins which package its DNA. It is unique also in transposing much more often than other transposons, providing nearly a hundred new copies of *Mu* during its hour-long life cycle.

Since phage *Mu* is a transposon, the terminals of its DNA are never free, instead they are embedded in other DNA of the bacterial host. Like other transposons, *Mu* has inverted repeats (IR) at and near the ends of its DNA and it makes a short duplication with 5 base pairs of the bacterial DNA into which it is inserted.

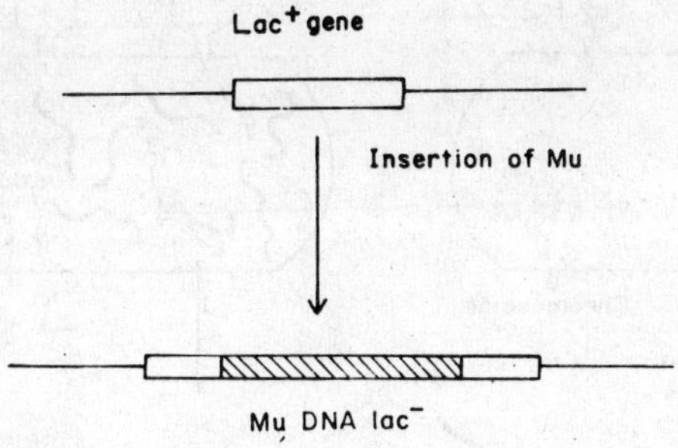

Fig. 10.6 Insertion of phage Mu DNA into a lac gene renders the gene inactive (lac)

10.3 PHAGE CROSSES

In normal conditions more than one phage particle at a time can grow in a single bacterium. If several particles adsorb at once, the chromosome from each phage particle enters the cell and duplicates to form a large number of copies. So long as the chromosomes exist free, i.e. are not enclosed by a protective coat, they can cross-over into similar chromosomes. This is shown by infecting cells having two or more genetically distinct phage particles and finding recombinant types among the progeny (Fig. 10.7). The frequency with which the recombinants appear depends upon the particular mutant used in the cross. Crosses between some pairs of markers give almost 50% recombinant phages; crosses between others give somewhat lower recombinant values, and sometimes almost no recombinants are found.

The genetic structure of phage chromosomes is similar to that of cellular chromosomes, but it is important to note a distinct feature of phage crosses. In meiosis each chromosome pairs just once, whereas in a viral cross, pairing and crossing over may occur repeatedly throughout the period when free chromosomes are present in the infected cell. Thus a given phage chromosome may participate in several pairing and crossing-over events. Another way in which a phage cross differs from recombination in conventional meiosis is that the phages are not separated into male and female particles. Phage particles do not have to fuse since crossing over occurs when two phage chromosomes of different genotypes enter a single host (bacterium). Moreover, in cells infected with several genetically distinct phages, crossing over can occur between both genetically identical and genetically different chromosomes. Thus, only a fraction of the cross-overs in a cell infected with different phages results in recombinant chromosomes of a new genotype.

10.4 GENE TRANSFER MECHANISMS

Three types of gene transfer mechanisms are known in bacteria:

(a) Conjugation,
(b) Transduction, and
(c) Transformation

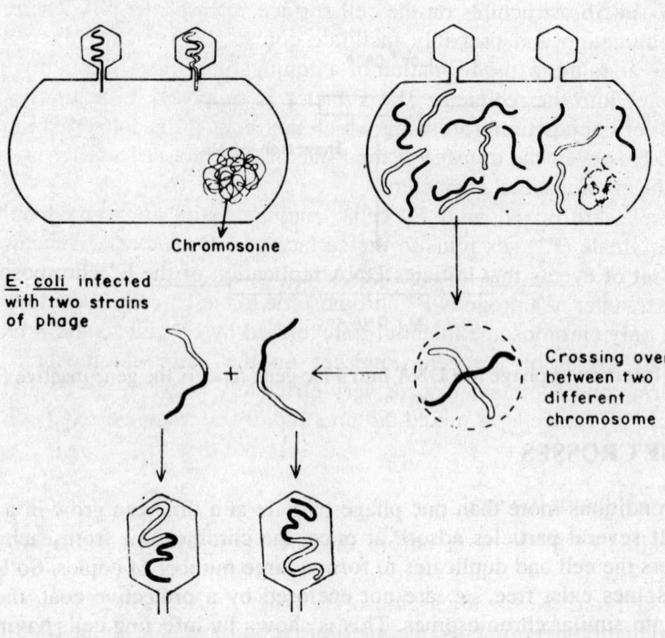

Fig. 10.7 Phage recombination in cells infected with two different strains of a phage

In all these processes, DNA is transferred from a donor bacterium to a recipient bacterium. The processes differ only in the manner in which the DNA is transported. The transfer process is immediately followed by DNA recombination in the recipient cell. In this the donor DNA is integrated into the DNA of recipient bacterium which is designated as a recombinant.

Conjugation

The transfer of genetic material from donor to the recipient by direct contact is called conjugation. In *E. coli* sexual cycle has distinctive features complicating conventional analysis. As in higher organisms, there are male and female cells, but these do not fuse completely. Instead, the transfer of genetic material is always unidirectional, with male chromosome moving into the recipient cells, the potential females. The converse movement of the female genes into male cells never occurs.

Male and female *E. coli* cells are distinguished by the presence of a distinct sex chromosome called the sex factor, fertility factor or the F factor, which is a transferable DNA segment. When it is present as a discrete body, the cells are male (F^+) and capable of transferring genes into female cells. In the absence of this fertility factor the *E. coli* cells are females (F^-) and act as recipients of genes. Like the main *E. coli* chromosome, the F-factor is a closed circular double-stranded DNA molecule present in one copy per cell. The relative particle weight of the molecule is 45×10^6 and it carries 94×10^3 base pairs which is 1/40th of the information content of the main chromosome. About one-third of F^+DNA consists of 19 transfer (*tra*) genes, which are specifically involved in the transfer of genetic material from males to females. Among these are those which are responsible for synthesis of sex-specific structures on the cell surface, such as the pili. These are essential for conjugation and probably facilitate contact between the donor and recipient cells so as to allow the formation of a conjugation bridge through which the DNA migrates into the recipient. The F factor is supposed to be transferred during the event of conjugation, following which the recipient is converted in to a potential donor. However, the transfer of the F factor does not necessarily involve the transfer of the chromosomal characters.

When F^+ cells are mixed with F^- cells, conjugal pairs are formed by the attachment of a male (F^+) sex pilus to the surface of a female cell. Attachment then triggers a set of events that initiates DNA replication of the F^+ chromosome and leads to the transfer of a progeny F^+ chromosome to the F^- cell. It is important to note that the only chromosome material transformed by F^+ cells is the F factor itself (Fig. 10.8). None of the genes present on the main chromosome are transferred. Pairing and F^+ transfer occur very efficiently between the F^+ and F^- cells. As a result, all the cells in mixed cultures rapidly become male (F^+) donor cells.

High Frequency Recombination

In a population of F^+ bacteria only a few are occasionally able to transfer chromosomal DNA. In such cells, through a crossing-over event the F factor becomes integrated into the single chromosome. When clones of such donor cells are used in conjugation experiments, the yield of recombinants is about 1000-fold higher than with ordinary F^+ strains. Such cells are called *Hfr* – high frequency recombinants (Fig.10.9). The integration of the F factor can occur in a limited number of sites on the bacterial chromosomes and in different orientations, which determine the beginning and the direction of chromosome transfer. Incorrect excision of the F factor from the chromosome gives rise to a F-factor containing a short piece of DNA which is called the F′ factor (Fig. 10.10). The cell containing an F′ factor is called a primary F′ cell.

The transfer process. When a population of *Hfr* cells is mixed with an excess of F^- cells, all the *Hfr* cells find F^- partners and conjugate. The conjugation–induced replication of the F DNA begins, and as the leading edge of the F^+ factor is now attached to the main chromosome, the transfer of the main chromosome follows (Fig. 10.11). Because DNA replication is involved, transfer is inherently a lengthy process requiring some 100 minutes at 37°C to bring a complete male genome into a female cell. However, complete transfer rarely occurs. The conjugal union between male and the female is very fragile, usually breaking and terminating transfer before it has been completed. Since the main chromosomal genes are inserted into the F^+ factor, only a portion of the F^+ genome is usually transferred into the recipient F^- cells. Consequently, the recipient cells usually remain female, whereas the $F^+ \times F^-$ matings invariably convert the recipient into a male F^+ cell.

Gene Transfer by F′ factors

The integration of the F factor is reversible. The F factor can be excised from the chromosome so that the *Hfr* cell reverts to a F^+ cell (Fig. 10.12). This excision occurs at about the same frequency as the frequency at which integration occurs. Correct excision depends on a break occurring on the chromosome at the same site as the integration site. In rare cases however, the break occurs at a neighbouring site so that on excision a neighbouring segment of DNA remains attached to the F factor (see Fig. 10.10). Such factors are called F′ factors and the cell containing an F′ factor is called a primary F′ cell. The DNA integrated into the F′ factor can be transferred to an F^- cell with the same frequency (100%) as the F factor from F^+ strains to F^- strains. The same piece of chromosomal DNA would be transferred from *Hfr* cells to F^- cells only with a frequency of 1%. Transfer of F′ factor from a primary F′ cell in which it originated, to a normal F^- cell results in a secondary F′ cell. In this cell, the segment of the bacterial chromosome is present twice, i.e. in the diploid state.

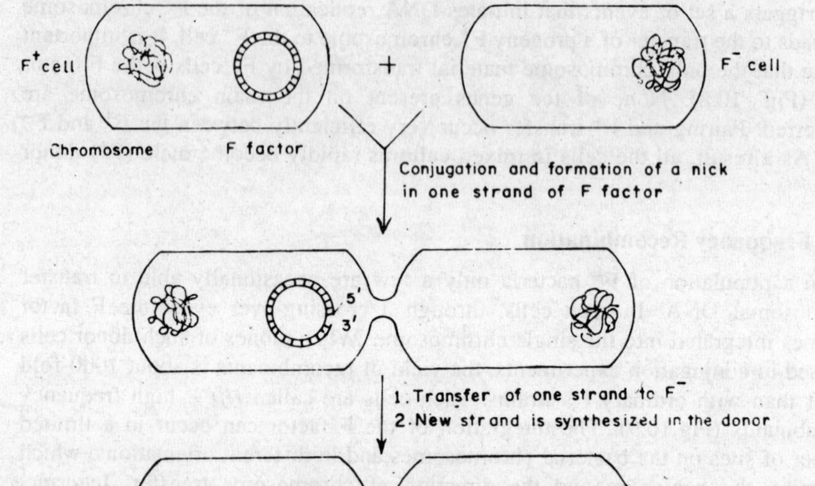

Fig. 10.8 Transfer of a F^+DNA to a F^- cell

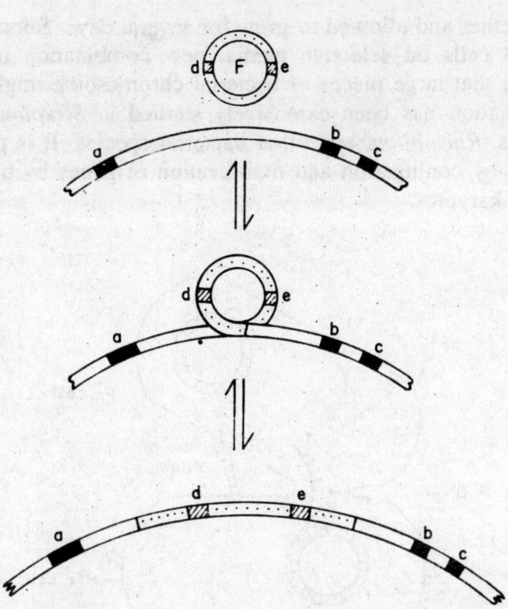

Fig. 10.9 Creation of an Hfr state by integration of the F factor into main bacterial chromosome

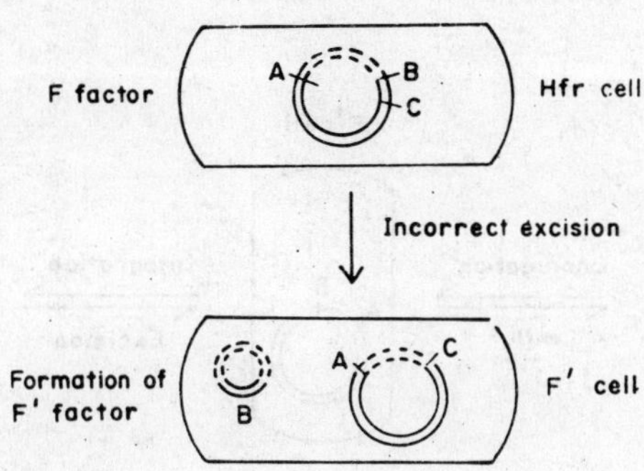

Fig. 10.10 Incorrect excision of the F factor from the chromosomal DNA

Conjugation in Other Groups of Bacteria

Genetic transfer by conjugation was discovered in *E. coli* and is very common in enterobacteria. Transfer of F factors from *E. coli* K 12 to *Salmonella* and *Shigella* has produced new genetic systems. Similar systems have been discovered in pseudomonads. The conjugation process in enterobacteria is highly developed and occurs readily when the partners are suspended in a liquid medium and allowed to stand for some time. In many other bacteria, conjugation takes place only when colonies of the two partners are intimately mixed in an agar medium streaked out together and allowed to grow for several days. Subsequently, by isolating individual cells on selective media, new combination of characters are found, indicating that large pieces of bacterial chromosome might have been exchanged. Conjugation has been extensively studied in *Streptomyces coelicolor*, *Nocardia* species, *Rhizobium*, and other bacterial species. It is probable that exchange of genes by conjugation and mobilization of genes by plasmid is widely distributed in prokaryotes.

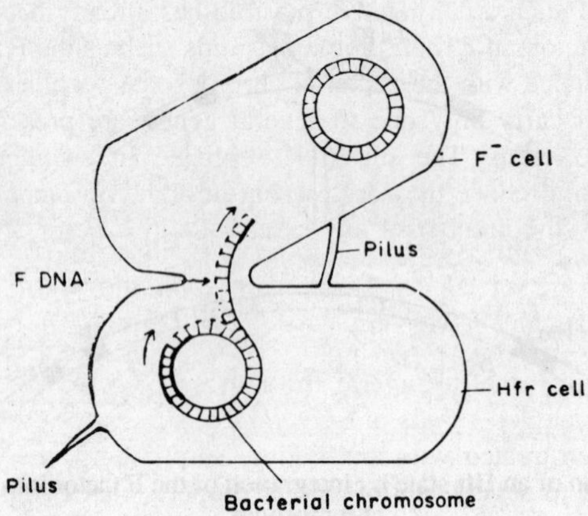

Fig. 10.11 Movement of the bacterial chromosome from *Hfr* cell to a F⁻ cell.

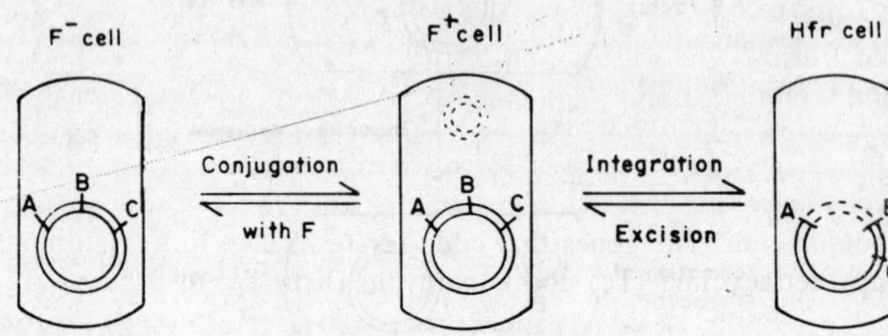

Fig. 10.12 Diagram showing reversible integration of the F factor into the bacterial chromosome

10.5 PLASMIDS

Many bacteria harbour extrachromosomal DNA elements, called plasmids. DNA molecules that possess their own origins of replication to self-duplicate are called *replicons*. The main bacterial chromosome generally circular, is a replicon. Plasmid DNAs are also circular, but much smaller molecules, and are capable of self-replicating. Plasmid DNAs usually encode no essential functions, and the bacteria lacking them multiply in a normal fashion. They are therefore, not essential for growth under normal conditions and plasmids in fact act as dispensable accessory sources of DNA which may impart unique functions to bacteria recognizable by some special characters conferred on the host cell. Some plasmids enable the host cells to conjugate with other cells. F factors, for example, are plasmids that encode the proteins which allow bacteria to conjugate and donate the genetic material which can be integrated with the host chromosome. They mobilize the host chromosome and mediate its transfer to another cell. In this way, further distribution of the plasmid by direct contact between cells is also facilitated.

A prototype of such a conjugative plasmid has already been presented in the fertility factor (F factor) of *E. coli*. Some plasmids such as the F factors are present in 1 : 1 correspondence with the bacterial chromosomes, while others, particularly the small ones that carry only one to several genes, are present in many copies (upto 50) per bacterium. The plasmids replicate independently of the main bacterial chromosome. Since they are present in high copy numbers, they provide a means to amplify the number of any essential single-copy gene that is inserted into their genome.

Resistance (R) Factors

Many bacteria resistant to several antibiotics were discovered in Japan during the 1950s. These included strains of *Shigella*, causing diarrhoea, isolated from patients who had been treated with antibiotics. Surprisingly, these bacteria showed multiple resistance to several antibiotics, a property that could be transferred to other bacteria, including *E. coli*. It is now known that resistance transfer factors (RTF) carry several genes that confer resistance to many kinds of antibiotics, such as ampicillin, streptomycin, chloramphenicol, sulphonamides and tetracyclins. Some resistance factors determine resistance to as many as eight antibiotics. Others confer resistance to toxic heavy metals (e.g. Hg, Ni, Cd, & Co). The R plasmids carry two groups of genes: (i) those which determine the transferability of the plasmid by conjugation (tra genes) and represent the resistance transfer factor; and (ii) the genes which determine the actual resistance, constituting a small part of the plasmid (Fig. 10.13). In the plasmid RP4, for example, two regions have been recognized which carry tra genes, a large and a small tra fragment. These are responsible for conjugation. The genes that code for resistance to the antibiotics ampicillin (Amp), tetracycline (Tc) and kanamycin (Km) are distributed all over the plasmid.

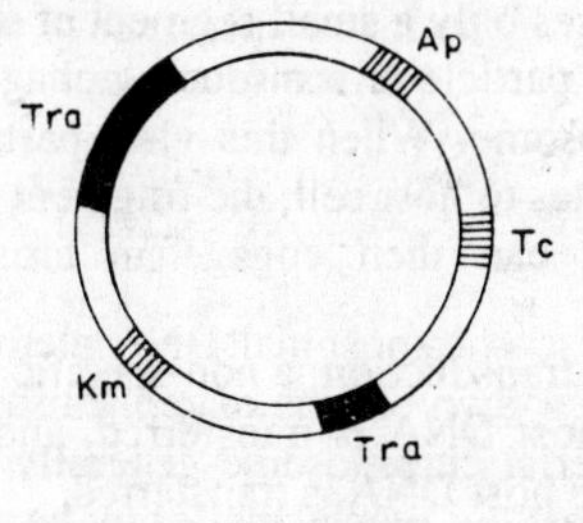

Fig. 10.13 Genetic map of the resistance plasmid RP4. Two regions carry tra genes which are responsible for conjugation.

The resistance transfer factor (RTF) includes all those genes that are responsible for the transfer of the R factor from one cell to another, usually by conjugation. The RTF region is actually found to have a large degree of structural homology with the F factors of *E. coli*. Some R factors are characterized by a wide host spectrum. They can be transferred between several bacterial genera. In some cases, the transfer of the R factor is accompanied by the transfer of chromosomal genes, evidently mobilized by the R factor. Since the R factor can also undergo genetic recombination, there is the possibility that new combinations and extensions of resistance properties may arise.

10.6 INSERTION OF PHAGE CHROMOSOMES

Some bacteriophages, such as lambda (λ), do not always multiply upon entering a host cell. Instead, their DNA is inserted in to the host chromosome. The phage chromosome is now an integral part of the host chromosome and replicates like the bacterial chromosome (Fig. 10.14). The phage chromosome becomes a *prophage*, sometimes called an episome. Bacteria containing prophages are called *lysogenic* bacteria and the phages are known as lysogenic phages. In contrast, some phages such as T_2 and T_7 always multiply when they enter a host cell and lyse the cell because of the large number of phage particles produced. These are called lytic phages. How the phage DNA is inserted into the host chromosome is best understood by the route followed by phage lambda. Integration of λ is achieved by crossing-over between *E. coli* chromosome and a circular form of the lambda chromosome. Prior to integration, the original lambda chromosome assumes a circular shape and attaches itself to a specific region of the *E. coli* chromosome. Both the host as well as the phage lambda chromosomes then break and rejoin in such a way that the broken ends of the phage chromosome join the broken ends of the *E. coli* chromosome, thereby inserting the phage DNA in the chromosome (Fig. 10.15). This prophage form of DNA detaches from the host chromosome by a reverse process. The prophage-host chromosome junctions pair prior to cross-over event that ejects the λ chromosome. The free λ genome can then begin to multiply as if it were the chromosome of a lytic virus.

10.7 TRANSDUCTION

Besides conjugation and transposition, the bacterial genes from one bacterium to another can be transferred passively by a process called *transduction*. Transduction is the transfer of DNA from a donor to a recipient bacterial cell through bacteriophages. In most cases only a small segment of the host DNA is transferred. This happens when a virus particle or transducing phage accidentally encapsulates sections of a host chromosome. When this viral particle, usually defective and biologically inactive, attaches to host cell, the fragment of bacterial chromosome is injected into the cell. It can then engage in crossing over with the host chromosome.

There are two types of transduction: a non-specific or generalized transduction in which any part of the host DNA is transferred, and a specific transduction in which specific region of the host DNA is transferred.

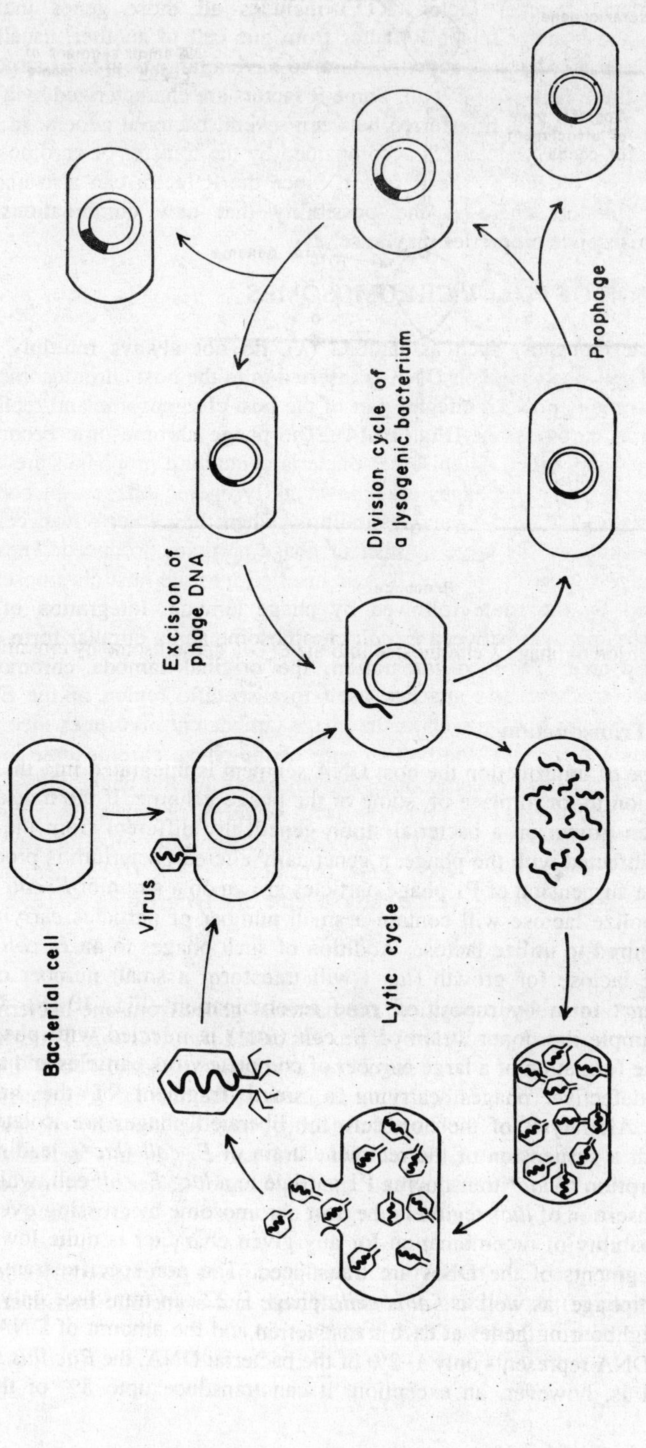

Fig. 10.14 The dual life style of a temperate phage. Infection of *E. coli* with phage can result in either phage multiplication and lysis or lysogenization of the bacterium.

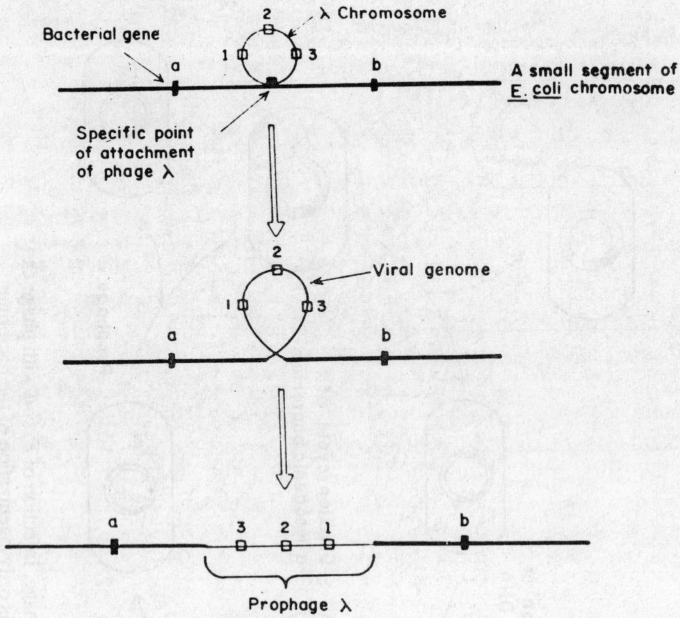

Fig. 10.15 Insertion of phage λ chromosme into the *E. coli* chromosome by crossing over

Non-specific Transduction

In this type of transduction the host DNA segment is integrated into the virus, either in addition to, or in place of, some of the phage genome. If the transducing phage has been grown on a bacterial strain genetically different from the strain subsequently infected with the phage, a genetically altered bacterium is produced. For example, a suspension of P1 phage particles grown on a strain of *E. coli* that is able to metabolize lactose will contain a small number of particles carrying the gene lac^+ required to utilize lactose. Addition of such phages to an *E. coli* strain unable to use lactose for growth (lac^-) will transform a small number of lac^- bacteria to lac^+ form by means of gene recombination (Fig. 10.16). In this particular example the donor strain of *E. coli* (lac^+) is infected with phage P1, resulting in the formation of a large number of complete virus particles and a small number of defective phages carrying a small fragment of the bacterial chromosome. After lysis of the host cell, the liberated phages are isolated and incubated with a suspension of the recipient strain of *E. coli* (lac^-), leading to a possible adsorption of lac^+ transducing P1 particle to a lac^- *E. coli* cell, which can result in the insertion of lac^+ region in the host chromosome by crossing-over.

The probability of recombination for any given character is quite low, since only small segments of the DNA are transduced. The non-specific transducing phage P1 (coliphage) as well as *Salmonella* phage P 22 can transduce only single or closely neighbouring genes at each transduction and the amount of DNA equal to the phage DNA represents only 1–2% of the bacterial DNA. the *Bacillus subtilis* phage PBS 1 is, however, an exception: it can transduce upto 8% of the host genome.

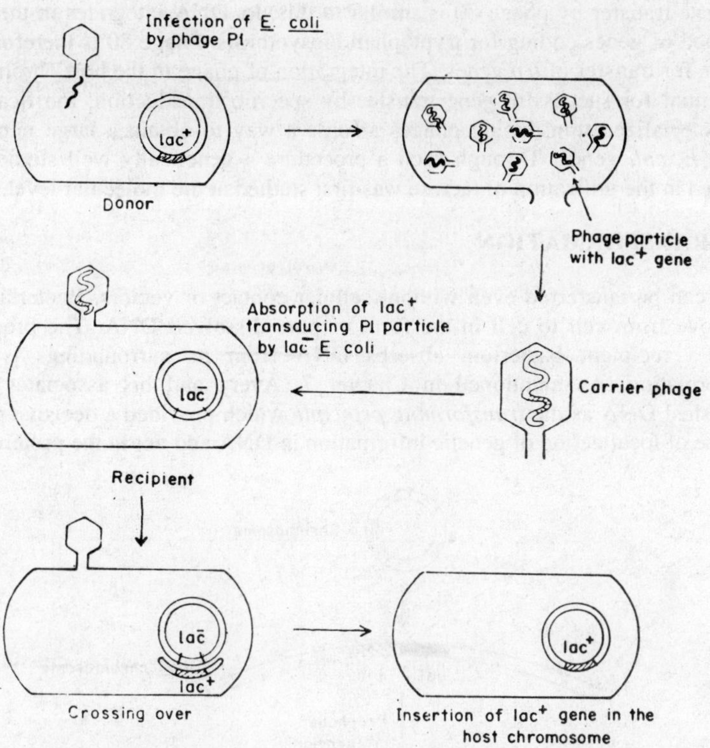

Fig. 10.16 Non-specific transduction, a mechanism for the passive transfer of genetic material from one bacterium (donor) to to a recipient by means of a carrier phage particle

Specific Transduction

In addition to transducing phages like P1, which randomly carries small sections of the *E. coli* chromosome, there are also specialized transducing phages that carry specific bacterial genes. The best known example of specific transduction is phage lambda. Certain defective phage particles possess hybrid chromosomes in which a section of the chromosome is linked to a fragment of the *E. coli* chromosome. Such hybrid chromosomes result when the lambda prophage improperly recombines or is excised out of the *E. coli* chromosome to generate a DNA fragment still attached to bacterial genes that lie immediately adjacent to the site where DNA inserts on the *E. coli* chromosome. For example á phage normally transduces only specific genes of the *gal* and *bio* operons. It integrates into the host genome during its transition to the prophage state site specifically between *gal* and *bio* operons. (Fig. 10.17). During the separation of the phage DNA from the host DNA, induced, for example, by UV irradiation, the phage DNA may not be precisely excised. It may leave a fragment in the host chromosome and acquire a neighbouring piece of the host DNA, which is then liberated along with the phage DNA. It is assumed that illegitimate recombination may be responsible for this event. When a transducing phage infects a recipient cell with a defect in the appropriate gene such as *gal*$^-$, the intact transduced gene can exchange with the defective host gene, and then the resulting recombinants are *gal*$^+$

Gene transfer by phage 80 is similar to this. Its DNA integrates in the neighbourhood of genes coding for tryptophan biosynthesis. Phage 80 is therefore most suitable for transfer of *trp* genes. The integration of phage in the host chromosome is essential for successful gene transfer by specific transduction. Purification of these specialized transducing phages affords a way to obtain a large number of certain *E. coli* genes. Through such a procedure a genetically well studied gene involved in the utilization of lactose was first studied at the molecular level.

10.8 TRANSFORMATION

Genes can be tansferred even without cellular contact or vectors. Bacterial genes can move from cell to cell in the form of pure dissolved DNA. The process by which a recipient bacterium absorbs DNA from its surroundings is called transformation. As mentioned in Chapter 7, Avery and his associates (1944) established DNA as the *transforming principle* which provided a decisive proof at the time of localization of genetic information in DNA and not in the protein.

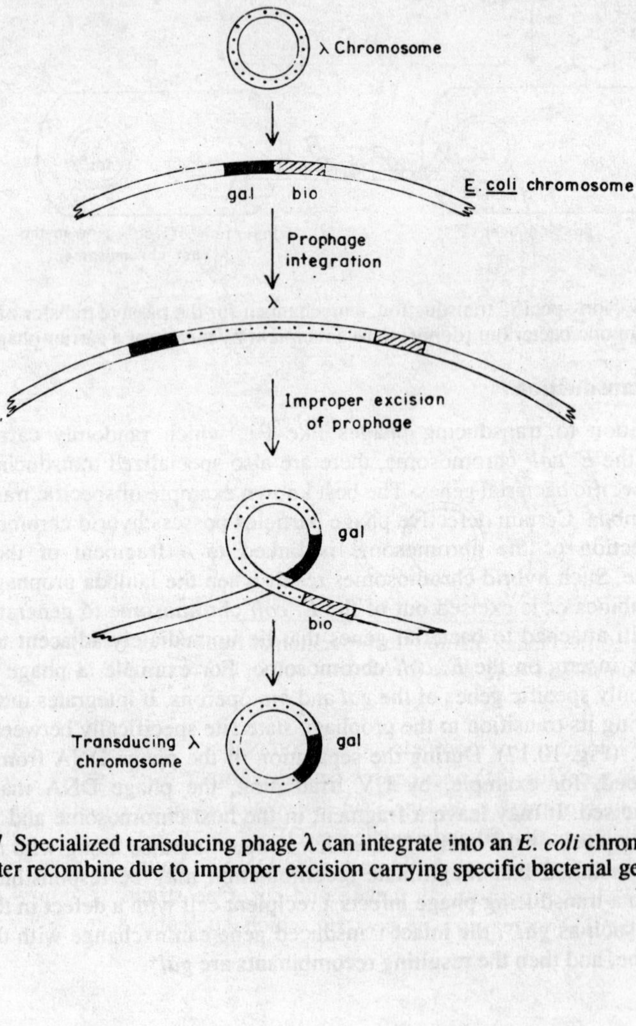

Fig. 10.17 Specialized transducing phage λ can integrate into an *E. coli* chromosome and later recombine due to improper excision carrying specific bacterial genes

The potential for transformation is apparently restricted to bacteria that are able to take up high molecular weight, double stranded, intact DNA. This ability is defined as competence. Although competent bacteria can take up any DNA, only the DNA of closely related species can lead to recombination, namely the exchange of homologous DNA fragments.

The potential to transfer genetic characters by isolated DNA has since been established in *Haemophilus influenzae, Bacillus subtilis, E. coli, Neisseria, Pseudomonas* and a number of other bacteria. Indeed some bacteria possess special proteins that mediate the entry of DNA into cells and that only recognize and take up DNA from related species (Fig. 10.18a and b). The mechanism of transformation can be explained by the following steps:

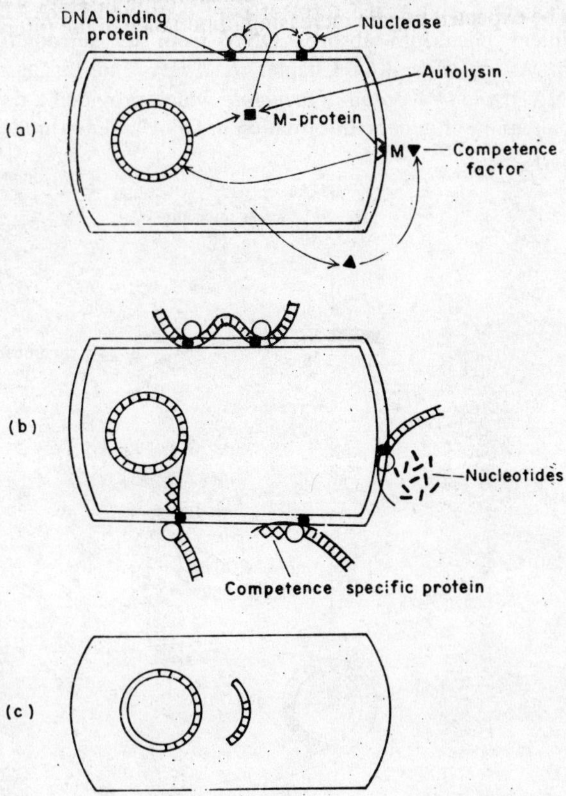

Fig, 10.18 Transformation: Incorporation of free DNA into a bacterial cell.
(a) Development of competence (b) Mode of entry of a naked DNA (c) Transformed cell

(a) Development of competence: A competence factor (▲) interacts with a cell surface receptor (M) and causes certain competence–specific proteins to be expressed. One of these is an autolysin (■), which exposes a membrane associated DNA-binding protein (●) and a nuclease (O),

(b) A long fragment of double-stranded DNA is bound at several sites on the cell surface. It is progressively nicked and cut. One strand of the cut segment is degraded by the nuclease, a process by which nucleotides are released into the medium. The other strand becomes associated with a competence-specific protein (◆). In this form it enters the cell and recombines.

Some organisms which were formerly regarded as non-competent such as *E. coli*, which do not retain the ability of taking up functionally active DNA, can be rendered competent by modification of the cell surface. A powerful trick now available for artificially introducing functional DNA is to treat cells with $CaCl_2$. It has been discovered that DNA added in the presence of high amounts of Ca^{++} can transform. This Ca^{++}– induced transformation works in *E. coli* only when the added DNA is a complete *replicon* capable of self-replication. Plasmid DNAs and intact viral chromosomes transform with high efficiency. As Ca^{++} treatment renders virtually all bacteria susceptible to transformation, gene transfer by this mechanism can be expected to gain increasing importance in the future

MOLECULAR BASIS OF MUTATIONS

The chromosome theory of inheritance which established the mechanism of inheritance through the transfer of hereditary determinants or genes from the parent to the offspring was established at the beginning of the twentieth century. The hereditary specificity is determined by the sequence of genes, of which the chromosomes are composed. Though the genetic information is faithfully transmitted from generation to generation, occasionally some mistakes occur in the genetic material which may be inherited through the generations. These errors impose variations in the organisms which bring about potential changes in the organization. Mistakes of this kind that alter the structure and functions of the genes are known as *mutations*, and are a source of biological diversity.

11.1 MUTATIONS ARE A SOURCE OF VARIATION

Mutations have been studied in a wide variety of organisms, but they were first studied seriously in *Drosophila melanogaster* by T. H. Morgan in 1912 for which he was awarded the Nobel Prize five years later. While working with the *Drosophila* fly, Morgan came across a white-eyed male fly as against the normal red-eyed character. Morgan performed a series of experiments in which he crossed the white-eyed male with a red-eyed female. Subsequently, he was able to show that the white eye character in *Drosophila* was a result of mutation at the gene level, caused by spontaneous or sudden change in the natural population. Thus mutations are sudden, heritable changes in the genetic material which are expressed in the phenotype of the organism, such as the one observed in *Drosophila* in the form of the white-eye character.

In chapter 9 it has been seen that the sequence of amino acids in a protein chain is determined by the base sequence of the DNA. Since a gene is a portion of the DNA that codes for single protein, any mutation or alteration in the base sequence could bring about an error in the translation process. Thus mutation induces a change at the molecular level by converting one base sequence to another or by deleting or adding base pairs to the genome. In a majority of the cases mutations may be fatal because they bring in biochemical imbalances in the organisms, but in rare cases they may survive and are then preserved through inheritance.

In order to study mutations and their impact on the characteristics of organisms, it is necessary to examine thousands of individuals over a number of generations. In the initial stages, *Drosophila* was the easiest available tool to study mutations, but it had a constraint of time because it takes about 15 days to grow one generation of flies. Biologists were trying to find suitable organisms in which mutations could be studied accurately and faster. This was possible by using bacteria and microorganisms. The advantages of using microorganisms for study of mutations stem from many considerations.

1. One of the best known organisms suitable as experimental material is the bacterium *Escherichia coli* which has a short life cycle and generation time. A genetic cross can be set up and analyzed in one day, hence a number of generations can be grown within a short span of time.

2. The micro-organisms can be grown on well defined media under controlled physiological conditions so that mutations can be characterized owing to specific biosynthetic processes.

3. A majority of micro-organisms are haploids, hence their life cycles can be easily studied. Haploidy is especially advantageous since the necessity of masking the recessive gene does not arise.

4. Micro-organisms are most suitable for mutational studies. Biochemically uniform and large populations can be obtained among which a specific mutation can be selectively studied. Loss of a gene in a particular mutant will not allow the organism to grow unless the product of that gene is supplied in the growth medium. Similarly, mutation rates can be more easily studied in a large population by using selective techniques. In *E. coli*, the frequency of mutations could be of the order of 1 in 10^4 cells.

Some Properties of Mutations

Mutations are sudden heritable changes in the genetic material, but mutations occurring in the somatic cells are not inheritable. During meiosis some genes may be lost or altered, these are chromosomal mutations. Replication of DNA is usually very accurate but sometimes alteration of one or more base pairs occurs to produce mutations. A mutation usually causes loss of a function, but in some cases, mutation in a single gene may affect more than one character. These are known as *pleiotropic mutations*. On the extreme side of these visible mutations are the lethal mutations which may cause death of an individual. However, mutations without any apparent effect are called *silent* mutations. Mutations that inactivate a gene represent forward mutations. Back or reverse mutations involve the conversion of a mutant to the original type. An exact reversal of the original mutation occurring at the same site is called a *true reversion*. If a mutation occurring at some other site compensates for the original mutation, it is called *second-site reversion*. Mutations occurring at some other site may also suppress the original mutation, this is called *extragenic suppression*.

Molecular Basis of Mutations

Since mutation means a hereditary change, the mutational event will cause a permanent alteration of a base pair or a few base pairs in the DNA molecule. Alteration of a single base pair is called *point mutation* and if it involves many bases, it is called multisite *mutation*. Point mutations may involve base substitution either by replacement of a base or by alteration of a base; by insertion of one or more bases or deletion of one or more bases from the DNA sequence. Base substitutions may be of two types.

Transitions: A transition mutation consists of a substitution of one pyrimidine with another (C – T; T – C) or of one purine with another (A – G; G – A). Nitrous acid causes oxidative deamination of cytosine and converts it into uracil. In the next replication cycle uracil pairs with adenine and therefore, a G – C pair is replaced by a T – A pair. Nitrous acid also deaminates adenine causing reverse mutation. Transition can also be brought about by base mispairing. The compound 5-bromouracil acts as a base analog of thymine and this results in replacement of an A – T pair by a G – C pair.

Transversion: This involves a mutation in which a purine is replaced by a pyrimidine or vice versa.

A – T —— C – G
A – T —— T – A
G – C —— C – G
G – C —— T – A

Thus an A – T pair is substituted by T – A or a C – G pair. Ethyl ethane sulphonate or ethyl methane sulphonate may cause transversion in some cases. Both these compounds have an ethyl group which can be donated to the nitrogen in the 7th position of the purine base. Alkylation of nitrogen results in the hydrolysis of the base-sugar linkage and gives rise to a gap in one DNA strand. Hence at the time of replication any one of the four bases may be inserted with an equal possibility.

11.2 HOW MUTATIONS ARE CAUSED

Mutations result in the loss of a gene function. Hence, mutations can be induced in nucleic acid by physical and chemical agents. Such studies have been carried out by using micro-organisms because they are haploid and permit mutations to be expressed phenotypically within a short period. A large number of mutations have been studied in the mould *Neurospora* that cause nutritional deficiencies.

Mutations by Ionizing Radiations

X-rays are a source of ionizing radiations. X-rays as well as radiations from radioactive substances have high energy levels which can change the atomic configurations. When such powerful radiations come in contact with an atomic structure, there is an electron loss accompanied by the formation of an ion, or sometimes, an electron pair from an inner orbit moves into the outer orbit. This results in bringing the atom in an excited state forming a free radical, which is in a highly reactive state. Such radiations are called ionizing radiations.

X-rays induce mutations in the DNA in several ways, but the magnitude of mutations depends upon the rate at which the atomic nuclei are bombarded. X-radiations cause serious structural damage to the genetic material. They frequently cause *deletions* or *translocations* of large segments of DNA strands by inducing breakage at one or more sites. As a consequence, a number of bases may be lost or undergo rearrangement, resulting in the cell's inability to synthesize proteins. Low doses of X-rays for a small duration may cause even point mutations.

Ionizing radiations are quite powerful and give rise to free radicals which not only change the chemical equilibrium of the cell, but also support the formation of strange molecular aggregates. Such changes may lead to cell death. If intact cells are exposed to ionizing radiations, chromosomes undergo fragmentation and the cell may ultimately die. In certain cases, delay in cell division is noticeable.

Mutations by Ultraviolet Light

Ultraviolet light is a form of nonionizing radiation which has a short wavelength in the non-visible range of the spectrum. It has sufficient energy to cause alterations in the atomic structures, so that mutations can be induced in biological molecules. Ultraviolet light penetrates the tissues very slightly, but it is absorbed by nucleic acids to a greater extent. The most effective radiations lie in the wavelength region 260 nm for DNA molecules. Since ultraviolet radiations have a low energy level, they penetrate slowly and bring about displacement of an electron pair in one of the atoms to higher shell, causing atomic instability.

Nucleic acids are very stable and thus maintain cellular integrity. However, any change in their atomic structure will have a long range effect reflected through replication and faulty synthesis of proteins. Both DNA and RNA absorb UV light preferentially in the wavelength region around 260 nm, as a result of which the nitrogenous bases become free radicals and thus become highly reactive. In case an *m*RNA molecule is irradiated, the triplet code for a given amino acid may be changed and a faulty protein formed.

The main affect of UV radiations on a DNA molecule is to convert one base into another, i.e. a purine into another purine or a pyrimidine base to another pyrimidine (Fig. 11.1). Base substitution could be lethal if the DNA is affected at a site which produces *t*RNA or an enzyme for a specific amino acid. Alteration of a single base by UV light is called a *point mutation*. UV light can also cause deletion of a single or multiple bases. If multiple deletion is not a multiple of three, the entire reading frame will shift and an aberrant protein will be synthesized. If a long segment of DNA beyond single gene is deleted, the syn...esis of more than one protein will be affected.

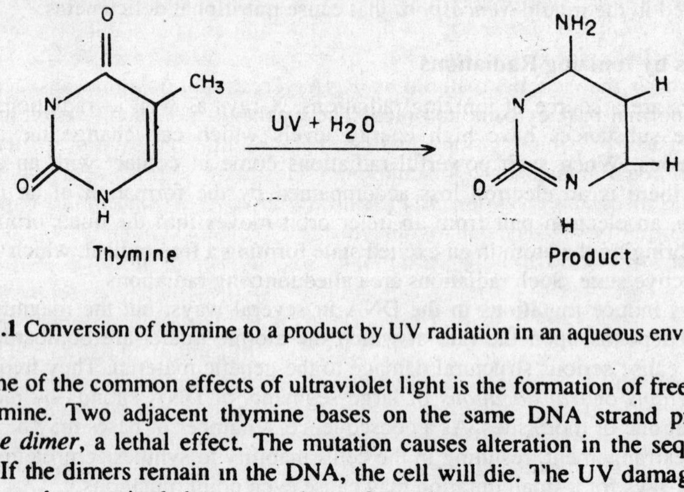

Fig. 11.1 Conversion of thymine to a product by UV radiation in an aqueous environment.

One of the common effects of ultraviolet light is the formation of free radicals of thymine. Two adjacent thymine bases on the same DNA strand produce a *thymine dimer*, a lethal effect. The mutation causes alteration in the sequence of bases. If the dimers remain in the DNA, the cell will die. The UV damage of this kind can be repaired by a number of mechanisms. One repair system is *photoreactivation*. If the UV-damaged cells are exposed to blue light, the repair enzyme becomes active and removes the thymine dimers from the DNA molecule. The excised material is replaced by the enzymes and the nick sealed. It has however, been found that prolonged exposure of DNA to UV light causes a permanent damage which is beyond repair (Fig.8.11).

Chemical Mutagens

There are many chemicals which cause mutations, hence called *mutagens*. In laboratory experiments mutations in DNA have been induced by a number of chemicals, which either affect the DNA directly or at the time of DNA synthesis. Some of these will be described here.

Base Analogues: The base analogues, such as 5-bromouracil and 2–aminopurine, are mutagenic. 5–bromouracil resemble thymine, and if during DNA synthesis thymine is not available, bromouracil will be incorporated in its place. 5-bromouracil causes transition mutation and since it is an analogue of thymine it normally pairs with adenine. However, occasionally it may pair with guanine instead of adenine which may lead to A T ↔ G C transitions. 2–aminopurine normally pairs with thymine, but can occasionally undergo tautomeric shift, i.e. sometimes the hydrogen of adenine involved in bonding with the complimentary strand shifts from amino group to the nitrogen in the ring. Thus 2–aminopurine can produce A T ↔ G C transitions. Such mutations arise spontaneously as copy errors during DNA duplication.

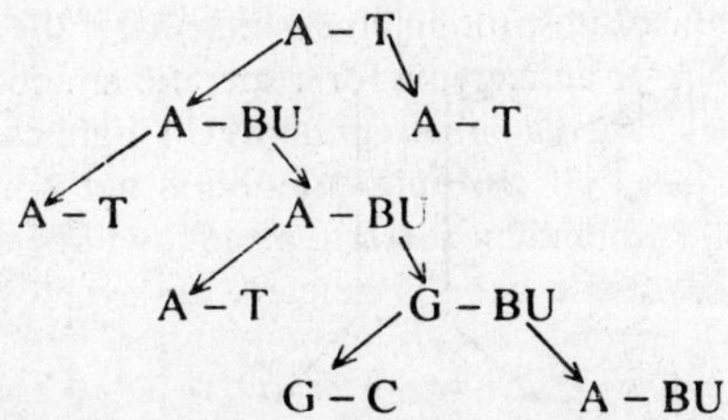

Nitrous Acid: Nitrous acid is a direct acting mutagen which chemically modifies the bases of DNA. The effect of nitrous acid has been studied in many viruses where the genetic material is RNA. Treatment with nitrous acid will change a normal base cytosine to uracil. The mechanism of modification involves the conversion of amino groups to hydroxyl groups. For example, adenine is oxidatively deaminated to hypoxanthine, cytosine to uracil, and guanine to xanthine. Consequently, hypoxanthine pairs with cytosine and uracil pairs with adenine. Similarly xanthine pairs with cytosine.

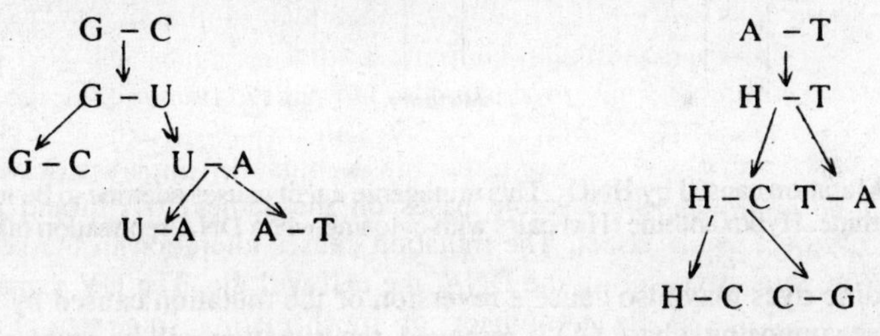

Thus nitrous acid causes A T ↔ CG transitions through cytosine-hypoxanthine base pairing (Fig. 11.2). Similarly, hydroxylamine (NH_2OH) is also a mutagen which reacts with cytosine to produce a modification that pairs with adenine rather than guanine.

Acridine Dyes: Acridine dyes are flat aromatic molecules which cause frame shift mutations by deleting or adding single or more base pairs from the DNA molecule. The mode of action of acridine dyes is interesting. These dyes slip in between the adjacent base pairs in the DNA double helix and by removing or adding base pairs in the DNA, alter the reading frame. For example:

ACA	AGC	AGA	GGA	GUA
ile	**ser**	**arg**	**gly**	**val**

Acridine might add a base in the sequence, hence

ACA	AAG	CAG	AGG	AGU	A...
ile	lys	gln	arg	ser	

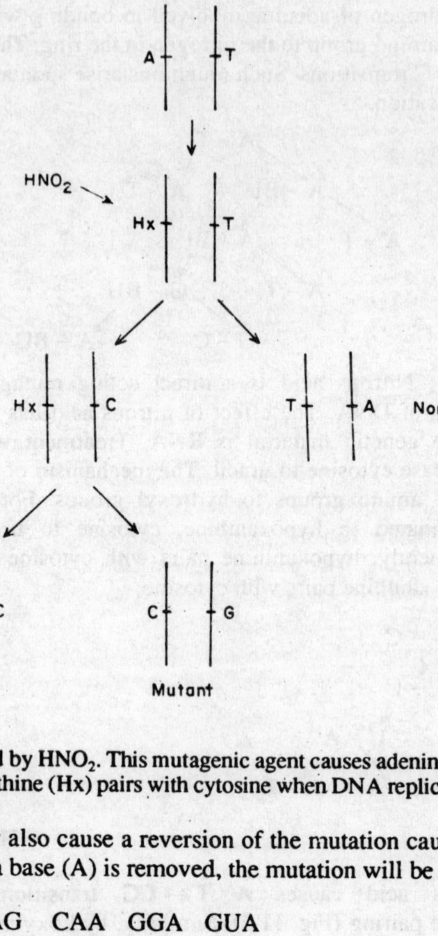

Fig. 11.2 Mutation caused by HNO_2. This mutagenic agent causes adenine to be replaced by hypoxanthine. Hypoxanthine (Hx) pairs with cytosine when DNA replication takes place.

Acridine dyes may also cause a reversion of the mutation caused by them and in that case supposing a base (A) is removed, the mutation will be reversed.

ACA	AAG	CAA	GGA	GUA
lie	ser	arg	gly	val

Alkylating Agents : The mutagens considered so far are found to be reactive during replication of the DNA. However, certain alkylating agents, such as ethyl ethanesulphonate or mustard gas, are mutagens which act on non-replicating DNA by introducing ethyl group to guanine. The action of alkylating agents shows a delayed effect. The mutagen introduces a copy error after several duplications of DNA have taken place. The alkylating agents add ethyl or methyl groups to guanine, making it analogous to adenine. In certain cases the alkylating agents cause *depurination* resulting in the loss of a base (Fig. 11.3). The loss of base leaves a gap in the DNA, which may be filled up by a wrong base.

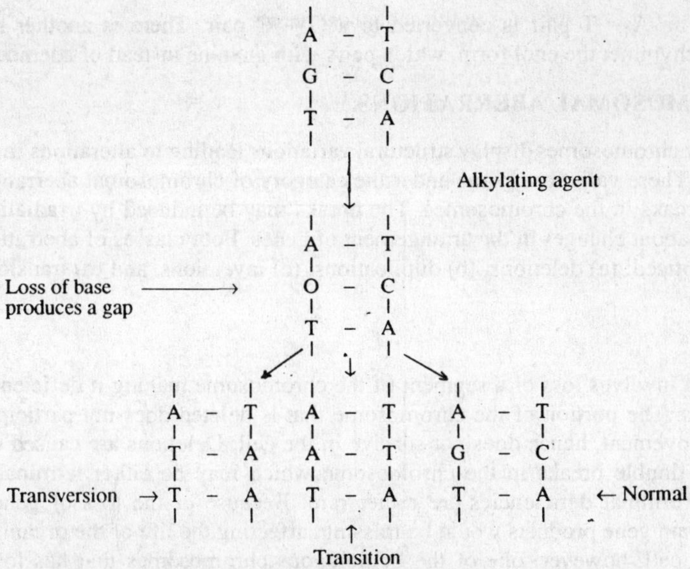

Fig. 11.3 Mutations caused by alkylating agents (e.g. ethyl methane sulphonate, ethyl ethane sulphonate).

Tautomerism : Certain chemical mutagens, such as hydroxylamine, give rise to tautomers or alternate forms of bases by arrangement of electrons and protons in the molecules. Changes in the electronic configuration of molecules bring about alteration in the normal base pairing pattern so that a purine may pair with another purine, or a pyrimidine pairs with another pyrimidine. These changes are described under the phenomenon tautomerism. In the normal way, adenine pairs with thymine, but the imino form of adenine pairs with cytosine. This rare combination is unstable and reverts to the normal base pair at the next replication cycle.

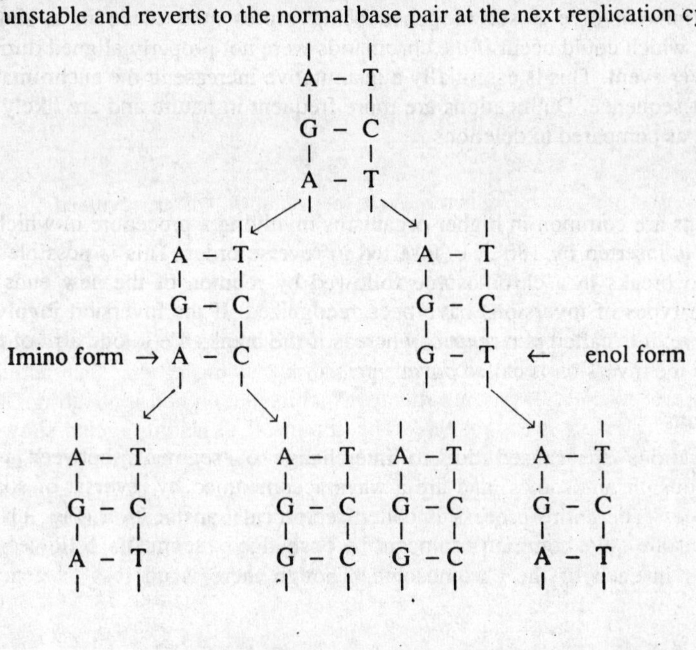

In this way an A – T pair is converted to a G – C pair. There is another rare tautomer of thymine, the enol form, which pairs with guanine instead of adenine.

11.3 CHROMOSOMAL ABERRATIONS

Occasionally chromosomes display structural variations leading to alterations in the phenotypes. These variations come under the category of chromosomal aberrations caused by breaks in the chromosomes. The breaks may be induced by irradiations which bring about changes in the arrangement of genes. Four classes of aberrations have been noticed: (a) deletions, (b) duplications, (c) inversions, and (d) translocations.

Deletions

Deletion involves loss of a segment of the chromosome making it deficient in certain genes. The portion of the chromosome that is deleted does not participate in mitotic movement, hence does not survive in the cell. Deletions are caused due to single or double breaks in the chromosome which may be either terminal or interstitial. Terminal deficiencies are rather rare. Because of the loss of genetic material certain gene products would be missing, affecting the life of the organism. In a diploid cell, however, one of the homologous chromosomes that has lost a segment will not have much deleterious influence because its corresponding alleles are intact in the homologue. Nevertheless, if a large segment is lost, the effect would be lethal even in a heterozygous state. Interstitial deficiency takes place when two breaks occur in the chromosome (Fig. 11.4). In some cases this may result in the formation of a ring chromosome.

Deletions have been often associated with the development of abnormalities in animals and humans. Deficiencies in the heterozygous state have led to many congenital disorders, such as myeloid leukemia, retinoblastoma etc.

Duplications

Duplications involve a set of genes being represented twice in a single chromosome which could occur if the chromatids were not properly aligned during a crossing over event. This is essentially a quantitative increase in the euchromatin with a repeat sequence. Duplications are more frequent in nature and are likely to be less lethal as compared to deletions.

Inversions

Inversions are common in higher organisms involving a procedure in which a set of genes is inserted by 180°, i.e. inserted in reverse order. This is possible by initiating two breaks in a chromosome followed by reunion of the new ends in reverse. Two types of inversions have been recognized. If the inversion involves the centromere, it is called *pericentric,* whereas if the breaks are in one arm of the chromosome the inversion is called *paracentric.*

Translocations

Translocations are caused due to interchange of segments between two nonhomologous chromosomes, and are always accompanied by reversal of some gene sequences. The entire process is called reciprocal translocation (Fig. 11.5). Translocations are quite common and occur by exchange of segments, followed by single breaks in each of the chromosome. Though there is no loss of genetic material, yet two new chromosomes are produced with altered gene sequence. Such aberrations are quite common in many plant genera.

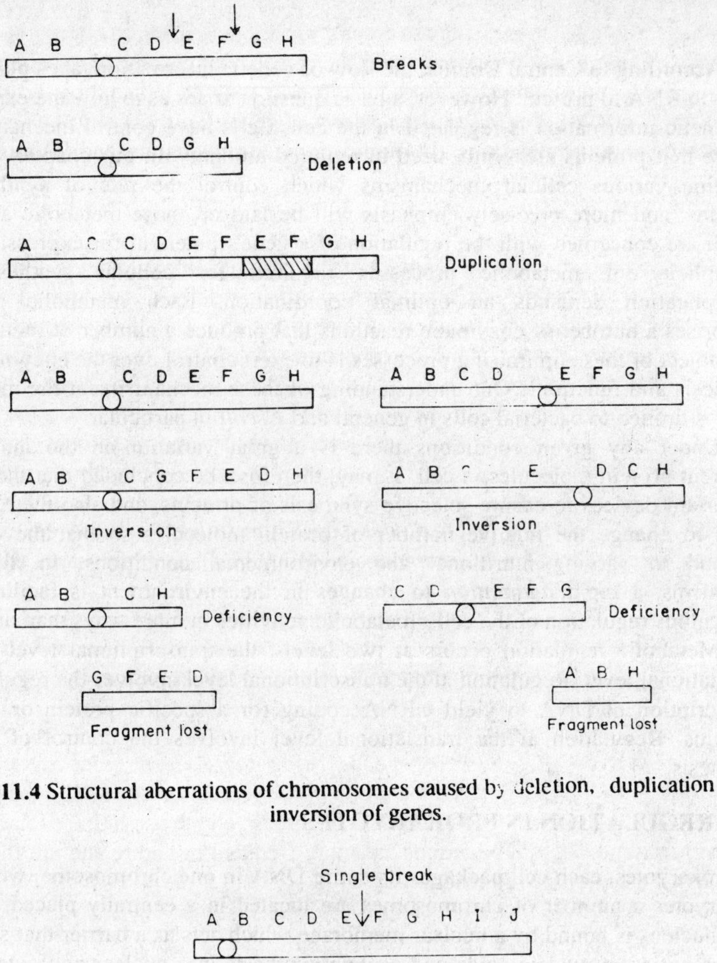

Fig 11.4 Structural aberrations of chromosomes caused by deletion, duplication and inversion of genes.

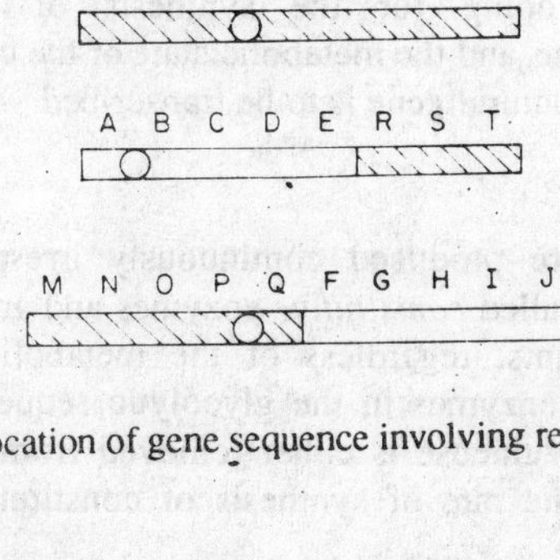

Fig. 11.5 Translocation of gene sequence involving reciprocal exchanges.

GENES AND REGULATION OF METABOLISM

According to Central Dogma, the flow of genetic information takes place from DNA to RNA to protein. However, a basic question arises as to how the expression of genetic information is regulated in the cell. Cells have control mechanisms to ensure that proteins are synthesized in required amounts. In this chapter we shall examine various cellular mechanisms which control the rate of synthesis of proteins, and more precisely emphasis will be laid on those metabolic activities which are concerned with the regulation of a gene's potential for expression. The multiplicity of metabolic processes required for cellular synthesis and multiplication demands an optimal coordination. Each metabolic pathway comprises a number of enzymatic reactions that produce a number of metabolites. The object of these optimising processes is to exert control over the enzymes, their synthesis and functions. Our understanding of these mechanisms at the molecular level is limited to bacterial cells in general and *E. coli* in particular.

Under any given conditions there is a great variation in the number of different protein molecules in cell. It may, therefore, be concluded that there must be certain devices to ensure selective synthesis of proteins, and also there may be ways to change the relative number of protein molecules so that the cell can respond to varying nutritional and environmental conditions. In all living organisms, a rapid *adaptation* to changes in the environment is facilitated by meticulous regulation of the cell's metabolic activities in more ways than one.

Metabolic regulation occurs at two levels, the transcriptional level and the translational level. Regulation at the transcriptional level involves the regulation of transcription of DNA to yield *m*RNA coding for a specific protein or a set of proteins. Regulation at the translational level involves the control of enzyme synthesis.

12.1 REGULATION IN PROKARYOTES

In prokaryotes, each cell packages its entire DNA in one chromosome, whereas in eukaryotes a number of chromosomes are located in a centrally placed nucleus. The nucleus is bound by a nuclear membrane which acts as a barrier that separates the activities of nucleic acids and protein synthesis into nuclear and cytoplasmic compartments. It has been observed that in prokaryotes, gene expression is mostly regulated at the transcriptional level. Transcription is subject to regulation for the majority of genes coding for the synthesis of polypeptide chains and environmental conditions and the metabolic state of the cell determine whether and at what frequency a structural gene is to be transcribed.

Enzyme Induction

Many enzymes are produced continuously irrespective of environmental conditions. These are called *constitutive* enzymes and are formed at constant rates and in constant amounts, regardless of the metabolic state of the organism. Concentrations of the enzymes in the glycolytic sequence, for example, do not radically change when glucose is either removed from, or added to, the growth medium of *E. coli*. The rate of synthesis of constitutive protein is genetically determined by : (a) the natural efficiency of the promoter from which its messenger is made, (b) the rate at which ribosomes read the messenger, and (c) the sensitivity of the messenger to degradation by nucleases.

The other class consists of *inducible* enzymes which are either completely absent or present as traces. Their concentration can quickly increase a thousand-fold or more when their substrate is present in the medium, particularly when the substrate is the only carbon source of the cell. Protein synthesis requires energy, hence it is not advantageous for the cell to synthesize all the enzymes all the time. The cell has therefore evolved a system of control to synthesize inducible enzymes in the presence of a substrate or an *inducer*. The existence of cellular control mechanisms was proposed at the Pasteur Institute in Paris by Jacob and Monod. They concentrated on the enzymes that utilize lactose in *E. coli*.

Induction of β-galactosidase

The most extensively studied example of enzyme induction is the utilization of lactose by *E. coli*. Lactose is a disaccharide which is hydrolysed by β-galactosidase to glucose plus galactose before it can enter the catabolic pathway. Wild type *E. coli* cells growing on glucose do not utilize lactose and have barely demonstrable β-galactosidase activity. Such cells contain only about 5 copies of β-galactosidase per cell. However, if such cells are grown on a culture medium containing lactose as the sole carbon source, they are unable to utilize it to start with but within a couple of minutes their β-galactosidase level increases a thousand-fold. Enzyme production reaches its maximum rate within 3 minutes of addition of lactose at 37°C. Jacob and Monod proposed that compounds which stimulated enzyme synthesis be known as *inducers*. Lactose is both an inducer and a substrate of β-galactosidase. If induced cells are transferred to a medium containing glucose but no lactose, synthesis of β-galactosidase stops immediately and its concentration drops rapidly to an extremely low level.

Normally lactose is a natural substrate and not an inducer *in vivo*. Rather, allolactose, an intermediate formed during lactose metabolism, is believed to be the natural inducer. Allolactose is an isomer of D-lactose in which the D-galactose and D-glucose residue are joined in a 1,6-linkage. Analysis of the regulatory mechanism was facilitated by the use of substrate analogues, such as isopropyl-β-D-thiogalactoside (IPTG) and O-nitrophenyl-galactoside (ONPG). IPTG is a good inducer of the enzyme but does not act as a substrate, therefore addition of thiogalactoside results in *gratuitous induction* when the enzyme is produced but it cannot hydrolyze the inducer. In contrast, ONPG is an excellent substrate but does not induce the enzyme system. The structures of lactose, IPTG and ONPG are given in Fig.12.1.

Regulation of Enzyme Synthesis

Synthesis of many catabolic enzymes is regulated by enzyme induction. It is economically advantageous for the cell that the enzymes required for the utilization of a substrate and for the entry of its catabolic products in the metabolic pathway are produced only when the substrate is present in the medium. In other words, the catabolic enzymes that the cell is potentially able to produce are not synthesized when they are not required.

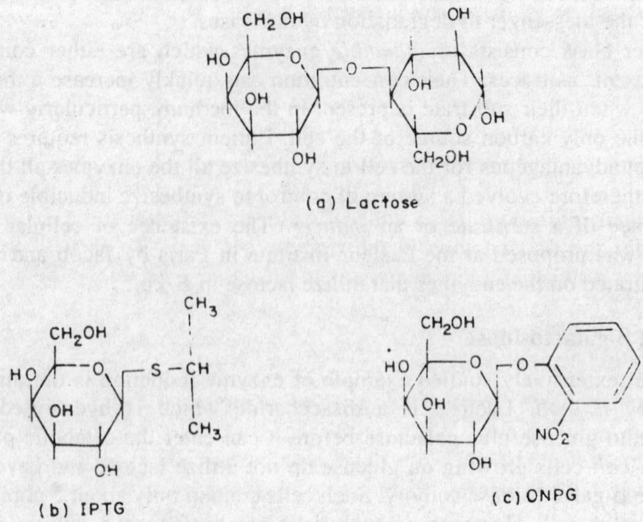

Fig. 12.1 Chemical structures of lactose, IPTG and ONPG

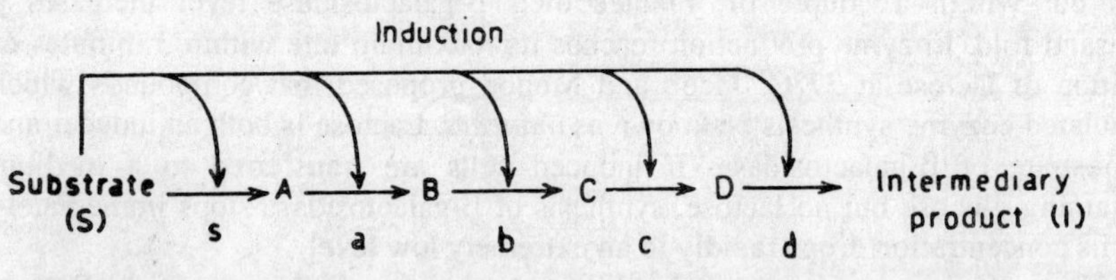

The above scheme shows the conversion of S into a product I through induction, where s, a, b, c, and d are inducible enzymes.

The synthesis of many *anabolic enzymes* is regulated by enzyme repression. Again, metabolic economy is promoted if the enzymes of a biosynthetic pathway are not produced when the respective end-product is present in reasonable amounts. Accumulation of an end-product thus leads to *repression* of all enzymes that are specific components of the biosynthetic pathway.

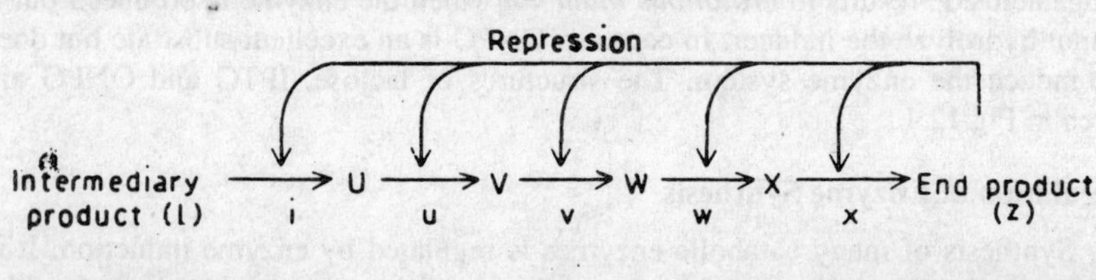

In the above scheme i, u, v, w and x are repressor enzymes. Normally the enzymes for the synthesis of building blocks are continually produced, but their formation is repressed when the end-product accumulates in the system. This is called *end-product repression*. The effects of both types of regulations, i. e. enzyme induction and repression on metabolism are rather slow and are regarded as "coarse tuning".

Another type of regulation exists at the level of enzyme activity, which usually concerns key enzymes in cellular metabolism. In this case, the catalytic activity of a particular enzyme in a specific metabolic pathway can undergo variation; it can be increased by a positive effector and decreased by a negative effector.

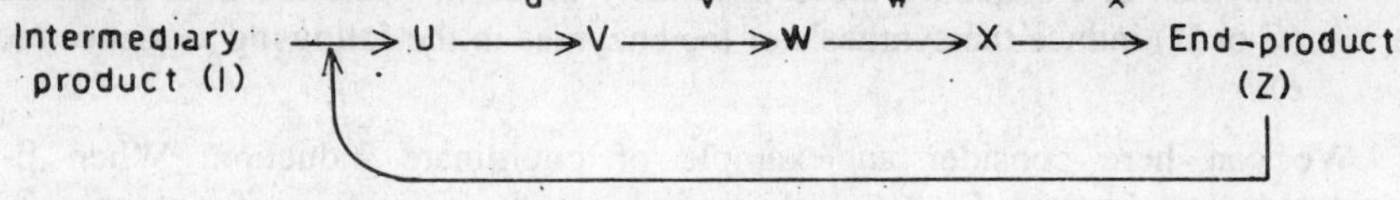

End-product inhibition

In end-product inhibition (also called feed-back inhibition), the end-product is inhibitory for the first enzyme in the reaction sequence. In such inhibitions the change in the activity of a key enzyme is immediately effective and is regarded as "fine tuning".

Product Induction

Some catabolic enzymes are induced by the product of the first or subsequent reactions in the pathway. When tryptophan breaks down, for example, it is converted in to catechol via formyl kynurenine, kynurenine and anthranilate. The first enzyme in the pathway is induced by kynurenine. This kind of enzyme induction requires that the basal level of the enzyme converting tryptophan to kynurenine is sufficiently high to yield at least traces of kynurenine in the presence of tryptophan.

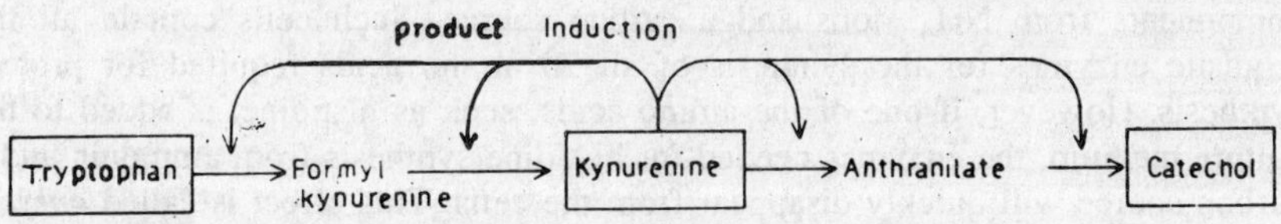

Product induction by tryptophan catabolism can be regarded as a protective mechanism to prevent the induction of catabolic enzymes by endogenously produced tryptophan which is required for protein synthesis. Tryptophan can be catabolized only when it is present in high concentration in the cell.

Sequential and Coordinated Induction

When a substrate is degraded via a number of intermediates by enzymes a, b, c, etc., induction of enzymes can occur according to several theoretical possibilities.

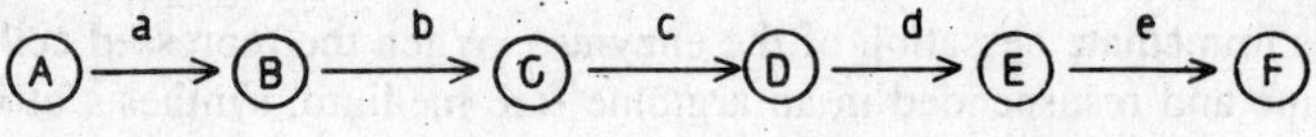

(i) *Strictly sequential*. Enzyme synthesis can occur sequentially or in steps, each succeeding enzyme is induced by the product of the preceding reaction.

(ii) *Strictly Coordinate.* All the participating enzymes are coordinately induced, i.e. the substrate A induces all enzymes a to e.

(iii) *Sequential induction of coordinately regulated enzyme groups.* Individual enzymes of a sequence are coordinately induced, while the later products (C or D) induce the synthesis of the enzymes in the following sequence (d, e).

We can here consider an example of coordinate induction. When β-galactoside is added to *E. coli* cells, it induces the formation of not only β-galactosidase but also two functionally related enzymes: β-galactosidase permease and β-thiogalactoside acetyl transferase (transacetylase). The former is a protein in the membrane of *E. coli* which can promote the passage of β-galactosides into the cell against the concentration gradient. The latter is presumably used in the utilization of thiogalactosides in the cell. The three proteins are coordinately induced by the same inducer. Induction of a group of related proteins or enzymes by a single inducing agent is known as *coordinate induction.* This kind of induction is advantageous for the cell because it enables it to react rapidly in the presence of a substrate.

Enzyme Repression

Two types of enzyme repressor mechanisms have been recognized. *Coordinate Repression.* When *E. coli* cells are grown in a medium containing an ammonium salt as the sole nitrogen source, they make all their nitrogenous components from NH_4^+ ions and a carbon source. Such cells contain all the requisite enzymes for the synthesis of all 20 amino acids required for protein synthesis. However, if one of the amino acids, such as histidine, is added to the culture medium, the enzymes needed for histidine synthesis from ammonia and a carbon source will quickly disappear from the cells. This effect is called *enzyme repression.*

Arginine biosynthesis is another example which can be used to show end-product inhibition of enzymes. Arginine synthesis starts from glutamate and leads via ornithine, citrulline and arginosuccinate.

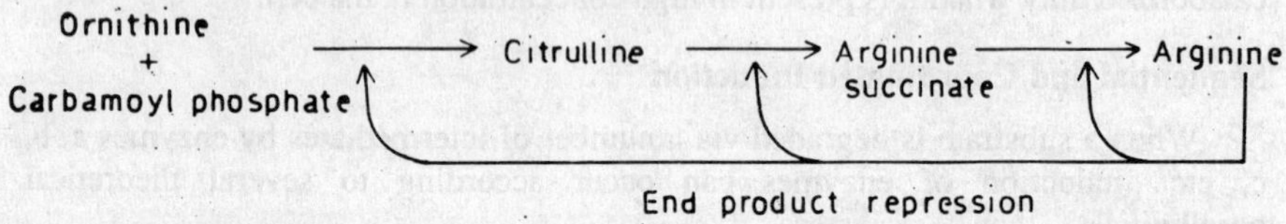

If *E. coli* is grown in a minimal medium, the enzymes of arginine synthetic pathway are found in normal amounts, but addition of arginine to the medium produces immediate cessation of the enzymes. When the repressed cells are freed of arginine and resuspended in an arginine-free medium, synthesis of enzymes is derepressed and the enzyme concentration rises rapidly. Repression of a group of enzymes catalyzing a sequence of constitutive biosynthetic reactions is called end-product or coordinate repression.

Catabolite Repression : The enzymes of the degradative pathways in bacteria are repressed by another mechanism known as catabolite repression. It has been observed that the presence of glucose as fuel represses some of the enzymes of respiration and electron transport because of the presence of a readily fermentable fuel supply, which can be preferentially used by the bacteria through the most common catabolic pathway, i.e. glycolysis, turning all other energy-yielding pathways off. Since elevation of the level of glucose or its catabolic products within the cell provides the signal of such an effect, the process is called catabolic repression or glucose effect.

Regulation of Lactose Utilization

The molecular and genetic relationships between enzyme induction and enzyme repression were first propounded by J. Monod and F. Jacob, who studied the induction of β-galactosidase activity in *E. coli*. They concluded that there are three different loci in the genetic map of *E. coli* that influence the formation of β-galactosidase, designated as *z*, *i* and *o* (Fig.12.2). The *z* locus specifies the amino acid sequence of the β-galactosidase molecule. Mutations in this locus lead to synthesis of a faulty or an inactive enzyme molecule. This gene is called a structural gene. It can also be defined as a gene that carries the message for the amino acid sequence of a specific protein such as β-galactosidase. The second locus *i* functions as an inhibitory gene and determines whether the structural gene for β-galactosidase will be transcribed. Such genes are called regulatory genes. When these genes undergo mutation and become defective, they can no longer inhibit the transcription of the structural gene. In such cases β-galactosidase is synthesized regardless of the inducer.

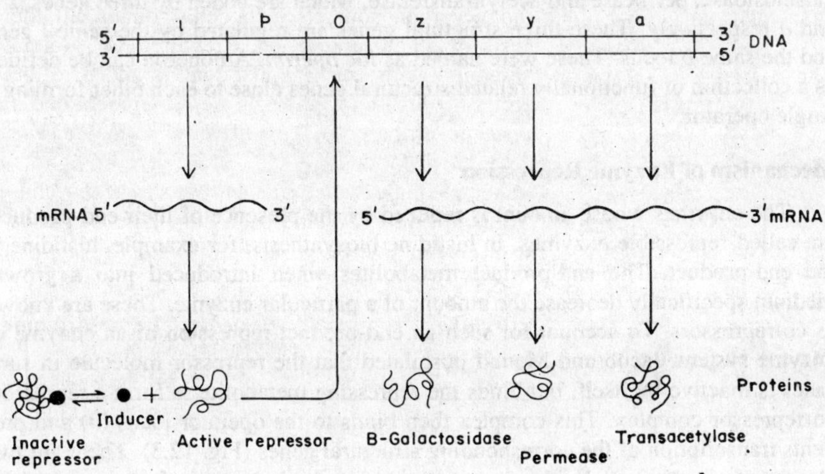

Fig. 12.2 The lac operon of *E. coli* and its regulation. The regulator gene (i) encodes a repressor that binds to the operator locus (o), thereby preventing transcription of mRNA from the structural genes z, y, a. Inducer metabolite combines with the repressor so that it no longer binds to the operator locus. The structural genes are then free to be transcribed to yield a polycistronic mRNA, which is translated into three proteins-the enzymes. P is the promoter gene.

Jacob and Monod also proposed that the regulatory gene codes for an amino acid sequence of a specific protein called the *repressor*. The repressor protein physically binds a specific site on the DNA of the cell next to or near the structural gene whose product it controls. When the repressor is bound in this manner it prevents the transcription of the structural gene coding for the enzyme. The site at which the repressor molecule binds is called the *operator* or the *o* locus. The operator is located close enough to the structural gene for β–galactosidase.

The Operon Model

This is a mechanism of coordinate induction in prokaryotes. According to Jacob and Monod, in the absence of an inducer the repressor protein is made by the cell following transcription from the *i* gene. In the absence of an inducer the repressor occurs in its active state and combines with the operator (*o*) locus, thus preventing the transcription of structural gene for β-galactosidase (Fig. 12.2). When the inducer is added, it combines with the repressor protein to form a repressor-inducer complex, rendering the repressor incapable of binding to the *o* locus. The structural gene for β-galactosidase thus becomes free for transcription to yield *m*RNA and the enzyme is then synthesized. The interaction between the repressor and the inducer is reversible and accounts for the fact that when the inducer is removed from the medium or is used up by the enzyme, the repressor reverts to its inhibitory form and binds to the operator, preventing the synthesis of enzyme.

Jacob and Monod extended their hypothesis for the regulation of protein synthesis to provide a mechanism for coordinate induction in which a sequence of enzymes can be induced as a group by a single inducer. We have seen that β-galactosidase coordinately induce a set of three enzyme proteins, namely β-galactosidase, permease and acetyltransferase, which are coded by three genes, *z, y* and *a* respectively. These three structural genes are regulated by the same *i* gene and the same *o* locus. These were named as *lac operon.* An operon can be defined as a collection of functionally related structural genes close to each other forming a single operator.

Mechanism of Enzyme Repression

The enzymes whose amount is reduced by the presence of their end-products are called repressible enzymes. In histidine biosynthesis, for example, histidine is the end-product. The end-product metabolites when introduced into a growth medium specifically decrease the amount of a particular enzyme. These are known as co-repressors. To account for such an end-product repression of an enzyme or enzyme system, Jacob and Monod postulated that the repressor molecule in such cases is inactive by itself, but binds the repressing metabolite to form a repressor-co-repressor complex. This complex then binds to the operator locus (*o*) and prevents transcription of the corresponding structural genes (Fig. 12.3). There are two classes of repressor molecules, one operative in the induction of enzyme activity and the other in repression of the end-product. Induction and repression are thus based on similar principles and no covalent bond is formed between repressors and their inducers or corepressors. A portion of each repressor molecule is complementary in shape to a specific portion of its inducer or corepressor, and the two are joined by weak bonds. These weak bonds are readily made and broken, allowing the active or inactive repressor to adjust quickly to the physiological need.

The Promoter

The promotor represents a short sequence of bases, normally less than 100 nucleotides, which is recognized by a DNA-dependent RNA polymerase. It has been identified by genetic and biochemical studies that the promoter lies between the regulatory gene (*i*) and the operater (*o*) of the lac operon. The promoter-RNA polymerase binding initiates the transcription of the neighbouring gene. The promoter is also the binding site for another specific type of protein, the cyclic AMP receptor protein (CRP) or catabolite-gene activator protein (CAP).

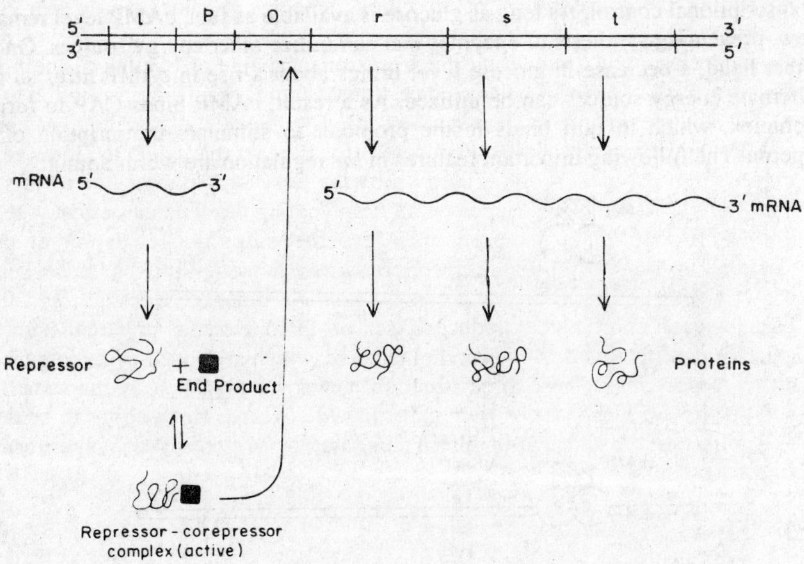

Fig. 12.3 Operon model of Jacob and Monod. I-regulator gene; P-promotor gene; o-operator locus; r, s and t-structural genes.

Catabolite-gene Activator Protein

For many years regulation of lac operon transcription was thought to reside solely in a competition between repressor and RNA polymerase for access to the same controlling element. However, it has recently become apparent that the lac operon is also under positive control by a dimeric regulator protein, CAP. More recently, it has been seen in *E. coli* that the glucose effect is brought about by two molecular intermediaries, namely CAP and cyclic AMP (*c*AMP). CAP is sensitive to intracellular levels of *c*AMP and has no influence on transcription until *c*AMP binds with it. It then acquires the ability to bind specific sites on the DNA, and by doing so increases the rate of transcription of adjacent operons. Thus CAP is the positive control element for all the glucose-sensitive operons and the main purpose of this system is to allow preferential catabolism of glucose since it is the main carbon source.

cAMP and Transcription of Operons

When glucose is added to the growth medium of growing *E. coli* cells, it depresses the levels of *c*AMP. Addition of *c*AMP to the suspension of *E. coli* overcomes the normal catabolite repression of β-galactosidase synthesis by glucose. Experiments have shown that *c*AMP and CAP are components which initiate transcription of the lac operon. But the question arises as to how *c*AMP and CAP enhance this transcription. Observations indicate that CAP functions as an activator, stimulating lac transcription, and that *c*AMP acts as an effector in conjunction with CAP to enhance its activity to recognize the start signal by binding RNA polymerase to the promoter site (Fig. 12.4). This is called positive transcriptional control. As long as glucose is available as fuel, *c*AMP level remains low, preventing synthesis of enzymes that can utilize other energy sources. On the other hand, a decrease in glucose level brings about a rise in *c*AMP titres so that alternate energy sources can be utilized. As a result, cAMP binds CAP to form a complex, which in turn binds to the promoter to stimulate transcription of its operon. The following important features in *lac* regulation are worth noting:

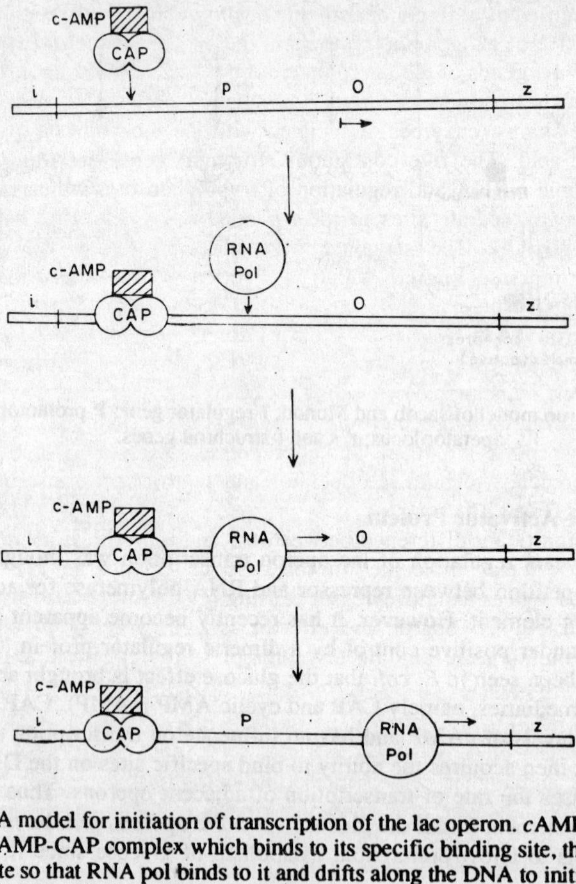

Fig. 12.4 A model for initiation of transcription of the lac operon. *c*AMP binds to CAP forming a *c*AMP-CAP complex which binds to its specific binding site, thus destabilizing the entry site so that RNA pol binds to it and drifts along the DNA to initiate the process of transcription. Glucose inhibits the formation of *c*AMP, hence the transcription of lac operon.

1. The operon is under negative control. When it interacts with the *lac* repressor, operon transcription is inhibited. The repressor is specified by *lac i* regulatory gene.

2. The *lac* repressor binds to *lac* operator to prevent transcription. When an effector, viz. lactose or a lactose analogue, is added, the repressor leaves the operator and *lac* transcription commences.

3. The operon is also under positive control. When it interacts with a regulator protein, such as CAP, operon trnascription is enhanced. A rise in glucose concentration causes CAP to leave its controlling element and thus *lac* transcription drops significantly.

Regulation of Tryptophan Biosynthesis

Biosynthesis of each amino acid is carried out by a series of enzymes and is regulated at the level of transcription through modulation of certain enzymes. Here we shall discuss the transcriptional control of *m*RNA involved in tryptophan biosynthesis. The tryptophan pathway has five distinct enzyme-catalyzed steps, all of which are specified by a single operon, the *tryptophan operon*.

In the presence of exogenous tryptophan, the synthesis of enzymes is sharply reduced because the genes of the operon are not transcribed. However, the absence of tryptophan in the medium selectively stimulates synthesis of *m*RNA and the five polypeptide chains (enzymes) associated with the biosynthesis of tryptophan from chorisimic acid. The five contiguous structural genes are transcribed as a single polycistronic *m*RNA and regulation of tryptophan biosynthesis at the gene level occurs at two separate sites in the *trp* operon, the first one being the *trp* operator locus (*trp o*). The sequence of nucleotides in the *trp* operator is recognized by a repressor protein which is encoded by *trp R* gene that maps far away from the *trp* operon.

The *trp* operon is repressible, i.e. transcription of its structural genes proceeds as long as tryptophan levels are low, and falls when this concentration rises. The *trp* repressor alone is incapable of binding to the *trp* operator, hence it is known as an *corepressor*. Binding can occur only when tryptophan acting as an effector (or corepressor) associates to form an aporepressor-corepressor complex. This complex then proceeds to bind the operator and inhibits *trp* transcription in the wild-type strains (Fig.12.5).

The most significant difference between *trp* and *lac* repressors relates to the function of the effector molecule. In the case of *lac* operon, the effector is allolactose which acts as an antirepressor causing release of the repressor from the operator, whereas in *trp* operon the effector is L-tryptophan which acts as a corepressor, stimulating the binding of repressor to the operator. This system is therefore a negative one.

Tanslational Control in Prokaryotes

In bacterial and higher cells a transcriptional control is believed to operate predominantly. We however, now know that ribosome binding to initiation sites on *m*RNA can also be regulated. Translational control mechanisms operate after a gene transcript has left the template. Translation in most cases appears to follow automatically once a *m*RNA is formed. In fact, there is a tight coupling of transcription and translation. The *m*RNA is translated by ribosomes as it is being transcribed by the RNA polymerase.

Translational control mechanisms govern the rate at which ribosomes initiate translation of the message, the rate at which the message is read, stability of the *m*RNA and so on. Regulation at the translational level is also believed to occur under some special circumstances, such as the regulation of the drop-off of ribosomes from the polycistronic *m*RNA. In such a case the product of different genes is not made in equimolar amounts. One gene product of an operon may be made in hundreds of copies, while another coded by some polycistronic *m*RNA may be made in a few copies only. In such situations the ribosomes may drop off the *m*RNA before completing translation of all the structural genes of the operon. As mentioned, the translational control is mediated by devices that determine the availability of the initiation site of *m*RNA to the ribosomes. Ribosome binding sites may be trapped in the secondary structure of RNA. If this happens, the ribosomes cannot initiate protein synthesis until the structure is broken. Such a translation-dependent exposure of an enclosed ribosome binding site can be a specific regulatory device linked to a control signal. This mechanism allows cells to express a gene specifying resistance to erythromycin that binds to the ribosomes of sensitive cells to inhibit protein synthesis.

Ribosomal proteins, moreover, are their own translational repressors which act by blocking the ribosome initiating sites of their own *m*RNAs when they accumulate in excess. It is advantageous to the cell when a protein represses its own translation, for it allows quick and precise adjustments of expression.

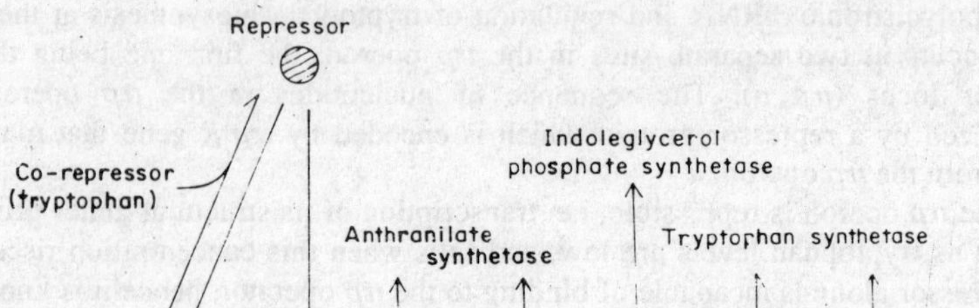

Fig. 12.5 The tryptophan operon of *E. coli*.

12.2 REGULATION IN EUKARYOTES

The molecular basis of regulation of gene expression in eukaryotes is more complex than in prokaryotes. Besides transcriptional control, gene expression may be regulated by the processing of *m*RNA in the nucleus, control at the level of ribosomes in cytoplasm or by post-translational modifications of the polypeptide products. Transcriptional control appears to be the major regulatory process nevertheless. In order to conserve energy and metabolites, nearly all regulatory mechanisms function at or near the beginning of a process rather than at the end.

Two classes of the regulatory phenomenon can be recognized in eukaryotes viz., short-term or reversible regulation and long-term or irreversible regulation. Short-term regulation represents a cell's response to fluctuations in the environment, especially those involving modulation of enzyme activities in response to changes in enzyme or hormonal levels. Long-term regulation is associated with determination, differentiation or development at the level of the entire organism. Some examples may be considered here.

Regulation in *Neurospora*

All the enzymes required for the synthesis of aromatic amino acids (e.g. tyrosine and phenylalanine) are present in a *Neurospora* cell. When the levels of these amino acids in the cells become high, synthesis of several enzymes is induced that help in the breakdown of these amino acids into non-aromatic compounds. Induction of three such *scavenger enzymes* is accomplished by an elevated level of the aromatic metabolite, quinic acid. Once quinic acid is catabolized, synthesis of the scavenger enzymes is turned off. The scavenger enzymes include a dehydrogenase, a dehydroquinase and a dehydrase, the genes of which are present in a cluster, called the *qa gene cluster*. The genes are named as *qa*-3, *qa*-2 and *qa*-4 respectively (Fig. 12.6). These *qa* genes of *Neurospora crassa* appear to be under positive control. The inducer quinic acid is thought to combine with a regulator, the product of a closely linked gene qa-1, to turn on the expression of the qua gene cluster.

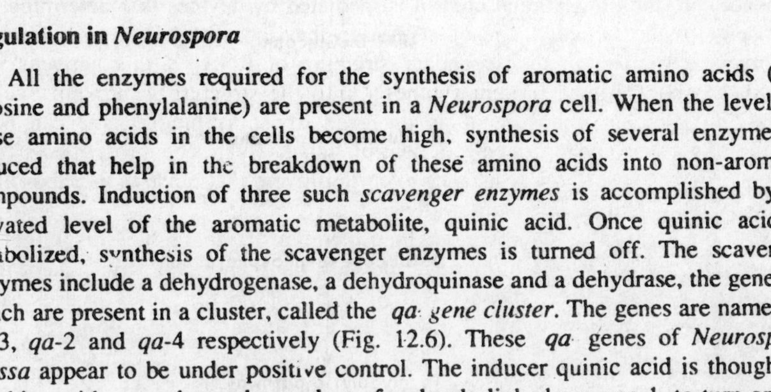

Fig. 12.6 Qua gene cluster in *Neurospora crassa* and the enzymes specified by them in the quinic acid catabolic pathway. QDH-quinic acid dehydrogenase; DHQ-dehydroquinase; DH-dehydrase.

Regulation in Yeast

Regulation of the galactose utilizing genes in yeast has been well investigated. In *Saccharomyces cervisiae* there are several regulatory genes that code for three enzymes involved in fermentation of galactose (Fig. 12.7). The enzymes are galactokinase specified by GAL-1 gene, galactose transferase—a product of GAL-7, and galactose epimerase encoded by GAL-10 gene (Fig.12.8). Together these enzymes transform galactose into glucose-1-phosphate, which is isomerized to glucose-6-phosphate before entering the glycolytic pathway. The enzymes of the GAL gene cluster are produced only when the yeast cells are grown in the presence of galactose.

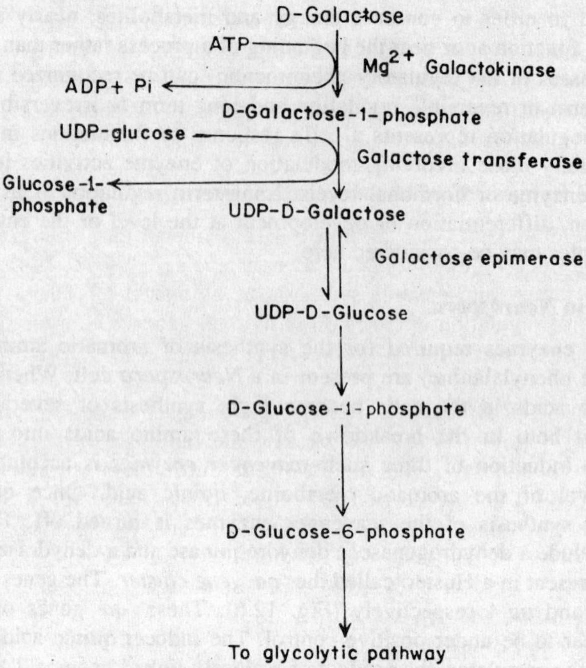

Fig. 12.7 Galactose utilization in yeast cells.

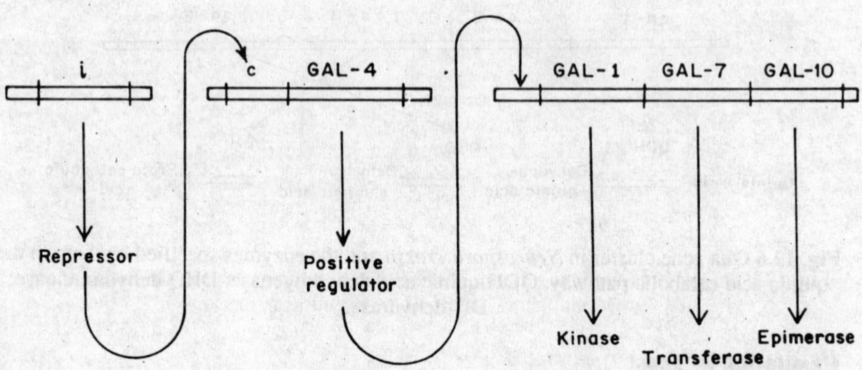

Fig.12.8 Regulation of galactose fermentation in yeast

Although the GAL genes are located close to each other, they do not comprise an operon, instead, they are controlled by the product of an unlinked gene which is the repressor *i*. The GAL-1 and GAL-10 *m*RNAs are transcribed from a common promoter, whereas a separate promoter functions for the GAL-7 gene. These three genes are in turn regulated by another gene GAL-4 which is shown to be a positive regulator. The repressor *i* acts on GAL-4 and not on the GAL cluster. Associated with GAL-4 is a region called *c* which behaves like an operater locus. The GAL-4 acts as a positive regulator and activates the transcription of GAL-1, GAL-7, and GAL-10 genes. In the presence of galactose, transcription of galactose utilizing enzymes is turned on, whereas in its absence through a repressor protein the GAL-4 product gets inactivated, switching off the transcription of the three GAL genes.

Differential Expression of Haemoglobin Genes

Haemoglobin genes are arranged in a sequence and are expressed during development. The regulation occurs at two levels. The first one operates on the entire gene family and ensures synthesis of globins in erythroid cell lines. The second type of regulation operates on individual genes within the family of genes to regulate the transcription of globin genes during an appropriate developmental stage.

The gene requirement for slightly different haemoglobin proteins at various stages of development is fulfilled by a whole family of related α and β-globin genes. Form the embryo to the foetus until its birth, the expression of these genes is turned on an off in coordination and sequentially. The α-globin genes in humans reside on chromosome 16, while β-globin genes are clustered on chromosome 11 (Fig.12.9), arranged in the order of their expression at various times during development, i.e. first the embryonic genes are switched on, followed by the foetal and then the adult genes. At all times during development, two α- and two β-polypeptide chains are assembled to form the functional haemoglobin tetramer.

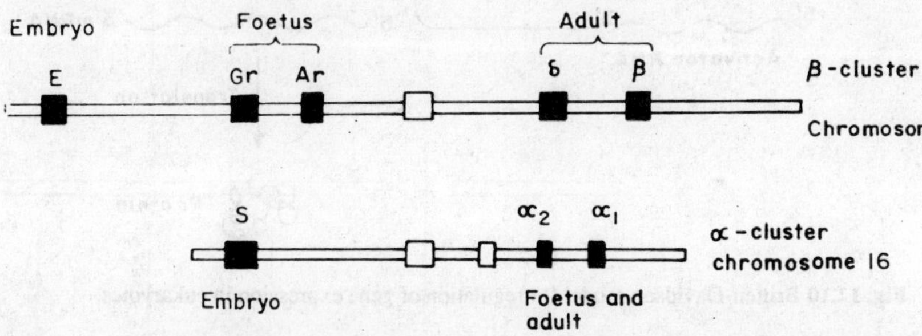

Fig. 12.9 Arrangement and differential expression of human α and β-globin genes. ■ are functional genes, while ❑ are pseudogene regions.

Britten and Davidson Model

R.J. Britten and E.H. Davidson have proposed a model explaining the gene regulation in eukaryotes. This model does not envisage the existence of operons consisting of adjacent genes rather, it accounts for the regulation of related genes in different chromosomes on an individual basis. This is because the genome of eukaryotes is very large. Besides, since eukaryotes have highly differentiated cells, a large part of the genome is permanently or perhaps irreversibly repressed. For example, a kidney cell cannot be transformed into a liver cell or a muscle cell, though each may contain structural genes characteristic of the other. Thus a small fraction of the total genome can be induced or repressed reversibly.

The model proposes that the eukaryotic genome contains a large number of sensor sites, which recognize various signalling agents (Fig. 12.10). The signalling agents may include hormone-receptor complexes or substrates of specific enzymes. Adjacent to each sensor site is an integrator gene region, while the structural genes are located elsewhere in the genome, and are in a repressed state unless activated. When a specific substrate binds the sensor, it becomes activated and allows transcription of the adjacent integrator gene to produce an activator RNA. Subsequently, the activator RNA binds the receptor locus which is located either on the same or on an other chromosome. Binding of activator RNA to the receptor results in the transcription of an adjacent structural gene to yield *m*RNA, which is then translated.

An important feature of the Britten and Davidson model is that the structural genes make up a small part of the large genome, and much of the DNA functions in regulation in the form of sensor, integrator and receptor loci.

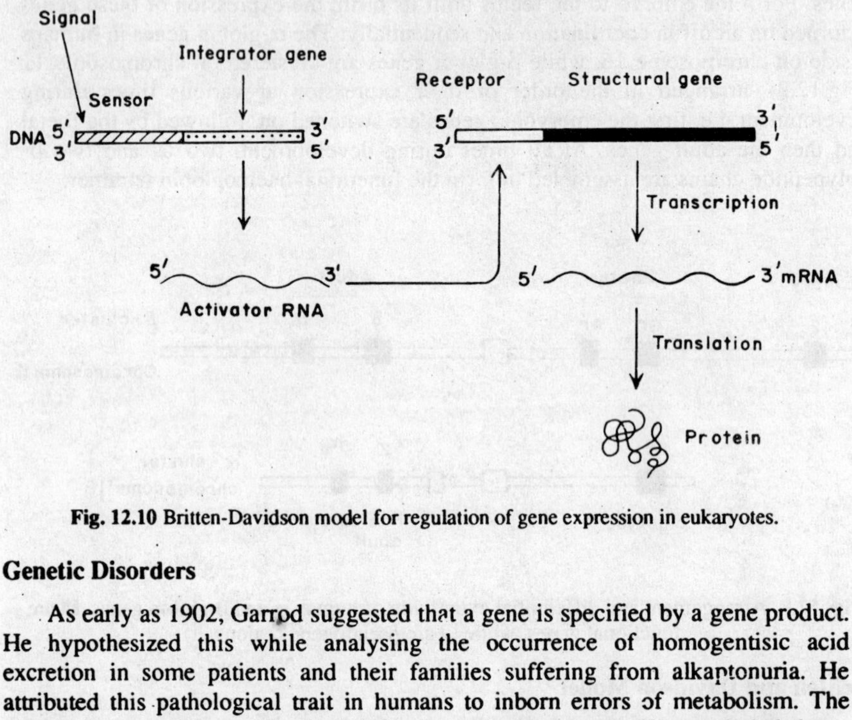

Fig. 12.10 Britten-Davidson model for regulation of gene expression in eukaryotes.

Genetic Disorders

As early as 1902, Garrod suggested that a gene is specified by a gene product. He hypothesized this while analysing the occurrence of homogentisic acid excretion in some patients and their families suffering from alkaptonuria. He attributed this pathological trait in humans to inborn errors of metabolism. The disease is caused by an inherited defect in a single enzyme. He concluded that a genetically controlled enzyme deficiency was the cause of this error. In this disease, the urine of affected individuals turns black on exposure to air. In normal individuals tyrosine and phenylalanine are catabolized to homogentisic acid, which is then oxidized to fumaric acid and acetoacetic acid. Alkaptonurics lack the enzyme homogentisic acid oxidase, and homogentisic acid is excreted in urine which turns black since it is oxidized and polymerized to a black melanin pigment.

In patients with alkaptonuria abnormal pigmentation causes darkening of the skin and they also develop arthritis.

Many inborn metabolic disorders have since been described, such as albinism, phenylketonuria etc. Phenylketonuria is produced by a defect in phenylalanine hydroxylase which normally converts phenylalanine to tyrosine. In phenylketonurics, however, phenylalanine is converted to phenylpyruvate which can be detected in the urine. Accumulation of phenyl compounds produces irreversible and acute mental retardation and other neurological disorders. Normally individuals subject-to such genetic disorders are homozygous for a recessive gene which codes for a particular defective enzyme.

CHAPTER 13

RECOMBINANT DNA TECHNOLOGY: GENETIC ENGINEERING

The first successful attempt by Nirenberg and Matthaei in 1961 to break the genetic code proved to be a landmark in the history of molecular biology. In 1970 Khorana and his colleagues succeeded in synthesizing a gene in the test tube by assembling its constituent base pairs and provided the foundation for recombinant DNA technology. The real story, however, began with the discovery of a group of enzymes by Hamilton Smith in the 1970s, called restriction endonucleases type II. These enzymes were obtained from micro-organisms, and resulted in the development of techniques for transformation of *E. coli*, joining DNA molecules and monitoring the cutting and joining reactions.

Effective experimentation on the molecular features of the genome is only possible through the employment of recombinant DNA procedures. To understand how the genes work, they are first isolated in pure form through gene cloning procedures and then reintroduced into cells where their expression can be studied.

In real terms genetic engineering or gene cloning is not a discipline, but rather a very powerful tool that has helped revolutionize our concept of genetics. To be more specific, genetic engineering can be defined as a deliberate manipulation of genes within or between species for the purpose of genetic analysis or strain improvement. Recombinant DNA technology is a more specialized category of genetic engineering. This area of study involves *in vitro* re-arrangement of genetic material by enzymic manipulation, producing recombinant molecules composed of at least two different fragments of DNA. These recombinants have no significance until they are introduced into a living host where they can be maintained and propagated indefinitely.

13.1 IMPORTANT GENERALIZATIONS

It is convenient to consider the recombinant part of the technology under four headings:

(i) the generation of DNA fragments using restriction endonucleases,
(ii) incorporation of these fragments into a suitable vector, i.e. formation of an artificial recombinant,
(iii) introduction of the artificial recombinant into a suitable host where it can be replicated and subsequently grown in a culture to produce clones with multiple copies of the incorporated DNA fragment, and
(iv) identification, selection and characterization of clones of recipient cells that have acquired the recombinant for which a number of ingenious techniques have been devised.

To explain the significance of gene cloning, we must first consider the essential requirements for a successful gene manipulation procedure. A typical cloning experiment (Fig. 13.1) requires:

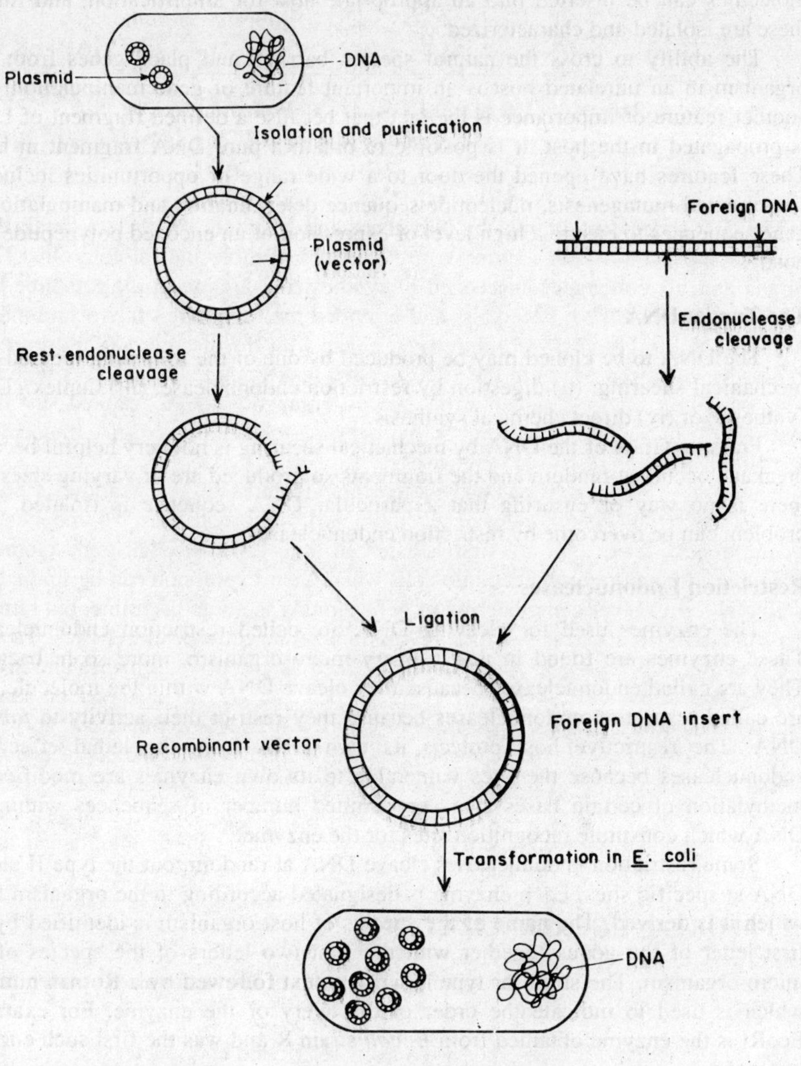

Fig. 13.1 A typical recombinant DNA experiment depicting the cloning of eukaryotic DNA fragments in *E. coli*

(i) DNA of interest , also known as foreign, insert or target DNA,
(ii) a cloning vector,
(iii) restriction endonucleases, the enzymes for cutting DNA molecules,
(iv) DNA ligase or enzymes for joining DNA fragments, and
(v) a prokaryotic or a eukaryotic cell to serve as host.

Once the foreign DNA fragment has been produced and a vector isolated, they are treated with the same restriction endonuclease to produce site-specific scissions in the DNA, complementary to each other, and annealed; thus producing a recombinant molecule. The recombinant molecule is made permanent only when all gaps are sealed by the DNA ligase. Once this is accomplished the recombinant molecules can be inserted into an appropriate host for amplification, and finally these are isolated and characterized.

The ability to cross the natural species barriers and place genes from any organism in an unrelated host is an important feature of gene manipulation. Yet another feature of importance is the fact that because a defined fragment of DNA is propagated in the host, it is possible to obtain a pure DNA fragment in bulk. These features have opened the door to a wide range of opportunities including site-directed mutagenesis, nucleotide sequence determination and manipulation of gene sequences to ensure a high level of expression of an encoded polypeptide in a host.

The Target DNA

The DNA to be cloned may be produced by one of the following methods: (i) mechanical shearing, (ii) digestion by restriction endonuclease, (iii) duplex *c*DNA synthesis, or (iv) direct chemical synthesis.

Fragmentation of the DNA by mechanical shearing is not very helpful because breakage occurs at random and the fragments so produced are of varying sizes and there is no way of ensuring that a particular DNA sequence is isolated. This problem can be overcome by restriction endonucleases.

Restriction Endonucleases

The enzymes used for cleaving DNA are called restriction endonucleases. These enzymes are found in nearly every micro-organism, more so in bacteria. They are called endonucleases because they cleave DNA within the molecule, and are called restriction endonucleases because they restrict their activity to *foreign* DNA. The restrictive host protects its own DNA from the lethal effects of endonucleases because the sites vulnerable to its own enzymes are modified by methylation of certain bases at a very limited number of sequences within the DNA which constitute recognition sites for the enzyme.

Some restriction endonucleases cleave DNA at random, but the type II cleave DNA at specific sites. Each enzyme is designated according to the organism from which it is derived. The name of the species of host organism is identified by the first letter of the genus together with the first two letters of the species of the micro-organism. The strain or type is written next followed by a Roman numeral which is used to indicate the order of discovery of the enzyme. For example, EcoRI is the enzyme obtained from *E. coli* strain R and was the first such enzyme isolated.

Type II restriction endonucleases recognize and break DNA at specific sites within hexanucleotides, but some recognize tetra or penta nucleotides. Restriction endonucleases create double-stranded breaks in the DNA to yield restriction fragments. Different types of ends that are produced have important implications. The enzyme EcoRI for example, cuts between G and A as indicated by arrows in the sequence and thus produces staggered ends with termini bearing 5′-phosphate and 3′-hydroxyl groups.

5′ — G A A T T C ⟶3′
3′ ⟵ C T T A A G —5′

```
               ↓
5′ —G      A A T T C →3′
3′ ←C T T A A      G —5′
                   ↑
```

The staggered ends with complementary bases are called *sticky ends* because they can combine with similar sequences produced by the same enzyme on a suitable vector DNA.

Other enzymes, for example, SmaI derived from *Serratia marcescens*, produces blunt-ended fragments cutting between C and G, on a specific hexanucleotide sequence on both strands.

```
5′ —C C C G G G →3′
3′ ←G G G C C C —5′

5′ —C C C | G G G →3′
3′ ←G G G ↓ C C C —5′
```

To-date around 475 types of restriction enzymes with more than 100 specificities have been isolated (Table 13.1). Thus, with such a large variety of enzymes a whole genome can be subdivided into smaller, discrete pieces so that genes can be isolated, cloned and characterized by nucleotide sequence analysis.

13.2 CLONING VECTORS

Cloning vectors play an important role in recombinant DNA experiments (see Fig. 13.1). In early 1970s, biologists first used the term vector to describe a vehicle for cloning DNA. Three types are in common use: (i) plasmids, (ii) bacteriophage, (iii) cosmids. All of these replicate independently of their host cell, and are therefore called *replicons*. Since most DNA fragments are incapable of self-replication, an autonomously replicating segment of DNA must be used to accomplish replication of the target DNA fragment.

Complementary DNA (cDNA)

Cloning with eukaryotic *c*DNAs have special advantages. An important fact is that they lack introns which are usually present in the corresponding genomic DNA and have only those regions which generate precursor of RNA not excised during transcription. This forms *m*RNA and specifies the primary structure of the gene product. These regions are called `exons'. On the other hand, `introns' are non-coding sequences which usually occur within eukaryotic gene sequences. They can either be present within the coding sequence or may occur in 5′ or 3′ non-translating regions of the gene. In any case, they are copied into RNA by RNA polymerase during transcription. Therefore, when *c*DNA is derived from *m*RNA, it lacks introns because the primary transcript, a precursor of *m*RNA, undergoes specific changes during transcription, which include the removal of introns and joining of exons by a process called splicing.

Table 13.1 Some restriction endonucleases, their recognition sequences and target sites. (By convention the recognition sequences are written in 5′ → 3′ direction and are given for one DNA strand; N = any base)

Enzyme	*Organism*	*Reognition sequence 5′ → 3′ and cleavage site (*)*
		Tetranucleotides
Hpa II	Haemophilus parainfluenzae	C* C G G
Alu I	Arthobacter luteus	A G* C T
		Pentanucleotides
EcoR II	Escherichia coli R245	*C C (T) G G
Hinf I	Haemophilus influenzae Rf	G* A N T C
		Hexanucleotides
Pst I	Providencia stuartii	C T G C A* G
Bam H I	Bacillus amyloliquifaciens H	G* G A T C C
Eco R I	Escherichia coli RY 13	G* A A T T C
Hind III	Haemophilus influenzae Rd	A* A G C T T
Hsu	Haemophilus suis	A* A G C T T
Sma I	Serratia marcescens	C C C* G G G
Xma I	Xanthomonas malvacearum	C* C C G G G
		Heptanucleotides
Mst II	Microcoleus species	C C* T N A G G

cDNA Synthesis

Initially *m*RNA has to be isolated from a relevant tissue. Suppose a copy of a globin gene is required. Then *m*RNA is isolated from the reticulocytes and purified. The *m*RNA is then exposed to reverse transcriptase to produce a single-stranded DNA copy (Fig. 13.2). The RNA-DNA hybrid is subsequently treated with alkali which destroys the RNA strand, leaving the DNA strand intact. The *c*DNA has a hair-pin loop at the 3′ end which acts as a primer for synthesis of a *c*DNA strand in the presence of DNA polymerase. The single-stranded hair-pin is then removed with S1 nuclease which specifically degrades single-stranded DNA.

Plasmids as Vectors

Plasmids are widely used as cloning vehicles. Though several plasmids have now been specially constructed for recombinant DNA work, they also occur naturally in yeast and various bacteria. These are extra-chromosomal replicons and are inherited. These are small double-stranded circular DNA molecules, possessing a limited number of restriction sites. Their most important attribute is that they confer antibiotic resistance on the bacterial host. One way of categorizing plasmids can be on the basis of the number of copies per cell:

(i) Relaxed plasmids, which are normally maintained as multiple copies per cell, and (ii) Stringent plasmids, which have a limited number of copies per cell. Plasmids can also be classified as conjugative and non-conjugative, depending on whether or not they carry a set of transfer genes called the *tra* genes that promote bacterial conjugation as discussed in Chapter 10. Generally, conjugative plasmids are of high molecular weight and are present as 1-3 copies per cell, whereas non-conjugative plasmids have low molecular weights and are present in multiple copies, i.e., 20-25 copies per cell.

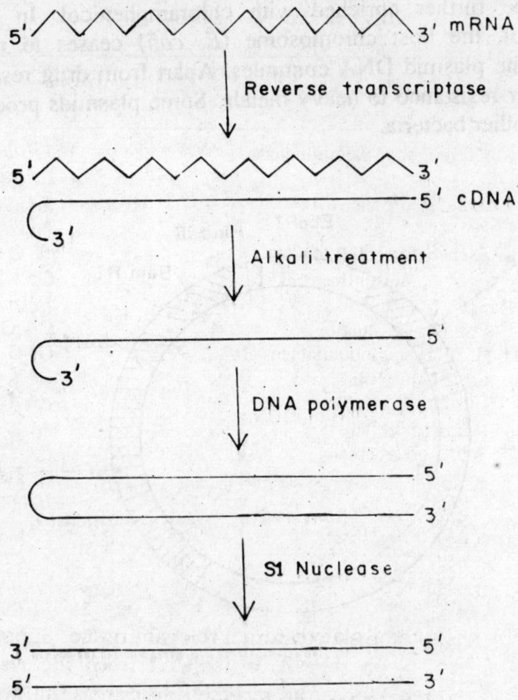

Fig. 13.2 Synthesis of cDNA from *m*RNA using reverse transcriptase

Criteria for Plasmid Cloning

The following criteria are used to design an ideal cloning vector:

(i) A plasmid vector should be small and should contribute as little as possible to the overall size of the recombinant molecule.

(ii) It should be well characterized with respect to single cleavage sites for a large number of restriction endonucleases located in one or the other genetic marker.

(iii) The plasmid should be easily propagated in the desired host so that the number of recombinant DNA molecules can be amplified.

(iv) It should have a large number of copies.

(v) It should have the ability to confer readily selectable marker (phenotypic trait) on the host cells that can be used to distinguish the transformed cells from the non-transformed ones.

(vi) A plasmid vector should have an additional genetic marker that can be inactivated by insertion of the foreign DNA, so that the inactivated gene may help in distinguishing the cells harbouring recombinant molecules on the basis of a readily altered phenotype.

These properties provide maximum flexibility in terms of cloning different types of restriction fragments. A commonly used plasmid pBR 322 (Fig. 13.3), is tailor-made as far as these attributes are concerned. Plasmids with a high copy number can be further enriched with chloramphenicol. In the presence of chloramphenicol, the host chromosome (*E. coli*) ceases to replicate but the replication of the plasmid DNA continues. Apart from drug resistance, plasmids may also confer resistance to heavy metals. Some plasmids produce bacteriocins which can kill other bacteria.

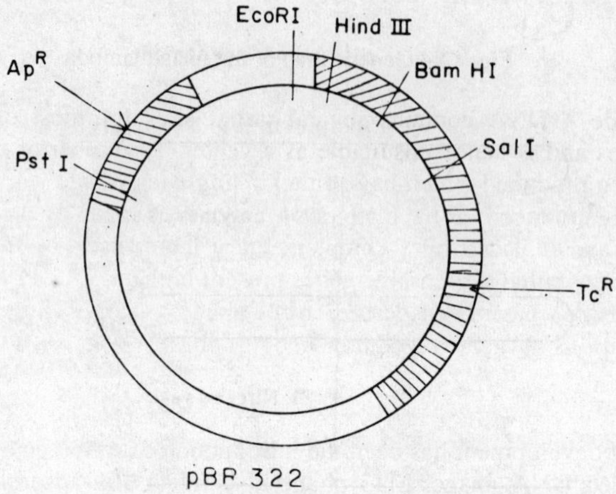

Fig. 13.3 A pBR 322 Plasmid vector

Bacteriophages as Vectors

Bacteriophages are viruses which infect bacteria. The bacteriophage lambda (λ), a temperate phage is one of the best understood elements in bacterial genetics. Upon entering *E. coli* , its DNA circularizes by the joining together of its sticky or cohesive ends, consisting of 12 complementary bases at each end of the linear DNA molecule. At this stage it replicates either by the lytic or the lysogenic path. In the lysogenic phase, the infecting viral genome integrates with the host chromosome in a harmless fashion, whereas in the lytic mode the viral DNA replicates to produce a large number of progeny particles which are released when the host cell bursts. The DNA of phage lambda is a linear duplex molecule of about 48.5 kb pairs possessing some 60 genes. Functionally related genes are clustered together on the map. Genes on the left of the linear map code for head and tail proteins of the phage (Fig.13.4). In the middle is a cluster of genes involved in the lysogenic phase, i.e. concerned with recombination.

An important feature of the phage lambda is that it replicates in the lytic phase. Much of the central region, including the genes for lysogeny, are not essential for phage growth and can be deleted or substituted without impairing the infectous growth cycle. Some of the important genes to the right of the central region are concerned with regulation and prophage immunity to infection, DNA synthesis and host cell lysis.

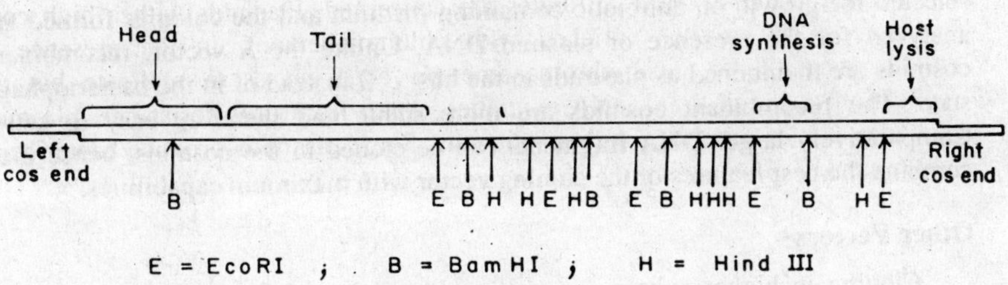

Fig. 13.4 Genetic map of the phage lambda

Wild type λ DNA contains several target sites for most of the restriction endonucleases and is hence unsuitable as a vector. A number of modifications of this have been produced which have either a single target site where a small insert DNA can be introduced, or have a pair of target sites spanning a segment that can be removed and replaced by a complementary large insert of foreign DNA. By replacing the so-called dispensable genes present in the middle of the genome over a hundred vectors have been derived from phage lambda. These include charon phage and various gt (generalized transducing) phages.

M -13 Phages

A recent development has been the introduction of M13 vectors-a filamentous single-stranded DNA phage. M13 phage also infects by adsorbing on the male bacterial cell through proteinaceous filaments called 'pili'. Once inside the host, their DNA is replicated, transcribed and translated into viral gene products. The pilus (+) strand of the virus is used to synthesize a complementary (–) strand giving rise to a double-stranded copy of M13, known as the parental replication form (RF). This serves as template from which several i.e. 100-200 copies of RF molecules are synthesized. The single-stranded DNA phages do not kill the bacteria, instead the phage particles are released continuously from the cell without any harm to it. However, the infectious process does retard the cell growth. No packaging of viral DNA is required with this vector system.

When M13 particles continue to infect the surrounding male bacteria, a *plaque* of slow-growing cells is formed. Therefore, with the M13 system, as with phage, recombinant phages can be distinguished from the non-recombinants on the basis of plaque colour.

Cosmid Vectors

Making use of the *in vitro* packaging system, the most novel modification of the λ vector cloning system was done for an artificial construction of cosmids (cos + plasmids), containing the plasmid DNA with a fragment of λ DNA including the *cos* site packaged in a phage particle. Cosmids consist essentially of a circular duplex DNA with antibiotic resistant genes, several restriction sites and cos sites of phage lambda.

The cosmid HC79, for example, consists of a 1.7 kb DNA fragment containing cos region of λ inserted into a non-essential region of pBR 322. Foreign DNA fragments can be cloned into any one of the restriction sites on pBR 322 and the product after ligation is packaged in vitro. The cosmid infects the host which is selected for growth on antibiotic-containing medium and the colonies formed are analyzed for the presence of plasmid DNA. Unlike the λ vector, recombinant cosmids are maintained as plasmids in the host cell instead of in the bacteriophage state. The recombinant cosmids are more stable and therefore easy to store. Comparatively larger DNA fragments can be cloned in the cosmids, hence they combine the best features of the cloning vector with maximum capabilities.

Other Vectors

Cloning in higher eukaryotes, particularly plants, is in a primitive stage. In plants vectors are being developed from DNA viruses e.g. the cauliflower mosaic virus CaMV and Ti plasmids of *Agrobacterium tumefaciens*. In the early 70s studies were undertaken to identify a `tumour inducing principle' in a wide variety of gymnosperms and angiosperms suffering from Crown gall disease. The tumour could be induced by inoculation of a soil bacterium, *Agrobacterium tumefaciens* (Agt) at the wound site and later it was discovered that the tumour-inducing factor was a Ti plasmid present in the virulent strain of the bacterium. The Ti plasmid has proved to be a natural vector for engineering plant cells because it can transfer its T-DNA segment from the bacterium to the plant which gets inserted into the host-cell DNA. Agt, however, infects only dicotyledons and many of the world's major food crops are cereal or monocotyledonous crops (rice, barley, maize, wheat etc.). For such plants an alternate strategy was followed involving the use of viral vectors, such as double-stranded DNA virus-CaMV.

The development of cloning systems in animal cells has assumed greater importance. Several approaches are followed to introduce manipulated genes into animal cells where their regulation can be assayed by several methods. The most widely used animal viruses employed as vectors are Simian virus 40 (SV 40) and adenovirus, which are DNA viruses.

13.3 LIGATION OF DNA

We must now consider the ways in which DNA fragments can be joined to create recombinant DNA molecules. The construction of recombinant DNA is dependent on the ability of the DNA ligase to covalently seal nicks in the single-stranded DNA. When a restriction enzyme produces sticky ends, the cohesive ends will come together and will be held by hydrogen bonding between the complementary bases. This process in which two separate strands of nucleic acid interact to form a duplex molecule by interaction between complementary base pairs present in the individual strands, is often termed annealing. The two molecules are then sealed and stabilized by the joining enzyme DNA ligase. This enzyme catalyzes the formation of a phosphodiester bond between 3′-hydroxyl and the 5′-phosphate of two adjacent nucleotides, establishing the structural continuity of the DNA strand.

DNA ligase plays a critical role in recombination, DNA replication and repair of the damaged DNA. DNA ligases have been isolated from prokaryotes, eukaryotes and their viruses. Though they catalyze the same reactions, their cofactor requirement varies. DNA ligase from *E. coli* requires NAD^+, while T4 DNA ligase requires ATP. In both cases an enzyme-adenylate complex is formed which binds to the nick bearing a 3′-hydroxyl and 5′-phosphate groups (Fig. 13.5). However, T4 DNA ligase has the unique ability to join the blunt-end DNA fragments (Fig.13.6) and because of this property T4 DNA ligase is preferred for in vitro ligation of blunt ends of DNA fragments. Joining of cohesive ends produced by restriction endonucleases is accomplished by *E. coli* DNA ligase (Fig. 13.6). Cohesive end ligation is essentially a nick-sealing reaction which proceeds 100 times faster than the blunt-end ligation. Moreover, in vitro ligation of DNA molecules reduces the susceptibility of DNA to nucleolytic degradation inside the cell and generally increases transformation efficiency.

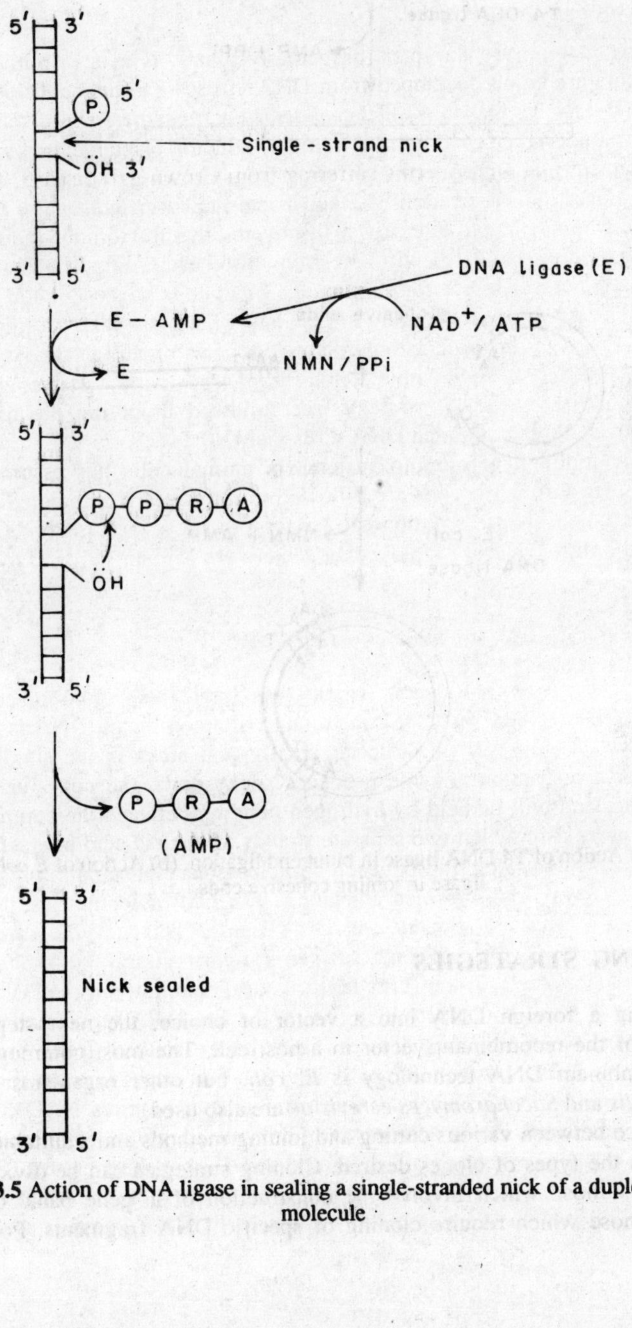

Fig. 13.5 Action of DNA ligase in sealing a single-stranded nick of a duplex DNA molecule

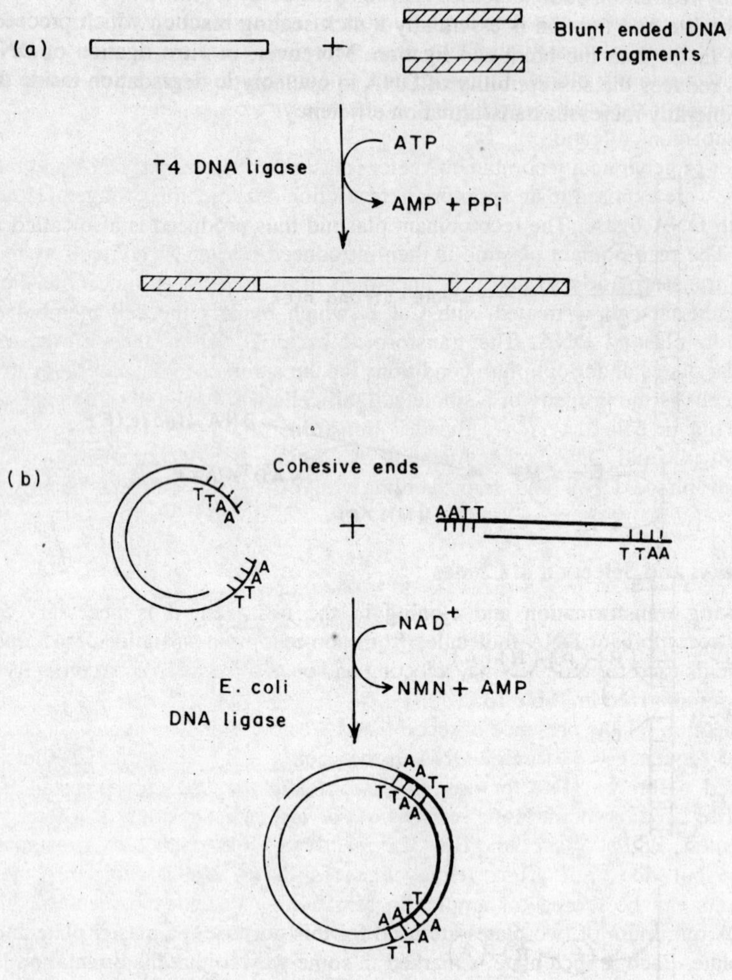

Fig. 13.6 (a) Action of T4 DNA ligase in blunt end ligation, (b) Action of *E.coli* DNA ligase in joining cohesive ends

13.4 CLONING STRATEGIES

After inserting a foreign DNA into a vector of choice, the next step is the introduction of the recombinant vector in a host cell. The most commonly used host in recombinant DNA technology is *E. coli*, but other organisms such as *Bacillus subtilis* and *Saccharomyces cerevisiae* are also used.

The choice between various cutting and joining methods and a suitable vector depends upon the types of clones desired. Cloning strategies can be divided into two categories–those which involve the construction of a 'gene bank' or 'gene library' and those which require cloning of specific DNA fragments. For larger genomes of higher eukaryotes it is necessary to clone the maximum number of different restriction fragments to obtain the gene of interest at a reasonable frequency. Thus we obtain a very large number of recombinants which contain a complete collection of nearly all DNA sequences in the entire genome. Such a collection of randomly cloned fragments that encompass the entire genome of a given species is called the gene library. Sometimes it is also known as a clone bank or a shot-gun collection..

In order to generate a recombinant vector (plasmid), the plasmid DNA and the foreign DNA are exposed to an appropriate restriction enzyme, mixed together and treated with DNA ligase. The recombinant plasmid thus produced is also called a *chimaera*. The recombinant plasmid is then introduced into an *E. coli* cell by the method of transformation so that it is integrated in to the host chromosome. For this, the bacterial cell is treated with $CaCl_2$ which renders the cell membrane permeable to plasmid DNA. The transformed bacterial cell is then grown on nutrient agar plates under optimum conditions for the colonies to develop. As each colony of cells is the progeny of a single cell, all cells will have the same genetic makeup. This is called a *clone*. Besides transformation of *E. coli* cells with recombinant plasmid DNA, other methods of cloning include transfection with recombinant phage DNA and transduction with either recombinant phage or cosmid.

Identification and Selection of Clones

Following transformation and cloning in the host cell, it is necessary to distinguish recombinant DNA molecules from non-recombinant molecules. Some of the methods used for indentifying, selecting and characterization of recombinant clones are summarized in Table 13.2.

(1) Selection for the presence of recombinant vector: The next step is to detect the inserted sequences i.e., foreign DNA in the vector, or to determine whether it has acquired a foreign DNA or not. In case of a plasmid vector this can be recognized by an altered antibiotic resistance. For instance, insertion of a foreign DNA fragment in Pst I site of pBR 322 inactivates the ampicillin resistance mechanism but does not affect tetracycline resistance. Tetracycline resistant transformants can be scored for ampicillin sensitivity to detect the recombinant plasmids. A minimum of two plates are used for this purpose—a master plate and a replica plate. Each replica plate is marked in some way so that the orientation is known and therefore reference can be made to the original master plate. The phenotype of each transformant is determined in terms of its growth on the replica plate. The master plate serves to verify the original phenotype. In the above referred example colonies will not grow in the presence of ampicillin though they will remain resistant to tetracycline, and in this way by referring back to the original master plate the recombinant clones can be identified, separated and cultured. In case of phage vectors, advantage is taken of the fact that certain strains produce blue plaques when plated in the presence of chromogenic substrate (X-gal), but in case of successful insertion in the gene region responsible for this colour change, colourless plaques are formed.

2. Selection for the presence of recombinant clones: Having identified the colonies in which recombinants have been formed, the next step is to screen and characterize recombinants which contain a specific DNA sequence. For selection and screening several approaches are possible (Table 13.2).

Table 13.2 Some methods used for identification, characterization and selection of the recombinant clones

	Methods
1. Selection of clones with a vector	(a) Antibiotic resistance (b) Plaque formation
2. Selection of clones with a recombinant vector	(a) Antibiotic resistance (b) X-gal test
3. Selection of recombinant clones with a specific DNA sequence	(a) Genetic (b) Immunochemical (c) Nucleic acid hybridization (d) Recombinational

(i) *Genetic Methods*: Genetic selection is a powerful tool since it can be applied to large populations. All useful vectors carry a selectable genetic marker. Plasmid and cosmid vectors carry drug resistance nutritional markers and for phages plaque formation itself is a selection property. The genetic methods depend on the inserted foreign DNA being expressed. The inserted gene may complement a metabolic defect in a strain of *E. coli* which has a requirement for a particular growth factor which the inserted DNA can synthesize. Transformants can thus be selected for using an appropriate culture medium.

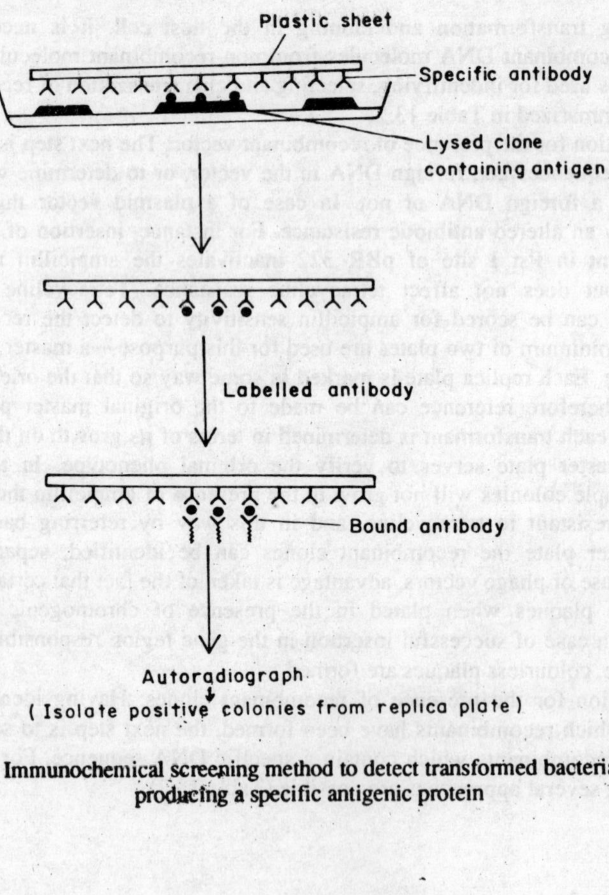

Fig. 13.7 Immunochemical screening method to detect transformed bacterial colonies producing a specific antigenic protein

(ii) *Immunochemical Methods*: These methods rely on antigen-antibody reaction, if specific antibodies are available to the gene product (antigen) of the inserted DNA. Such immunochemical detection of clones synthesizing a foreign protein has been found to be successful in cases where the inserted gene sequence is expressed. Antibodies adsorb strongly to plastic and if a plastic sheet is coated with a specific antibody and then gently applied to the surface of the lysed bacterial colonies on a replica plate, those colonies synthesizing the particular antigen will bind to the antibody. The immobilized antigen is then localized using a second antibody which is radioactively labelled and then the plastic sheet is subjected to autoradiography (Fig. 13.7). By referring back to the master plate relevant colonies synthesizing the particular antigen can be identified.

(iii) *Hybridization Method*: It is also possible to detect recombinant clones employing hybridization with DNA isolated and purified from the transformed cells. Thus clones containing a specific foreign DNA sequence can be identified by

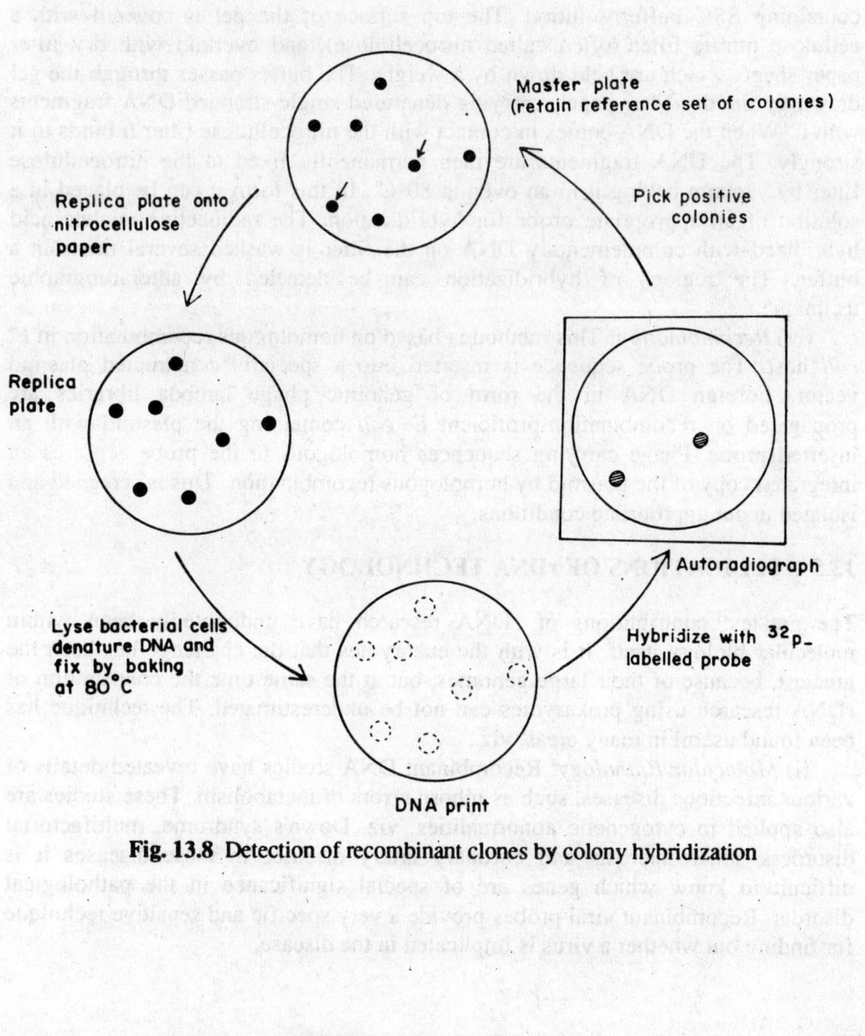

Fig. 13.8 Detection of recombinant clones by colony hybridization

hybridization with an appropriate probe. A probe is a single-stranded RNA or DNA segment which has been labelled either radioactively or with some biochemical marker so that it can search out or locate complementary DNA sequences in the presence of a large amount of non-complementary DNA.

Hybridization is carried out on a replica made on a nitrocellulose filter on which the cells have been lysed, and the DNA denatured and finally fixed to the filter by baking (Fig. 13.8). Colonies found to hybridize with a specific probe can be identified by autoradiography and finally located on the master plate.

Hybridization of DNA extracted from a clone can be carried out on electrophoresis gel using Southern Blotting technique (Fig. 13.9) developed by Southern in 1975. This is done by slicing the gel, eluting the DNA and hybridizing it with a radiolabelled probe DNA or RNA. The extracted DNA is first exposed to appropriate restriction endonuclease and the resulting fragments are subjected to electrophoresis on agarose gel. The gel is then transferred to a dish containing sodium chloride/sodium hydroxide (SSC) solution in order to denature the DNA. The gel is then rinsed and neutralized, and placed on a filter paper circle in a bath containing SSC buffer solution. The top surface of the gel is covered with a cellulose nitrate filter (often called nitrocellulose), and overlaid with dry filter paper sheets which are held down by a weight. The buffer passes through the gel drawn by the dry filter paper, carrying denatured single-stranded DNA fragments with it. When the DNA comes in contact with the nitrocellulose filter it binds to it strongly. The DNA fragments are then permanently fixed to the nitrocellulose filter by vacuum baking it in an oven at 80°C . In this form it can be placed in a solution of an appropriate probe for hybridization. The radioactive nucleic acid hybridized with complementary DNA on the filter is washed several times in a buffer. The regions of hybridization can be detected by autoradiographic technique.

(iv) *Recombination*: This method is based on homologous recombination in *E. coli* host. The probe sequence is inserted into a specially constructed plasmid vector. Foreign DNA in the form of genomic phage lambda libraries are propagated on recombination-proficient *E. coli* containing the plasmid with an inserted probe. Phage carrying sequences homologous to the probe acquires an integrated copy of the plasmid by homologous recombination. This is screened and isolated under appropriate conditions.

13.5 APPLICATIONS OF rDNA TECHNOLOGY

The greatest contributions of *r*DNA research have undoubtedly been within molecular biology itself. It is with the eukaryotes that the challenge has been the greatest, because of their large genomes, but at the same time the contribution of *r*DNA research using prokaryotes can not be underestimated. The technique has been found useful in many areas. viz.,

(i) *Molecular Pathology*: Recombinant DNA studies have revealed details of various infectious diseases, such as inborn errors of metabolism. These studies are also applied to cytogenetic abnormalities, viz. Down's syndrome, multifactorial disorders, atherosclerosis and coronary artery disease. In these diseases it is difficult to know which genes are of special significance in the pathological disorder. Recombinant viral probes provide a very specific and sensitive technique for finding out whether a virus is implicated in the disease.

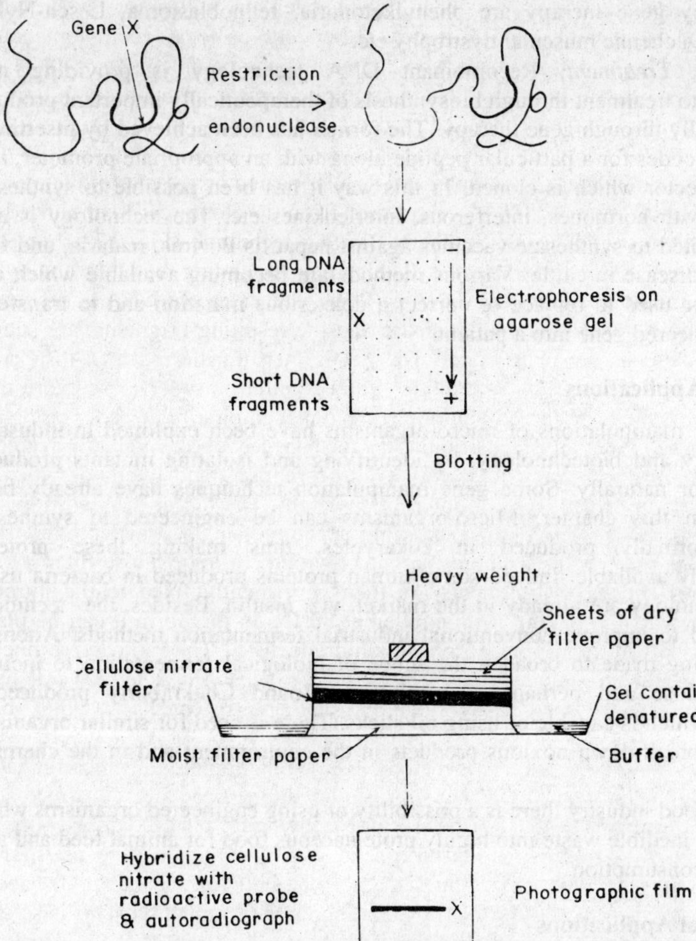

Fig. 13.9 Southern Blot technique. X indicates the gene or sequence of insert

In case of common cancers, *r*DNA technology has already revealed a great deal about their molecular pathology. It is now known that oncogenes occur normally and become activated in neoplastic tissues. Two areas of considerable interest are the mechanisms of oncogene activation and the nature of oncogene-encoded proteins.

(ii) *Prevention of Genetic Disorders*: Most genetic diseases are serious and as yet treatments are known for very few. Currently the most important practical application of *r*DNA technology is in the prevention of genetic diseases by providing novel and precise methods of preventive diagnosis. There are hundreds of genetic diseases afflicing mankind which arise from single recessive mutations. Among such diseases, inherited haemoglobin disorders have been extensively studied at the gene level. Some of the disorders that have been diagnosed and prevented by gene therapy are phenylketonuria, retinoblastoma, Lesch-Nyhan syndrome, Duchenne muscular dystrophy etc.

(iii) *In Treatment*: Recombinant DNA technology is providing new approaches to treatment through biosynthesis of therapeutically important products and eventually through gene therapy. The former has been achieved by inserting a gene which codes for a particular peptide along with an appropriate promoter, into a suitable vector which is cloned. In this way it has been possible to synthesize insulin, growth hormones, interferons, interleukines etc. The technology is also being exploited to synthesize vaccines against hepatitis B virus, malaria, and foot and mouth disease in cattle. Various methods are becoming available which one day could be used to replace or correct a deleterious mutation and to transfer a cloned engineered gene into a patient.

Industrial Applications

Genetic manipulations of micro-organisms have been exploited in industrial microbiology and biotechnology, by identifying and isolating mutants produced artificially or naturally. Some gene manipulation techniques have already been described in this chapter. Micro-organisms can be engineered to synthesize proteins normally produced in eukaryotes, thus making these proteins commercially available. Indeed some human proteins produced in bacteria using *r*DNA technology are already in the market, viz. insulin. Besides, the technique can be used to improve conventional industrial fermentation methods. Attempts are also being made to broaden the scope of biological fermentation to include wood cellulose and perhaps even plastics. Anand Chakraborty produced a bacterium which is capable of using oil slicks. There is need for similar organisms which can break down noxious products in the environment and in the chemical industry.

In the food industry there is a possibility of using engineered organisms which can convert inedible waste into highly proteinaceous food for animal feed and also for human consumption.

Agricultural Applications

In agriculture, perhaps the most important applications are in attempting to transfer genes for nitrogen fixation to cereal crops. Certain plants, notably legumes, are able to fix atmospheric nitrogen. This ability is due to the bacterium *Rhizobium* which lives symbiotically in the root nodules of leguminous plants. Transfer of this technique to cereal crops would reduce dependance on nitrogenous fertilizers. The Ti plasmid of *Agrobacterium* has been exploited as a means of transferring foreign genes into a host plant. Unfortunately *Agrobacterium* infects dicotyledons only and the world's major food crops are monocotyledons, namely, rice, maize, barley, wheat etc. As an alternative strategy the possibility of using viral vectors is being considered to overcome the problem.

Other Applications

Recombinant DNA technology has been of much use in animal husbandry through production of vaccines and transfer of commercially important gene-traits to improve milk yields. Other areas in which this technology is expected to benefit farming is in developing methods for transferring genes to improve photosynthesis, and ultimately the yield; resistance to pests, pathogens, herbicides and tolerance to adverse conditions such as frost, drought and increased salinity.

There are also possibilities of producing microorganisms with specific affinity for metals like uranium for use in mineral leaching and recovery. Biology, medicine, industrial microbiology and agricultural and pharmaceutical industries are likely to share a special relationship with genetic engineering in the future.

CHAPTER 14

THE ORIGIN OF LIFE

Life in the world, as it exists today, is the result of a long process of evolution. How and when did life originate on planet earth? In what form? These are difficult questions, which can be answered on the basis of certain assumptions, hypotheses and deductions. Whatever may have been the early form of life, it was certainly a primitive form which must have evolved through complex interaction of the forces of evolution. These forces are, mutation, genetic recombination, natural selection and reproductive isolation. Let us now examine the possible mechanisms which may have given origin to life.

According to the *Big-Bang* hypothesis, the planet earth originated about fifteen billion years ago, and it was immensely hot to begin with. When it cooled sufficiently, neutrons and protons fused together by random collisions to form atomic nuclei. Further cooling took about 7×10^5 years to allow electrons to bind the atomic nuclei. Only hydrogen and helium were produced in the early phases of earth's evolutionary history, followed by the formation of heavier elements after the stars were formed. The stars were formed from cold and turbulent solar dust clouds under the influence of gravitational forces. As they increased in density, the stars got heated up so that thermonuclear interactions started leading to the synthesis of heavier elements. The formation of planets perhaps began with the collision and clumping of tiny particles into larger particles. After the earth was fully formed, its atmosphere consisted of H_2, N_2, CH_4, NH_3, H_2S and water vapour, but no oxygen.

The liquid atmosphere of the earth was established soon after its formation had been completed. Water may have originated from the condensation of the earth's inner core as well as torrential rains, and existed or the surface as liquid due to ambient surface temperature. Sufficient primeval records are available from which the nature of the atmosphere and the hydrosphere in which these chemical reactions took place can be visualized.

14.1 EVOLUTION OF ORGANIC MOLECULES

Since life originated *de novo* on earth, one should understand the origin of complex molecules. It is argued that living systems evolved from a chaos of molecules. A plethora of evidence suggests that organic molecules may have been synthesized abiotically on the primeval earth. At the time of abiotic synthesis of organic molecules, the earth's atmosphere was probably non-oxidizing due to lack of oxygen and violent with volcanic eruptions, torrential rains and lightening. It contained H_2, N_2, H_2S NH_3 CO_2, CH_4, and water, but not oxygen. The small molecules acted as the precursors for organic molecules; and solar radiation, lightning and thermal energy produced by volcanic eruptions were the energy sources. Once the organic molecules were formed, they accumulated in water thus giving rise to an organic 'soup'. The best evidence for abiotic synthesis of organic molecules comes from Stanley Miller's experiment conducted in 1953, in which he simulated conditions prevalent on the primeval earth. The results of this experiment were significant, for the limited number of inorganic molecules gave rise to a large number of organic molecules (Fig. 14.1). The table 14.1 gives a list of biologically important organic molecules that were synthesized in the experiment.

Table 14.1 List of some organic molecules synthesized in the apparatus designed by Miller

1.	Formaldehyde	HCHO
2.	Formic acid	HCOOH
3.	Acetic acid	CH_3COOH
4.	Propionic acid	C_2H_5COOH
5.	Glycolic acid	$HO.CH_2.COOH$
6.	Lactic acid	$CHOH(CH_3)COOH$
7.	∝-hydrobutyric acid	$HO{-}CH(C_2H_5)COOH$
8.	Succinic acid	$HOOC(CH_2)_2COOH$
9.	Urea	$H_2N{-}\underset{\underset{O}{\Vert}}{C}{-}NH_2$
10.	Glycine	$H_2N{-}CH_2{-}COOH$
11.	N-methyl urea	$H_2N{-}CO{-}NH{-}CH_3$
12.	Alanine	$H_2N{-}CH(CH_3)COOH$
13.	Aspartic acid	$H_2N{-}CH(CH_2{-}COOH)COOH$
14.	Glutamic acid	$H_2N{-}CH(CH_2CH_2{-}COOH)COOH$
15.	Aminobutyric acid	$H_2N{-}CH(C_2H_5)COOH$

The amino acids were shown to be produced by simple condensation reaction by HCN, NH_3 and R. CHO (aldehyde) in aqueous medium as:

$$HCN + NH_3 + R.CHO \longrightarrow R\,\underset{\underset{NH_2}{|}}{CH}{-}CN \xrightarrow[2H_2O]{\uparrow} R.\ CH.COOH + NH_3 \quad \text{amino acid}$$

During the last 30 years, a large number of workers have synthesized practically all biological molecules abiotically in the laboratory.

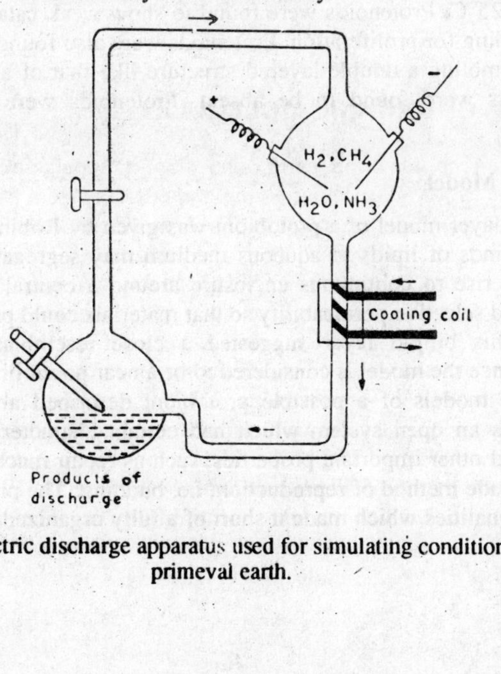

Fig. 14.1 Miller's electric discharge apparatus used for simulating conditions prevailing on primeval earth.

14.2 ORIGIN OF PROTOBIONT

Though there is general agreement on the abiotic synthesis of organic molecules, no consensus has been reached as to the manner in which these molecules organized themselves into living entities. A primitive form of life, *protobiont* might have appeared in the primeval sea, which was capable of showing activities so conspicuous of life. Several models of protobiont have been proposed, out of which three models have gained support. These are: (1) *Coacervate* model of Oparin, (2) Protenoid model of Sydney Fox, and (3) Lipid bilayer model of Richard Goldacre.

Coacervate Model

A coacervate is a droplet that is spontaneously formed when polar water molecules are oriented around an electrically charged molecule. The water boundary acts as a barrier between the interior of the coacervate and its unorganized surroundings. Several kinds of coacervates could be formed in the same medium, some of which may even fuse to form complex coacervates capable of concentrating organic molecules from their surroundings. Oparin described these coacervates as extremely diverse systems which can increase in mass and show complex interactions. They contained enzymes which could synthesize and hydrolyze starch, and release maltose to the surroundings. According to Oparin, this exchange of matter and energy is characteristic of open systems. A growing coacervate may fragment itself into smaller droplets or may give rise to small buds. They could also undergo destruction in media of unfavourable pH.

Protenoids

In 1965, Sydney Fox suggested that *protenoid* could be a likely candidate for a protobiont. Protenoids are protein-like molecules which could be synthesized by plymerization of amino acids when heated at temperatures between 130 to 160°C. Protenoids are spherical particles which are unstable at high temperatures, but stable at 0 to 25°C. Protenoids were found to show weak catalytic properties and exhibited budding for proliferation. Protenoids were also found to possess an outer boundary resembling a double layered structure like that of a plasma membrane, although lipids were found to be absent. Protenoids were also christened as *microspheres*.

Bilipid Layer Model

A bilipid-layer model of a protobiont was given by Robinson, who suggested that various kinds of lipids in aqueous medium may segregate to form bilayers, and thus give rise to continuous enclosure around a central space. This bilipid layer possessed selective permeability so that materials could pass in and out of the membrane. This bilipid layer suggested a close resemblance to the plasma membrane, hence the model is considered to be a near possibility.

The three models of a possible protobiont described above suggested that protobiont was an open system which had certain characteristics of the living. They also fulfil other important properties, such as repair machanisms, increase in mass, and a crude method of reproduction, i.e. budding. The protobiont admittedly lacked a few qualities which made it short of a fully organized cell. The formation of a protobiont in abiotic conditions was nevertheless a unique event and might have been a forerunner of a truly living cell.

14.3 ORIGIN OF A EUBIONT

For a cellular organization, three requirements have to be fulfilled. These are: a selectively permeable boundary separating the cellular components from the outside, a metabolic system that would allow a series of enzyme catalyzed reactions, and a genetic information system that would store and transmit hereditary information to future generations.

We have seen that protobionts, though they exhibited certain qualities of life, did not possess genetic system. However, if we could imagine that at some stage, the protobiont might have incorporated a genetic system by a chance event, the requirements of a true living cell would be fulfilled. It appears logical to think that at some stage of evolution, genetic material was evolved which became incorporated with the protein molecules and developed a boundary of lipid molecules. Let us now examine the evolution of polypeptides and polynucleotides.

Evolution of Proteins and Nucleic Acids

Proteins are polypeptides which are synthesized by polymerization of amino acids. This was described by Sydney Fox while describing the formation of protenoids. Formation of proteins may be considered as a major step in the evolution of genetic information. Similarly, nucleic acids or polynucleotides may have been formed by polymerization of organic compounds aided by catalytic activities of polyphosphates.

Once these two types of polymers were formed, they could have catalyzed the formation of other polymers. The polynucleotides have the capability of acting as templates to direct synthesis of complementary nucleotides. In fact, complementarity of a newly synthesized polynucleotide on the template must have been an important event in the evolution of living systems. This perhaps gave rise to a mechanism for transmission of genetic information. It has been postulated that polyribonucleotides or RNA were the earliest molecules which had the self-replicating property. The RNA molecules have two important characteristics: (a) information transfer through nucleotide sequence of replication, and (b) a uniquely folded structure that determines its function in response to environmental stimuli. These features are considered as prerequisites for the evolutionary process.

Information in RNA molecules is encoded in the form of triplet codes, but how did this code actually evolve ? Among the existing codes, those specifying the same amino acids have the first two bases common, but the third base may be different. The doublets formed by four types of bases could code for only 16 types (4^2) of amino acids which may have served the purpose for synthesis of primitive proteins on abiotic earth. But it appears that the doublet code could not stabilize, and hence a triplet code may have evolved. A triplet code was perhaps favoured and through natural selection, got established. More details about the triplet code have already been discussed in chapter 9. There are several theories to explain the origin of the genetic code.

Froze Accident Theory: This theory was proposed by Francis Crick in 1967, who suggested that the development of a genetic code was the outcome of a series of random interactions over time and space, and once the genetic code was stabilized, it remained so because any change in the codon assignment would have disturbed the sequence of proteins, which would have been lethal.

Vocabulary Expansion Theory: This theory is supported by two hypotheses. The first assumption, as proposed by Jukes in 1967, is that sixtyfour codons initially coded for a small number of amino acids, and in the event of a new amino acid entering the scheme, it would necessarily share the existing amino acid codons. The second assumption was made by Hartmann in 1978, according to whom the initial specificity of codons was for a few amino acids. This specificity was mediated by triplet codons. According to vocabulary expansion theory, the proteins assembled by limited specificity of codons were primitive in structural and functional complexities. Expansion of new vocabulary to produce more complicated proteins occurred when new and related amino acids were formed. In this process some of the original codons retained their specificity, but expansion of the code occurred by addition of additional information in the triplet.

Direct Interaction Theory: Which was proposed by Gamow in 1959 received much attention. The theory assumes that there exists a sort of complementarity between amino acids and nucleic acids. There seems to be differential affinity between amino acids and their coding sequence due to differences in their physicochemical properties. According to some authors, the genetic code originated due to interaction between nucleotide sequence in the anticodon and the amino acids.

The evolution of polypeptides probably occurred simultaneously with the polynucleotides. Polymerization of amino acids into polypeptides by heating is well established by Fox's experiment with protenoids. It has been found that polypeptide with random assembly of amino acids do possess some catalytic activity. This must have been followed by establishment of viable amino acid sequences by elimination of non-viable chains by natural selection. Once these polypeptide chains with specific amino acid sequence were formed, they might have interacted with the polynucleotide chains of specific nucleotide sequences. These interactions must have given rise to a definite relationship between amino acid sequences and nucleotide sequences, and an information coding system was thus evolved.

The information coding system probably evolved first in the RNA, since these molecules are unstable as their single strands are open to degradation by enzymes. Following this, a more stable system having complementary coding strands got evolved in the form of the DNA double helix. The RNA molecules in course of evolution under changing conditions on earth must have undergone modification to produce DNA by reduction of ribose sugar into deoxyribose and uracil by more suitable modification into thymine. Subsequently, DNA molecules became diversified by storing information in a variety of proteins.

A variety of polynucleotide and polypeptide species were initially formed in large numbers with ample building blocks in the surrounding primeval soup. However, increasing number of these diverse polymers must have competed with each other, and in the process the polymers which had the capacity to form speedily with minimum errors, and ability to coordinate a proper sequence of events useful for life were retained and the rest were destroyed by the process of natural selection.

Evolution of the Membrane

DNA-directed protein synthesis on RNA templates was perhaps the most important event which resulted in the appearance of the first living cell. The next crucial step was the formation of a cell membrane to delimit the genetic information system from the surrounding chaos of molecules in the primeval sea. As we know, all living cells are surrounded by a phospholipid-protein membrane endowed with selective permeability. In vitro experiments have demonstrated that when phospholipids are mixed with water, bimolecular phospholipid membranes are formed. Under appropriate conditions such membranes round up into closed vesicles of 1 to 10 nm diameter. If such vesicles enclose proteins and nucleic acids, they can function as structurally isolated units.

Coming back to protein and nucleic acid synthesis under primordial conditions on earth, the protein synthesized by RNA or DNA species could be of no use unless the synthesized proteins could remain in close association with, or near, the synthesizing nucleic acid. As long as these proteins remained free to diffuse among the replicating nucleic acid molecules, they could equally benefit any RNA species competing with its synthesizer. Whenever a superior RNA variant produced an equally superior enzyme, the latter could not help survival of that superior RNA species in its competition with others until some barrier evolved to contain the enzymes. It may be surmised that during stabilization of protein synthesis by RNA in the primordial soup on earth, a chance trapping of an RNA species along with its protein into a bilipid layer vesicle might have occurred to generate a first self-sustaining and self-replicating unit of life—the *eubiont*. Wolken (1984) described biological molecules as liquid crystals which are capable of aligning into various structures around other liquid crystals with an ordered organization as well as mobility. Such an organization fits into the features of a cell.

14.4 ORIGIN OF A PRIMITIVE CELL

It has been postulated that the first cell evolved in the primordial soup when phospholipid molecules organized themselves into membranous vesicles enclosing self-replicating RNA and protein molecules. Our discussion so far has been speculative, nevertheless, experimental evidence also points out that the story narrated so far appears to be correct. It is logical to assume that a chance collision of RNA, protein and phospholipid molecules spontaneously resulted in the first truly living primitive cell. In course of evolution and in an ever-changing primeval environment, RNA molecules must have been transformed into more suitable and stable DNA molecules. All these events might have produced the first cell about three or four billion years ago.

14.5 ORIGIN OF A PROCARYOTIC ORGANIZATION

If we compare the organization of the primitive cell with that of the *mycoplasma*—the simplest of the present day living cell, we find many similarities. An average mycoplasma cell has a diameter of about 0.3μm, much smaller than lipid bilayer vesicles which could be generated experimentally. Nucleic acids of the mycoplasma can synthesize about 750 types of proteins, the minimum number required by cells to survive. In mycoplasma both RNA and DNA are parts of the genetic information system, and thus they bear a close resemblance to the first evolved cell. Once the primordial cell was formed, it competed with others and survived by evolving its own metabolic regulating and energy trapping mechanisms from the surroundings. It also may have developed a primitive method of reproduction by cell division.

Evolution of Metabolic Activities

The primitive cell after having established its existence must have elaborated its internal organization by developing cytoplasm and genetic system to become more complex. There must have been perhaps, severe demands on the cell in an ever changing environment, and as a consequence mycoplasma-like cells evolved themselves into procaryotes which include bacteria and blue-green algae.

The procaryotic organization was perhaps more efficient in making use of matter and energy available in the environment, and so was able to evolve a more complex organization to carry out biochemical functions. In present-day environment also bacteria are available which utilize organic compounds like sugars, amino acids, fatty acids and other polymers. Some bacteria even draw their carbon atoms directly from CO_2 and nitrogen from atmospheric nitrogen. Bacteria and mycoplasma are found in our present-day environment and their ability to divide rapidly enables them to adapt to a changing environment. In order to utilize organic compounds, chemical reactions require a number of enzymes so as to sustain specific metabolic pathways, and hence an elaborate metabolic machinery might have developed.

The procaryote as outlined above was a complicated little creature, which was generated spontaneously since it could utilize monomers from its environment and transform them into polymeric sequences. When life began on earth, there was little need for elaborate metabolic pathways. The primitive cells could maintain themselves on the primordial soup of organic molecules. As the natural resources of the directly utilizable molecules started declining due to increasing number of cells, cells which could develop new enzymes to use more organic molecules were favoured. Thus a kind of natural selection was at work. The enzyme pool of the survivors and the most adapted ones gradually increased, generating diverse metabolic pathways. It has been suggested that enzymes evolved before the DNA code for protein structure, because of the abundance of nitrogen derivatives, vitamins and inorganic ions in the early seas. These might have acted as coenzymes which accelerated the enzyme activity. The evolution of enzyme diversity could have been possible by two possible mechanisms, one by the increasing variety of enzymes consequent upon depletion of useful organic matter and their precursors in the environment, and two the complete exhaustion of utilizable molecules with simultaneous availability of another type of molecules which could be converted into utilizable molecules.

Let us assume that a primitive cell used a molecule *D* for its maintenance from a pool of four related compounds *A*, *B*, *C* and *D*. In course of time *D* was depleted, and the cell evolved a new enzyme which could convert *B* into *D*. Similarly, when *B* became scarce, the cell evolved yet another enzyme to convert *C* into *B* and so on, thus a variety of enzymes were evolved. Suppose the cells originally used *A* for a specific purpose, and the concentration of *A* declined considerably, another related compound *D* is used for another work and is available in high concentration. Under such a situation, the cell developed an enzyme to convert *D* into *C*, which is more related to *A*. Thus by evolving another enzyme the cell was able to convert *C* into *B*. Since *B* was very close to *A*, the cell produced another enzyme to convert *B* into *A*, ensuring the supply of *A*. Thus development of a new enzyme system led to back-tracking of the original reaction.

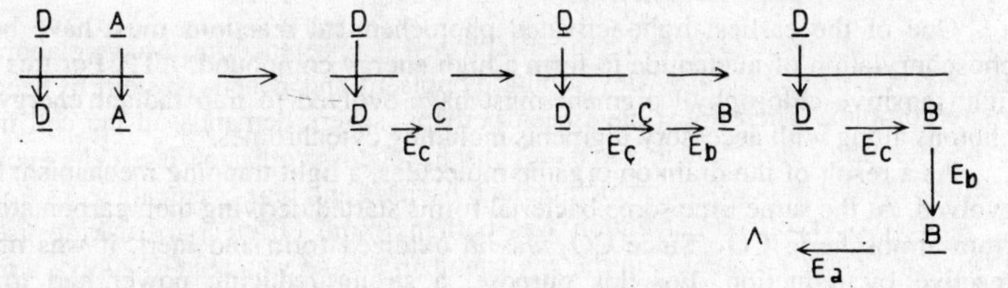

To explain the evolution of the primitive cell, Horowitz in 1945 proposed a hypothesis called "evolution backward". According to the postulates of this hypothesis, when utilizable organic molecules became scarce, relatively simple cells having the capacity to make the original utilizable molecules from the available ones held the selective advantage in the depleting environment. New metabolic pathways with coupled reactions and common intermediates involved in energy transfer must have evolved in a fashion to imprint these processes in the present day cells.

As discussed, eubionts have undergone changes in the genetic material so that their genome increased from a few genes to a few thousand. The genome expansion was perhaps non-random and definite. According to Francis Jacob, evolution of diverse enzymes has occurred mainly by *tinkering* with the pre-existing genetic system, rather than by evolving a new one. Modification of nucleotide sequence and regulation of gene expression have been the features for diversification of metabolism in present day cells.

Evolution of Photosynthetic Activity

Present day organisms have three modes of nutrition. Those drawing their food from plants and animals called *heterotrophs*, largely obtain energy through oxidation of organic molecules. Another category of organisms, called *chemotrophs*, derive their energy from inorganic molecules and synthesize their own food chemosynthetically. Many bacteria fall under this category. The third type of organisms are called *phototrophs*, which synthesize their food materials in the presence of sunlight and green chlorophyll. All green plants, algae and photosynthetic bacteria come under this category.

The primitive cells were heterotrophic and also anaerobic and drew their nutrition from the organic soup in the form of organic molecules. As the nutrients became scarce owing to the increasing number of cells, the primitive cell had to evolve by modifying its genome so that more enzymes could be produced to utilize the sun's energy for the synthesis of polymers. A photosynthetic cell requires more organizational complexity and has more enzyme systems than does a chemotroph. The selection pressure imposed by the environment had been working on the cell and electron transport pathways developed in three directions:

(1) Some cells evolved a photosystem to trap solar energy in which water was the electron donor,
(2) Some cells developed a system to release a large amount of molecular oxygen into air and water, and
(3) Some cells developed a respiratory electron transport chain in which molecular oxygen served as the ultimate electron acceptor.

One of the earliest light-activated photochemical reactions must have been phosphorylation of nucleotide to form a high-energy compound, ATP. For this the light-sensitive chlorophyll pigment must have evolved to trap radiant energy in photons along with accessory pigments including cytochromes.

As a result of the drain on organic molecules, a light trapping mechanism had evolved. At the same time some bacterial forms started deriving their carbon atoms from atmospheric CO_2. Since CO_2 was in oxidized form and inert, it was made reactive by reduction. For this purpose, a strong reducing power had to be generated. To accomplish this, the then existing photosystem was elaborated by addition of photosystem II to generate a continuous supply of reducing agents. Initially, the source to generate the reducing agent was H_2S, a fact that is supported by the sulphur bacteria which are anaerobes and use H_2S as an electron source. This system so developed could work well in a non-oxidizing environment. In the early photosynthetic process, electrons from the chlorophyll were transferred to NAD or NADP to generate a reducing power in the form of NADH or NADPH, and the electron deficiency was made up by pigments in the photosystem II. The electron deficient photosystem II was replenished with electrons extracted from H_2S, releasing elemental sulphur as a by-product.

The ever-growing competition to avail of H_2S from the environment by increasing photosynthetic bacteria resulted in scarcity of H_2S. Stress on H_2S led some groups of photosynthetic bacteria to modify their photosystem to use H_2O, which was plentiful in the environment. However, utilization of water as an electron source resulted in the evolution of a tremendous amount of molecular oxygen, changing a non-oxidizing environment into an oxidizing one, and oxygen started accumulating in the atmosphere.

Oxygen is highly reactive and was probably toxic to early heterotrophs. Some chemotrophs could not synthesize chlorophyll, but made use of molecular oxygen as the ultimate acceptor of electrons. In such cases, the electron transport system got modified into a non-photosynthesizing system, called the Respiratory Chain. Thus, the aerobic cell was born. However, photosynthetic cells also developed systems similar to the newly evolved aerobic cells, and photosynthetic aerobic systems evolved. Once the procaryote was developed, and the DNA-RNA-protein system established, the foundation was laid for the appearance of the eucaryotes.

14.6 ORIGIN OF EUCARYOTES

As the oxygen accumulated in the environment of the earth, a large section of anaerobic cells adopted an aerobic life-style, and the rest were subjected to three hard options: (1) either they had to perish, or (2) specialize to find a niche where oxygen supply was either too low or absent, or (3) take to aerobic life. A sizeable number of anaerobic heterotrophs perished due to oxygen toxicity, while some found niches with low oxygen supply, and continued with anaerobic mode of life. Quite a few cells started living in association with photosynthetic or aerobic cells.

According to "Symbiotic Theory of Eucaryotic Cell Evolution" of Margulis, this association resulted in the appearance of eucaryotes about 1.5 million years ago. Eucaryotes show certain basic differences when compared to procaryotes. These are:

(1) A mobile cell membrane instead of the rigid cell wall of procaryotes
(2) Presence of a membrane-bound nucleus
(3) Presence of a photosynthetic organelle
(4) Presence of mitochondria
(5) A membrane-bound compartmentalized cytoplasmic interior.

According to Sydney Fox, the efficiency of the primitive photosynthetic system was enhanced by development of membrane-bound lamellae as in case of cyanobacteria. These semi-autonomous lamellae permitted isolation of photolysis of water and nascent oxygen from the products such as NADPH and ATP or their equivalents. The primitive photosynthesizing cell coped with the emerging oxidizing atmosphere by the breaking up of internal photosynthetic membranes into small sacs. Stability of these small sacs was achieved through genomic competence and by their losing the ability to cleave the plasma membrane by binary fission. The loss of binary fission occurred possibly due to the toxic effect of oxygen.

The cells acquired multiple versions of these membranous sacs, and followed three evolutionary paths. One population of these cells developed chloroplasts, and a second group bagan to lose their photosynthetic ability but without losing their electron transport system, and developed into mitochondria. The third group lost their ability for electron transport reactions and transformed into nuclei by duplication of a single procaryotic DNA fibre to several copies of the chromosome. The normal cells required only one set of genome, hence the other sets were allowed to undergo mutation in response to changing environmental conditions. These mutations had either deleterious or beneficial effects on the survival of cell. The genomes with deleterious mutations perished, while the one having favourable mutations survived and adapted to new conditions. Presence of circular DNA, similar to that of procaryotes, in mitochondria and chloroplasts is indicative of preservation of some primitive characters.

The origin of the nucleus was proposed by Cavalier-Smith in 1978. He suggested that it is formed by invagination of the surface membrane which surrounds the DNA molecule to isolate it from the rest of the cytoplasm. Once the membrane-bound nucleus was formed, its circular DNA began to increase in mass to store more information. So, the circular DNA got broken up into linear pieces resulting in a new method of division to increase DNA copies. Cavalier-Smith proposed that invagination of the surface membrane seems to be a certainty as is evident from the mesosomes found in bacteria. Initially, invaginations might have originated to engulf solid food particles from the environment, and later some vesicles must have formed containing enzymes. Enzyme-containing vesicles became stable cytoplasmic compartments in the form of *lysosomes*. The mesosomal invaginations must have extended further into the cytoplasm to form tubular network to form Golgi complexes and endoplasmic reticulum. The development of these structures was an important landmark in the evolutionary process of the eucaryotes.

The appearance of mitochondria and in case of plants, chloroplasts in the cells was perhaps the most important event. The mitochondria show two basic similarities with procaryotes in possessing circular DNA and reproducing by fission. The eucaryotic cell does not possess any respiratory activity anywhere except in the mitochondria. Therefore, a suggestion has been made that the primitive anaerobic cells engulfed some aerobic procaryotes at some stage. This possibility is supported by a species of amoeba, *Pelomyxa palustris*, which lacks mitochondria but possesses aerobic bacteria in permanent symbiotic association. The chloroplasts also show these two similarities with procaryotes, which are shared by the mitochondria. In addition, they carry out photosynthesis in a fashion similar to that of cyanobacteria. All characteristics of chloroplasts suggest their origin from a cyanobacteria-like ancestor, which at some stage of evolution was engulfed by eucaryotes. This fact is borne out by two factors. The first evidence is the cyanobacteria-like structure of the chloroplast in moss cells, and the second and most definite evidence is the occurrence of cyanobacteria in certain present-day eucaryotic cells, such as *Cyanophora paradonai*. Once the eucaryotes were established, they were able to dominate the earth. Gradually they underwent modifications due to specialization of functions and became the successful survivors.

GLOSSARY

Acridine	—	dye used as a mutagen and as an inhibitor of the division of the kinetosome.
Activation energy	—	the energy in excess of the ground state that must be added to a molecular system to allow a chemical reaction to start.
Active site	—	Region of apoenzyme which is in direct physical contact with the substrate.
Active transport	—	movement of solutes against the concentration gradient across a membrane requiring energy input.
Actomyosin	—	the complex of actin and myosin formed when they interact.
Aerobic	—	needing the presence of air for growth.
Allele	—	one or more alternative forms of a gene concerned with the same trait or characteristic; one of a pair of a gene controlling the same trait lying at the identical loci on the homologous chromosomes.
Allosteric enzymes	—	enzymes having a binding and catalytic site for the substrate and a different site for a modulator (allosteric effector) to act.
Allosteric control	—	ability of an interaction at one side of a protein to influence the activity of another side.
Anaerobic	—	having the ability to live or grow in the absence of free oxygen.
Anaphase	—	mitotic stage during which the chromosomes move towards opposite ends of the spindle.
Antibodies	—	glycoprotein molecules produced by plasma cells capable of specifically binding with the antigen that triggered their production.
Anticodon	—	a group of three bases on a *t*RNA molecule that is complementary to the 3-base codon of *m*RNA.
Antigen	—	an agent that stimulates the production of antibodies; or the substance that is capable of generating immune response.
Asexual	—	lacking sex or functional sex organs.
Autotrophs	—	organisms that synthesize their own nutrition or organic compounds in the presence of sunlight, CO_2 and ATP.

Autoradiography	—	detects radioactive labelled molecules by their effect in creating an image on photographic film.
Autosomes	—	all chromosomes other than the sex chromosomes.
Avirulent	—	lacking virulence; a microorganism lacking the property that normally causes disease.
B cell	—	a differentiated lymphocyte involved in antibody-mediated immunity.
B-lymphocytes	—	white blood cells that are capable of producing specific immunoglobulins.
Bacteriophage	—	a virus whose host is a bacterium; viruses that infect bacteria.
Beta-galactosidase	—	an enzyme catalyzing the hydrolysis of β–linked galactose.
Binary fission	–	a process in which two cells of similar shape and size are formed by the division of one cell.
Biosynthesis	–	synthesis of chemical substances by the metabolic activities of living organisms.
Bivalent	–	structure containing all four chromatids as seen in meiosis.
Calvin cycle	—	pathway for carbondioxide fixation in photoautotrophs and chemolithotrophs.
Capsid	—	a protein coat of a virus enclosing a naked nucleic acid.
Carcinogen	—	cancer-causing agent.
Carotenoid pigments	—	pigments, usually yellow, orange, red, or purple, widely distributed among microorganisms.
Catabolism	—	metabolic reaction involving enzymic degradation of organic compounds to simpler compounds with the release of energy.
Cell-mediated immunity	—	specific acquired immunity involving T cells responsible for resistance to infectious diseases caused by pathogenic bacteria and viruses.
Chemical mutagens	—	chemicals that can modify nucleotide bases and increase mutation rate.
Chemiosmotic hypothesis	—	a theory put forward by Mitchell that establishes a proton and electrical gradient across a plasma membrane, and generates ATP.
Chemoautotrophs	—	microorganisms that obtain energy from the oxidation of inorganic compounds and carbon from CO_2.

Chiasma	—	site at which two homologous chromosomes appear to exchange material during meiosis.
Chloroplasts	—	membrane-bound organelles of photosynthetic eucaryotes where conversion of light energy to ATP occurs; the site of photosynthesis.
Chromatids	—	fibrils from a eucaryotic chromosome formed by replication prior to meiosis or mitosis.
Chromatin	—	complex of DNA and protein in the nucleus during interphase.
Chromosome	—	gene-containing structure in cells made up of DNA and protein.
Cilium	—	organelle of locomotion possessed by certain organisms.
Cistron	—	functional unit of inheritance; a segment of DNA that codes for a specific polypeptide chain.
Clone	—	a population of genetically identical cells derived asexually from a single cell.
Cloning vector	—	segment of DNA used for replication of foreign DNA fragments.
Codon	—	a triplet of adjacent bases in a polynucleotide chain of an *m*RNA molecule that codes for a specific amino acid.
Coenzymes	—	a non-protein component of enzymes with which they function and which act as acceptors or donors of electrons or functional groups during enzymatic reactions.
Complementary DNA (cDNA)	—	a single stranded DNA molecule that is complementary to the complete *m*RNA, used in cloning eucaryotic genes in bacteria.
Conjugation	—	the process in which genetic material is transferred from one micro-organism to another involving physical contact between two cells.
Conjugative plasmids	—	F and other plasmids that encode for self-transfer from one cell to another.
Constitutive enzymes	—	enzymes whose synthesis is not altered in response to changes in the environment and are therfore continuously synthesized.
Corepressor	—	a small molecule which triggers repression of transcription by binding with a regulator protein.
Cosmid	—	phage-plasmid artificial hybrids.

Covalent bond — a strong chemical bond formed by the sharing of electrons.

Cristae — infolded regions of mitochondrial inner membrane where electron transport and oxidative phosphorylation take place.

Crossing over — the process in which each of two adjacent DNA strands break and recombine so that there is an exchange of homologous regions of DNA.

Cyclic phosphorylation — a metabolic pathway involved in the conversion of light energy to chemical energy, with the generation of ATP.

Cytochromes — Haeme-containing proteins which transport electrons by way of iron atom valency changes.

Cytotoxic T cells — specialized class of T lymphocytes which are able to kill cells as part of the cell-mediated immune response.

Dalton — unit of mass (approximately equal to the weight of a hydrogen atom).

Dark reaction — light-independent photosynthetic reaction during which carbondioxide is reduced to carbohydrates.

Deficiencies — deletions of large number of base pairs that can result in the loss of genetic information for one or more complete genes.

Degenerate — describes redundancy inherent in the genetic code which occurs because there are several codons coding for the insertion of the same amino acid in the polypeptide chain.

Deletion mutation — mutation caused by the removal of one or more nucleotide pairs from the DNA.

Denaturation — of protein involves its conversion from physiological conformation to some other inactive conformations.

Diffusion — movement of molecules across the concentration gradient from the area of higher concentration to the area of lower concentration.

Diploid — cells having two sets of homologous chromosomes in the nucleus.

DNA polymerases — enzymes that catalyze the phosphodiester bonds in the formation of DNA.

Electron transport chain — a series of oxidation-reduction reactions in which electrons are transported from a substrate through a series of intermediate electron carriers to a final acceptor

		establishing an electrochemical gradient across a membrane that results in the formation of ATP.
Electrophoresis	—	the movement of charged particles suspended in a liquid under the influence of an applied electron field.
Endergonic	—	a chemical reaction requiring input of free energy; has a positive Δ G.
Endocytosis	—	movement of materials into the cell.
Enthalpy	—	the total heat of a system, Δ H.
Entropy	—	The portion of the energy of a system which cannot be converted into work, Δ S.
Enzymes	—	proteins that function as biological catalysts, increasing the rate of reaction without altering the equilibrium constant by lowering the energy of activation.
Episome	—	segment of DNA capable of existing in two alternate forms, one replicating autonomously in the cytoplasm, the other replicating as part of the bacterial chromosome.
Eucaryote	—	cellular organisms having a membrane-bound nucleus within which the genome of the cell is stored as chromosomes.
Exergonic	—	a reaction accomplished by liberation of energy.
Euchromatin	—	chromosomes that are not condensed and are considered metabolically active.
Exocytosis	—	movement of material out of a cell.
Exon	—	any segment of an interrupted gene that is represented in the mature RNA product.
Exonuclease	—	enzyme that progressively removes the terminal nucleotides of a polynucleotide chain.
F pilus	—	attachment structure that projects from cells of certain bacteria involved in mating; found on cells that donate DNA.
F plasmid	—	fertility plasmid coding for the donor strain that includes genes for the formation of F pilus.
Fab	—	antigen-binding fragment of immunoglobulin.
Facilitated diffusion	—	transport of molecules across the plasma membrane down a concentration gradient with the assistance of membrane proteins.
Fc	—	crystallizable fragment of immunoglobulin.
Feedback inhibition	—	a cellular control mechanism by which the end product of a series of metabolic

		reactions inhibits the activity of an earlier enzyme in the sequence of metabolic transformations.
Ferredoxin	—	small membrane-bound iron-sulphur protein which is the primary electron acceptor of photosystem I.
Free energy	—	the energy available to do work, particularly in causing chemical reactions.
Fluid-mosaic model	—	a model of cell membrane depicting proteins as floating in a bilayer of lipid molecules.
Freeze-fracture method	—	technique for preparation of cells for electron microscopy to *expose* surface structures.
Gene	—	a sequence of nucleotides that specifies a particular polypeptide chain or RNA sequence.
Generalized transduction	—	a form of recombination in which a phage carries bacterial DNA from a donor to a recipient cell, resulting in exchange of many homologous genes.
Genetic engineering	—	manipulation of genetic properties of an organism either through the selection of desirable traits, the introduction of new information on DNA, or both. The application of recombinant DNA technology.
Genetic code	—	code for specific amino acids formed by three sequential bases in *m*RNA.
Gene mapping	—	determination of the relative positions of genes in DNA or RNA.
Genome	—	the complete set of genetic information contained in a haploid set of chromosomes.
Gluconeogenesis	—	biosynthesis of glucose from noncarbohydrate sources.
Glycocalyx	—	specialized bacterial structure with an attachment function composed of a mass of tangled fibres of polysaccharides or branching sugar molecules.
Glycolysis	—	anaerobic oxidation of glucose to pyruvic acid.
Glycosidic bonds	—	bonds in disaccharides and polysaccharides formed by elimination of water.
Golgi complex	—	a membranous organelle of eucaryotes involved in the formation of secretory vesicles and synthesis of complex polysaccharides.

Grana	—	membranous unit formed by stacks of thylakoids.
Haploid	—	a single set of homologous chromosomes; having half of the normal diploid number of chromosomes
Hapten	—	a small molecule that acts as an antigen when joined to a protein.
Heterogeneous RNA	—	high molecular weight RNA formed by direct transcription in eucaryotes that is then processed enzymatically to form *m*RNA.
Heterochromatin	—	highly condensed chromosomes found in the interphase nucleus.
Heterotrophs	—	organisms requiring organic compounds for growth and reproduction, the organic compounds serve as sources of carbon and energy.
Histones	—	basic proteins found in chromosomes, rich in arginine and lysine.
Homozygous	—	individuals whose allelic forms of a gene are identical.
Hydrophylic	—	a substance showing affinity for water.
Hydrophobic	—	a substance lacking an affinity for water.
IgG	—	immunoglobulin of G type, the largest fraction of the body's immunoglobulins.
Immunity	—	relative unsusceptibility of a person or an animal to active infection by pathogenic micro-organisms or the harmful effects of certain toxins; resistance to disease by a living organism.
Inducible enzymes	—	enzymes that are synthesized in reponse to a particular substance in the environment.
Induction	—	an increase in the rate of synthesis of an enzyme; the turning on of enzyme synthesis in response to environmental conditions.
Insertions	—	the presence of an additional stretch of base pairs in DNA.
Insertion sequence	—	a transposable genetic element that can move around the bacterial chromosmes, and occurs at different locations on the chromosomes.
Intron	—	an intervening region of DNA of eucaryotes that does not code for any known protein or a regulatory function.
Isotope	—	an element that has the same atomic number as another element but has a different atomic weight.
Killer T cells	—	cytotoxic T cells.

Krebs cycle — the tricarboxylic acid cycle or the citric acid cycle in which pyruvate is converted into CO_2 and H_2O.

Lac operon — inducible enzyme system of *Escherichia coli* for utilization of lactose.

Ligases — a group of enzymes that catalyze reactions in which a bond is formed between two substrate molecules using energy obtained from the cleavage of a pyrophosphate bond.

Lymphocytes — lymph cells.

Lysis — rupture of the cells.

Lysogeny — the non-disruptive infection of a bacterium by a bacteriophage.

Lysosomes — membrane–bound cytoplasmic vesicles rich in hydrolytic enzymes.

Lytic phage — viruses that cause lysis of the cells within which they reproduce.

Macrophages — mononuclear phagocytes; large, actively phagocytic cells found in spleen, liver, lymph nodes, and blood.

Meiosis — cell division that results in reduction of chromosome number, i.e diploid to haploid.

Mesosomes — intracellular membranous structures observed as infoldings of bacterial cell membranes.

Metaphase — stage of mitosis during which chromosomes are highly coiled and aligned in the equatorial plane.

Mitochondrion — a semiautonomous organelle found in eucaryotes, the site of respiration and other cellular processes, and site of ATP synthesis.

Mitosis — the sequence of events resulting in the division of the nucleus into two genetically identical cells during asexual cell division, each of the daughter nuclei having the same number of chromosomes as the parent cell.

Monoclonal antibody — an antibody produced from a clone of cells making only that particular antibody.

Mutagen — an agent that promotes the occurrence of mutation.

Mutation — a stable, heritable change in the nucleotide sequence of the genetic material, resulting in an altered protein that gives the cell properties which are different from those of its parent.

Niche — the functional role of an organism within an ecosystem, the combined description of the physical habitat, functional role, and interactions of the microorganisms occurring at a given location.

Non-cyclic photophosphorylation — a metabolic pathway involved in the conversion of light energy for the generation of ATP in which an electron is transferred from an electron donor, normally water, by a series of electron carriers, with the eventual formation of a reduced coenzyme, NADPH.

Nucleoid — the compact body containing the genome in bacteria.

Nucleolus — an RNA-rich intranuclear body bound by a limiting membrane that is the site of *r*RNA synthesis in eucaryotes.

Nucleosome — structural unit of DNA in eucaryotes, having approximately 190 base pairs folded and held together by histones.

Okazaki fragments — short segments of DNA, about 1000-2000 bases, produced during discontinuous replication of the trailing strand of DNA.

Oncogene — the transforming gene of the retrovirus having a homologous sequence in the DNA of the normal organism.

Operon — a cluster of structural genes whose coordinated expression is controlled by a regulator gene.

Osmosis — the passage of a solvent through a membrane from a dilute solution into a more concentrated one.

Oxidative phosphorylation — a type of ATP synthesis coupled with the flow of electrons to molecular oxygen from a reduced substrate.

Packing ratio — ratio of the length of DNA to the unit length of the fibre containing it.

Palindrome — a word reading the same backward and forward.

Passive diffusion — unassisted movement of molecules from areas of high concentration to areas of low concentration.

Peptidoglycan — basic structural unit of the bacterial cell wall containing N-acetylglucosamine and peptide units, and N-acetylmuramic acid.

Permease — an enzyme that increases the rate of transport of a substance across the membrane.

Phage — a bacterial virus, also known as bacteriophage.

Phagocytosis — the process in which particulate matter is ingested by a cell, involving engulfment of that matter by the cell's membrane.

Phenotype — the observable structural and functional characteristic of an organism determined by its genotype.

Phosphodiester bond — the bonding of two moieties by a phosphate group; each moiety is held by an ester linkage.

Photolysis — liberation of oxygen by splitting of water during photosynthesis.

Pili — filamentous appendages that project from the cell surface of certain Gram-negative bacteria apparently involved in adsorption phenomena; involved in bacterial mating.

Pinocytosis — endocytosis involving the engulfment of soluble materials; cell drinking.

Plasmid — extrachromosomal genetic structure that can replicate independently in a bacterial cell.

Pribnow sequence — a sequence within nucleotide bases in DNA that determines the site of transcription initiation.

Prions — infectious proteins; substances that are infectious and reproduce within living systems but appear to be proteinaceous.

Procaryotes — cells whose genomes are not contained within a nucleus; e.g., bacteria and blue-green algae.

Promoter — a specific region of DNA where RNA polymerase enzyme binds for transcription on the DNA.

Prophage — the integrated phage genome when it becomes integrated with the bacterial chromosome and is replicated as part of the bacterial chromosome during subsequent cell divisions.

Protobiont — theoretically primitive, self-replicating, protein containing cell-like structures.

R plasmid — a plasmid enocoding for antibiotic resistance.

Rec genes — genes that code for enzymes involved in heteroduplex formation during homologous recombination.

Recombinant DNA — such molecules which contain DNA sequences derived from more than one species.

Recombinant DNA technology — see genetic engineering.

Regulatory genes — that serve as a regulatory function; genes that do not code for specific peptides but instead regulate the expression of structural genes.

Replica plating — a technique by which various types of mutants can be isolated from a population of bacteria grown under nonselective conditions.

Replication — multiplication of microrganisms; duplication of a nucleic acid from a template; the formation of a replica mold for viewing by electron microscopy.

Replication fork — the y-shaped region of a chromosome that is the growing point during replication of DNA.

Replicon — a nucleic acid molecule that possesses an origin and is therefore capable of initiating its own replication.

Restriction enzymes — enzymes capable of cutting DNA molecules.

Retrovirus — an RNA virus that propagates via conversion into duplex DNA.

Reverse transcription — mechanism for RNA synthesis in which the RNA viruses use their RNA genome as a template for an RNA-directed polymerase; RNA-directed synthesis of DNA.

Ribosomes — structures composed of *r*RNA and protein; the sites where protein synthesis occurs within cells.

RNA polymerase — an enzyme that catalyzes the formation of RNA on DNA templates.

S phase — part of the eucaryotic cell cycle during which synthesis of DNA occurs.

Semiconservative replication — replication of DNA involving separation of the duplex DNA strands, each then acting as a template for synthesis of a complementary strand.

Silent mutations — changes in the genotype that do not alter the phenotype; changes in the sequence of nucleotides which do not alter the amino acids in the gene product.

Splicing — the process of the removal of introns and joining of exons in the RNA.

Split genes — genes coded by noncontiguous segments of the DNA so that *m*RNA and the DNA for the protein product of that gene are not collinear; genes with intervening nucleotide sequences not involved in coding for the gene product.

Synapsis — association of the two pairs of sister chromatids representing homologous chromosomes that occurs at the start of meiosis.

Stem cells — formative cells; blood cells capable of producing various types of differentiated cells.

Structural gene — a gene whose product is an enzyme; structural protein, *t*RNA or *r*RNA, as opposed to a regulator gene whose product regulates the transcription of structural genes.

Svedberg unit — the unit in which the sedimentation coefficient of a particle is commonly expressed as S; when values are given in seconds, the basic unit is 10^{-13} seconds.

T cells — lymphocytes which are differentiated in thymus and are important in cell-mediated immunity, as well as in the modulation of antibody-mediated immunity.

Template — a pattern that acts as a guide for directing the synthesis of new macromolecules.

Thylakoids — flattened, membranous vesicles that occur in the chloroplasts where light reactions of photosynthesis occur.

Transcription — synthesis of *m*RNA, *r*RNA and *t*RNA from a DNA template.

Transduction — transfer of bacterial genes from one bacterium to another through a bacteriophage.

Transformation — a mode of genetic transfer in which a naked DNA fragment derived from one bacterium is taken up by another and subsequntly undergoes recombination with the recipient's chromosome.

Transfection — the process of incorporation of new genetic markers.

Transition — a point mutation in which one purine or one pyrimidine is replaced by another.

Translation — the process by which a particular base sequence in *m*RNA is used for the construction of a specific sequence of amino acids.

Translocation — nonhomologous recombination.

Transposons — translocatable genetic elements; genetic elements that move from one locus to

		another by nonhomologous recombination and thus move from one location to another.
Transversion	—	a point mutation in which a purine is replaced by a pyrimidine or vice versa.
Triplet code	—	the genetic code in which three sequential bases code for a specific amino acid.
Vectors	—	organisms that act as carriers of pathogens and are involved in the spread of disease; genetic elements that can carry foreign DNA.
Viroids	—	agents of certain diseases, resembling viruses in many ways but differing in apparent lack of viral organization.
Virus	—	a non-cellular entity that consists of a protein coat and a nucleic acid molecule enclosed in the coat, and that can replicate only after entry into specific types of living cells; it has no intrinsic metabolism of its own. Some viruses are infectious agents.
Xanthophyll	—	a pigment containing oxygen and derived form carotenes
Z scheme	—	the combination of cyclic and noncyclic photophosphorylation in photosynthesis which accounts for the trapping of sunlight, and generation of ATP, oxygen and NADPH.
Zygote	—	a diploid cell formed from two haploid male and female gametes during fertilization.

QUESTIONS

CHAPTER-1

1. Explain "The cell acts as an open system".
2. In what ways are the laws of thermodynamics applicable to living system? Discuss.
3. Explain that "the cells are self-regulating and self-replicating systems"

CHAPTER-2

1. Give the structure and functions of the following organelles found in the eukaryotes:
   (a) Mitochondria
   (b) Chloroplast
   (c) Golgi complex
2. What is a bacteriophage? What is its mode of infection? Explain the difference in the mode of infection of a plant and animal virus.
3. Differentiate between:
   (a) a prokaryote and a eukaryote
   (b) a prion and a viroid
   (c) a lytic and a lysogenic phage
4. Give a labelled diagram of an autonomous organelle found in a eukaryotic cell. Give its importance.
5. Give the structure of a typical prokaryote and list at least six characters which distinguish them from a eukaryote.
6. Differentiate between:
   (i) Lysosomes and peroxisomes
   (ii) Golgi and granular E R
7. Give structure and functions of the Nuclear organization.
8. Write briefly on cytoskeleton and its components.
9. Why are viruses called quasi-living particles?

CHAPTER-3

1. Explain the amphipathic nature of phospholipids.
2. Differentiate between RNA and DNA.
3. Give structures and names of the following compounds:
   a. a ribonucleotide
   b. a dipeptide

5.

6. Give structures of the following:
   a. Phosphatidyl choline
   b. Maltose
   c. Deoxyadenylic acid
   d. Ganglioside GM_2
   e. Cellobiose
   f. Amylopectin
7. What are gangliosides? Give their importance.
8. Differentiate between globular and fibrous proteins. Give atleast one example in each case.
9. Explain as best as you can:
   a. Sphingolipids, b. tRNA
10. Give the structures of the following:
    a. A sugar derivative
    b. A base present in RNA only
    c. A linear homopolysaccharide.

CHAPTER-4

1. Which energy-rich compound is not phosphorylated but is used to generate ATP from ADP? Give its importance.
2. Differentiate between exergonic and endergonic reactions.
3. What are energy-rich compounds? Explain their significance.
4. Describe the importance of electrochemical gradient across the inner mitochondrial membrane in ATP generation.
5. How does substrate level phosphorylation differ from oxidative phosphorylation? Give examples of substrate level phosphorylation during non-oxidative as well as oxidative respiration.
6. Explain the mechanism by which extra mitochondrial NADH generates ATP in eukaryotes.
7. In the reduction phase of photosynthesis, if one mole of carbondioxide is incorporated for the synthesis of a hexose, how many moles of ATP and NADPH + H^+ will be used?
8. A person is suffering from hypoglycemia (low blood glucose level). How the blood glucose level is restored to normal?
9. Give the reactions by which CO_2 reduction takes place in the tropical plants.
10. Give an account of the number of ATP molecules derived from a C-18 fatty acid molecule when it is completely metabolized in the mitochondria.
11. What does Hill Reaction prove?
12. Compare and contrast between oxidative and photophosphorylation.
13. Supposing the substance A is converted to product P through four enzyme catalyzed steps. After an hour the product P accumulates in the system resulting in the blocking of the whole sequence of reactions. Explain why and how does it happen?
14. Explain the non-cyclic photophosphorylation.
15. Explain chemi-osmosis in mitochondria.
16. How does an enzyme molecule maintain its structural integrity to perform catalysis?

17. How does the role of cytochromes in aerobic respiration compare with their role in photosynthesis?
18. Explain the reactions occuring in the stroma of the chloroplast. What are the raw materials and their sources for these reactions?

CHAPTER-5

1. Explain the mechanism by which a cell will internalize cholesterol or insulin from its environment.
2. Define action-potential.
3. What is the mechanism of glucose transport in an erythrocyte and an intestinal cell? Explain the mechanism in each case.
4. Explain the differences between active and passive transport mechanisms. Under what conditions active transport would cease?
5. Explain with diagram how antiport mechanism is involved in impulse propagation.
6. Name any two of the transport mechanisms where the energy is not involved and explain them.
7. Describe Natural Killer Cells.
8. Explain the molecular mechanisms involved in the generation of action potential and their importance in the conduction of the impulse through a myelinated nerve.
9. Differentiate between the following:
   a. Cell mediated and humoral immunity.
   b. Diffusion and osmosis.
10. A hormone acting on the cell membrane triggers a response resulting in the activation of phosohorylase kinase. Name the hormone, its mode of action and its effect on cell metabolism.

CHAPTER-6

1. Name the different phases of a cell-cycle and state the significance of each phase.
2. How humoral immunity is generated? Explain.
3. In a cross of $\frac{wfm}{wfm} \times \frac{+++}{+++}$, following gametes were produced:

Gametes	Nos.	Gametes	Nos.
wfm	268	+ fm	132
+++	268	wf+	33
w++	132	+f+	67
W+m	67	++m	33
			1000

Answer the following:-

a. Designate non-cross-overs, single cross-overs, and double cross-overs and calculate their percentages.
b. Determine the percentage of cross-overs between w and m and m and f.
c. Determine the order of genes and construct the gene map.

4. How the cell-mediated immune responses are generated? Explain.
5. What are linked genes? In an individual $\frac{++}{++}$ is crossed with $\frac{ab}{ab}$, what type of gametes will be produced and in what proportions? Assume that genes a and b show 20% recombination, what do you infer from the results?
6. Describe the molecular mechanism responsible for producing genetic variability in the formation of gametes.
7. Write briefly on the following:-
   a. G1 phase of cell cycle, b. Chromatin.
8. Seeds from red flowered plants following Mendelian traits, produced few white flowered plants. How will you account for this phenomenon?
9. An organism is heterozygous for four pairs of closely linked genes Aa, Bb, Cc and Dd. If AbCD are on the parental chromosome, show how ABCd and ABCD types of gametes will be produced?
10. In guinea pigs rough coat (R) is dominant over smooth (r) and black coat (B) is dominant over white (b). R and B are independent genes.
    a. Cross a homozygous rough black animal with a smooth white one. What will be the appearance of F1, of F2.
    b. What will be the appearance of the offspring of F1 crossed with the rough black parent, and with smooth white one?

CHAPTER-7

1. State the "Central Dogma". In case of certain viruses RNA is the genetic material. Explain the mechanism as to how RNA will be used to multiply itself.
2. Answer the following in brief:
   a. Plectonemic Property of DNA.
   b. Hypochromic effect.
   c. Palindrome
   d. Conclusion of Avery, McLeod and McCarty's Experiment.
3. Two double stranded RNA virus strains H & K have total number of 10^{12} and 10^{16} nucleotides respectively. Each strain has 10% guanine. Answer the following:
   a. Can total number of thymine bases be calculated for strain K? If yes, calculate this number.
   b. Calculate the number of all individual bases (except guanine) for strain H.
4. Answer briefly:
   a. Chargaff's rule
   b. Functions of DNA ligase enzyme
   c. Denatured DNA shows hyperchromicity. Why?
   d. A DNA molecule has 10^{12} nucleotides. Find out the mol. wt. and the number of helical turns in the molecule (Average mol. wt. of a nucleotide = 300).
5. How did Chargaff's equivalence rule helped Waston and Crick to bring out a model of double helix? Explain.

CHAPTER-8

Explain the repair of damage caused to DNA due to UV rays by the photo-reactivation mechanism.

Diagramatically depict the synthesis of RNA on DNA templates in *E. coli*.

Explain the enzymatic process for the correction of thymine dimer effect in a DNA molecule.

4. "In the intact living E. coli cells, DNA is replicated in a semi-conservative manner". Describe the experiment whose results demonstrated the semi-conservative replication of DNA.
5. A N^{15} labelled DNA molecule was allowed to multiply in N^{14} medium. Calculate the number of hybrid DNA molecules after 4 generations.
6. Enumerate the differentiating roles of DNA-dependent RNA polymerase and RNA-dependent DNA polymerase.
7. Why should there be a template specificity in transcription? Explain the role of sigma factor in transcription.
8. Write briefly on the following:
   a. Okazaki Fragments, b. Amino Acid Activation
9. A genome of *E. coli* containing 5×10^8 base pairs replicate 1000 base pairs/sec. at a single replication fork. Calculate the time taken for the replication of the entire genome.
10. Explain the following:
    a. Why only one strand of DNA duplex is normally transcribed?
    b. How heating and pH will effect DNA duplex?
    c. How replication will occur on the lagging strand?
11. Suppose you are given a total of 25 nucleotides having A, T, G, and C:
    a. Construct the sense strand of DNA with any random sequence.
    b. What will be its antisense strand?
    c. Construct the transcript of this DNA.

CHAPTER-9

1. Describe the step of amino-acid activation. What is the initiation complex and how is it formed during translation?
2. Give the mechanism of initiation of protein synthesis. What is the polarity of peptide synthesis during translation?
3. List the important components of the *E. Coli* protein synthesizing system.
4. What are the inhibitors of protein synthesis? Give the mechanism of puromycin termination of protein synthesis.
5. Describe the Genetic code. List briefly its important generalizations.
6. What do you understand by the post-translational modifications of proteins?

CHAPTER-10

1. Explain the process of conjugation in E. coli with the help of a suitable diagram and comment on the results of the following crosses:
   a. Hfr cell × F^- cell, b. Primary F′ × F^- cell

2. Describe the different mechanisms by which DNA is transferred to a bacterial cell and recombined with the recipient bacterial chromosome.
3. Write briefly on the Transposable Genetic Elements.
4. Describe the mechanism of gene transfer from a donor to a recipient bacterial cell through bacteriophages.
5. Explain the process of gene transfer without cellular contact or vectors.
6. What is the role of Rec A Proteins in homologous recombination?

CHAPTER-11

1. Define mutations. A DNA molecule exposed to nitrous acid was found to contain some defects. What are these and how are they caused?
2. An *E. coli* cell, which was already defective in the enzyme photolyase, when exposed to non-ionizing radiations was found to have a new mutation in its DNA. Name the mutation and the mechanism that can be used by the cell to repair it.
3. Differentiate between transition and transversion.
4. Why *E. coli* is considered as one of the best materials in mutation studies? Discuss transition mutations induced by 5-Bromouracil and Nitrous acid.
5. What do you understand by chromosomal aberrations? Describe their different types leading to altered phenotypes.

CHAPTER-12

1. Describe the 'lac-operon' of *E. coli*. How does it operate for the regulation of lactose utilization.
2. In an experiment, E. coli grown in glucose medium, were transferred to another medium containing lactose. According to observation, population of *E. coli* in lactose started only after 3 minutes, continued for another 12 minutes and thereafter stopped.
   a. Explain why a 3 min. lag period in the growth of *E. coli* is observed?
   b. What regulatory control mechanism is followed by *E. coli* in lactose medium?
   c. After 15 min, *E. coli* were again transferred to glucose, where they started growing immediately. Why? Explain with reasons.
3. What do you understand by Enzyme Repression? Explain the mechanism by giving an example.
4. Give the role of catabolite-gene Activator Protein (CAP) & cAMP for the initiation of transcription of lac operon.
5. Explain the following briefly:
   a. Enzyme Induction
   b. Product Induction
   c. Enzyme Repression
6. How does regulation of galactose utilization take place in yeast cell.
7. Describe the translational control in prokaryotes.
8. Explain Britten-Davidson model for regulation of gene-expression in eukaryotes.

CHAPTER-13

1. What should be the selection criteria of a cloning vector? Explain the steps to construct and identify a recombinant cloning vector.
2. Supposing you are given a mRNA fragment for insulin product. How do you proceed to clone the *E. coli* recombinants capable of producing insulin.
3. Name the four major steps of the recombinant DNA technology. Explain the mechanism by giving a typical recombinant DNA experiment depicting the cloning of an eukaryotic DNA fragment in *E. coli*.
4. How do you synthesize a c-DNA from mRNA?
5. What are restriction endonucleases? What is their role in Genetic Engineering?
6. What are plasmids? Explain their role in rDNA technology as cloning vehicles.
7. Describe the action of DNA ligase in sealing a ss-nick of a dsDNA molecule.
8. How would you do identification and selection of clones by colony hybridization and immunochemical methods?
9. Define the following terms:
   a. Vectors, b. Sticky end, c. Stringent plasmids,
   d. Cosmids, and e. Hybridization
10. Describe various applications of rDNA technology.

CHAPTER-14

1. Describe the experiment which simulated the conditions prevailing on primeval earth explaining the origin of life. What was its outcome?
2. Though the protobiont had many characteristics of the living, yet it lacked an information system. What progressive evolutionary changes it may have undergone to evolve into a eubiont?
3. Describe the evolution of metabolic and photosynthetic activities in a prokaryotic cell.
4. Give the origin of eukaryotes. What are the basic differences of eukaryotes and prokaryotes?

SELF EVALUATION

1. The bacterial cell is surrounded by a distinct cell wall made up ot
   A. Lipoproteins
   B. Peptidoglycans
   C. Hemicellulose
   D. Pectins and lignins
2. Blue-green algae is an example of
   A. Photosynthetic autotroph
   B. Chemosynthetic heterotroph
   C. Photoorganotroph
   D. Chemoorganotroph
3. Plasma membrane of prokaryotic cells is often thrown into small invaginations called
   A. Tonoplast
   B. Glycocalyx
   C. Mesosomes
   D. Flagellum
4. Communication between two adjacent cells is by
   A. Lamellosomes
   B. Plasmodesmata
   C. Vacuoles
   D. Glyoxysomes
5. Two glucose units linked together through a α 1–4 linkage make
   A. Cellobiose
   B. Maltose
   C. Sucrose
   D. Lactose
6. The organelle unbound by any membrane is
   A. Chloroplast
   B. Endplasmic reticulum
   C. Ribosome
   D. Mitochondria
7. Viroids are
   A. Naked DNA molecules
   B. Naked RNA molecules
   C. Naked proteins
   D. Naked nucleic acids
8. Phophatidyl choline is a derivative of
   A. Ganglioside
   B. Sphingosine
   C. Phosphatidic acid
   D. Ceramide

9. Sphingolipids are absent in
   A. Yeast cells
   B. Bacterial cells
   C. Animal cells
   D. Plant cells
10. NADP functions in photosynthesis in what way?
   A. As a water splitting enzyme
   B. As a maker of ATP
   C. As a hydrogen carrier
   D. As a chlorophyll stimulator
11. CO_2 is introduced into dark reaction by
   A. Combining with NADPH
   B. Absorbing the energy released from ATP
   C. Attaching to PGAL to form glucose
   D. Being added to a 5-carbon carrier
12. A major event in the cytochrome system is
   A. High production of ATP molecules
   B. Splitting of water to release oxygen
   C. Release of many CO_2 molecules
   D. All of the above
13. In respiration every hydrogen removal is accompanied by
   A. Addition of an oxygen
   B. Removal of an electron
   C. Production of much heat
   D. Depletion of an ATP
14. The two photosystems of the light reactions
   A. Carry out the same reactions but involve different chlorophylls
   B. Are never found in the same type of plant
   C. Allternate in producing ATP necessary for dark reactions
   D. Differ in their final products but interact in electron transport
15. DNA duplication requires the presence of
   A. DNA polymerase
   B. Messenger RNA
   C. Transfer RNA
   D. None of the above
16. The term ionizing radiations is used because
   A. Ions are source of radiation
   B. Various substances are ionized by radiation
   C. It ionizes only proteins
   D. None of the above
17. Radiation damage has hereditary effects if it
   A. Changes proteins
   B. Deactivates enzymes
   C. Alters DNA bases
   D. None of the above

18. Hypochromic effect is due to
 A. Stacking of bases in DNA duplex
 B. Denaturation of DNA
 C. Presence of larger no. of AT pairs in DNA
 D. None of the above

19. In eukaryotes DNA replication occurs
 A. Anywhere in the cell
 B. Within mitochondria
 C. In the nucleus
 D. Near the nucleus

20. The term linked genes refers to what?
 A. Genes on the same chromosome
 B. Genes linked by hydrogen bonds
 C. Genes found in chromosome maps
 D. None of the above

21. Practical application of crossing over is
 A. To produce strong offspring
 B. To form new hybrids
 C. To determine relative position of genes on chromosomes
 D. To develop pure-bred lines

22. Cell cycle is
 A. A part of cell division
 B. A modified form of cell division
 C. A type of division occurring in asexual organisms
 D. None of the above

23. Meiosis is responsible for what?
 A. Maintaining constancy of chromosome number
 B. Genetic variability
 C. To ensure purity of species
 D. None of the above

24. The equation that shows energy transfer from GTP to ADP is
 A. GTP + ADP = GDP + ATP
 B. GTP + ADP + Pi = GTP + ATP
 C. GDP + ADP + PPi = GTP + ATP
 D. GTP + ATP = GDP + ADP + PPi

25. The enzyme that transfers PO_4 group from ATP is
 A. Phosphorylase
 B. Kinase
 C. Isomerase
 D. Glycosyl transferase

26. Which of the following compounds is continuously recycled in Krebs cycle?
 A. Alpha ketoglutarate
 B. Oxaloacetate
 C. Citrate
 D. Acetyl Co-A

27. Initiation of DNA replication in eukaryotes takes place at
    A. A single point of origin
    B. Several points of origin
    C. Intervals determined by DNA Pol III
    D. A clockwise direction

28. Transcription requires
    A. RNA Pol and a primer
    B. DNA dependent RNA Pol and no primer
    C. Both DNA strands as templates
    D. Deoxyribonucleoside triphosphates

29. Where is messenger RNA synthesized in a cell?
    A. Nucleus
    B. Mitochondria
    C. Ribosomes
    D. Endoplasmic reticulum

30. What transfers amino acids to the growing plypeptide chain?
    A. tRNA
    B. Peptidyl transferase
    C. DNA Pol
    D. RNA Pol

31. Nirenberg's experiment listed in the text demonstrated
    A. That DNA is the genetic material
    B. That DNA functions only in the nucleus
    C. The meaning of triplet codes of mRNA
    D. That ribosomes and mRNA are involved in protein synthesis

32. The coding of amino acids by sets of nucleotides is termed what?
    A. Base pairing specificity
    B. Hydrogen bonding
    C. Triplet coding
    D. Chemical coding

33. What is the role of ribosomes in protein synthesis?
    A. They provide a source of amino acids
    B. They carry proteins to their site of action
    C. They translate the basic DNA code
    D. They attach tRNAs to mRNA

34. Antibodies are
    A. Foreign proteins that result in immune response
    B. Special proteins that destroy antigens
    C. A class of cells that engulf invading proteins
    D. Antigens that are partially inactivated?

35. Upon depolarization ions move through the membrane
    A. By active transport
    B. By polarity
    C. By diffusion
    D. By electrical attraction

36. Where is acetyl choline produced?
A. From the nerve cell body
B. At dendritic ends
C. At axon terminals
D. All along the axon

37. Recombination is an important event which occurs when
A. Egg and sperm unite in fertilization
B. DNA duplicates itself
C. Homologous chromosomes pair in meiosis
D. Two chromosomes are unrelated

38. To what does the chromosome theory refer?
A. Role of chromosomes in meiosis
B. Role of genes in heredity
C. Location of genes on chromosomes
D. Location of chromosomes in the nucleus

39. A hapten is
A. an epitope
B. a molecule that can react with antibodies
C. an antigen which cannot initiate an immune response
D. another name for H-antigen

40. Immunoglobulins are produced by plasma cells and their chemical nature is
A. Peptidoglycan
B. Glycolipids
C. Glycoprotein
D. Lipoprotein

41. Which of the following components make up the cytoskeleton of most eukaryotic cells, except.
A. Actin filaments
B. Micro filaments
C. Chromatin
D. Microtubles

42. In which of the following the double membrane structure is absent.
A. Mitochondria
B. Chloroplast
C. Nucleus
D. None of the above

43. Which of the following cell organelle has F_1 particle?
A. Ribosomes
B. Golgibodies
C. Mitochondria
D. Chloroplast

44. Which one of the following is used as sole carbon source by Autotrophs.

A. Glucose
B. Ethanol
C. Carbondioxide
D. None of above

45. In which of the following Centrosomes are rudimentary.
A. Animal cells
B. Plant cells
C. Bacteria
D. Virus

46. In viral invasion of plant cell
A. Only the DNA enters the cell
B. Involves entire virus
C. Cell is killed first
D. None of the above

47. Centrioles
A. are found in plants
B. are made up of nine tubules
C. play a role in protein synthesis
D. occur in two's in the cytoplasm

48. Grana are membranous areas where
A. Enzymes are synthesized
B. Lipids are stored
C. Energy is released
D. Chlorophyll is concentrated

49. Which functional group is present in glucosamine
A. –SH
B. $-NH_2$
C. –COCH
D. –CHO

50. Which of the following aminoacids has aromatic group?
A. Phenylalanine
B. Cystein
C. Isoleucine
D. Methionine

51. Which of the following in non elastic fibrous protein?
A. Actin
B. Myosin
C. Collagen
D. Lysine

52. Which of the following has the shortest half life?
A. tRNA
B. mRNA
C. HnRNA
D. rRNA

53. Neutral fats are accumulated in
A. Connective tissue
B. Adipose tissue
C. Nerve fibers
D. Brain

54. **Which of the linkages is present between two nucleotides?**
A. Phosphodiester bond
B. Peptide bond
C. Weak hydrogen bond
D. None of the above

55. **Which of the following polymers is present in chitin?**
A. Glucosamine
B. Acetyl glucosamine
C. Glucuronic acid
D. Hyaluronic acid

56. **Mucopolysaccharides in bacterial cell walls are made up of:**
A. Carbohydrate + lipid
B. Sugar + protein
C. Protein + lipid
D. Sugar + nucleotide

57. **Which of the following has branched polysaccharide chains?**
A. Starch
B. Amylopectin
C. Glycogen
D. All the above

58. **Which of the following proteins has four polypeptide chains?**
A. Myoglobulin
B. Insulin
C. Haemoglobin
D. None of the above

59. **Cholesterol is a precursor of:**
A. Vitamin D
B. Progesterone
C. Testosterone
D. All the above

60. **The flexibility & fluidity of the biological structure of fatty acids is due to**
A. C – C
B. C = C
C. C ≡ C
D. All the above

61. **Which of the following linkages exists between two glucose molecules?**
A. Peptide linkage
B. Phosphodiester linkage
C. Glycosidic linkage
D. All the above

62. Which of the following statements is correct?
A. Antibodies are firous in nature
B. Structural proteins are fibrous in nature
C. Functional proteins are fibrous in nature
D. Collagens are globular in nature

63. Which of the following has pyridine nucleotide?
A. FAD
B. NAD
C. AMP
D. GTP

64. Which of the following enzymes act in presence of water?
A. Lyases
B. Hydrolases
C. Dehydorgenases
D. Ligases

65. Kinases, belong to the class of enzymes called:
A. Isomerases
B. Oxidoreductases
C. Hydrolases
D. Transferases

66. Oxidative phosphorylation takes place in:
A. All prokaryotes & eukaryotes
B. In eukaryotes alone
C. In all aerobic cells
D. None of the above

67. Enzymes involved in oxidative phosphorylation are located:
A. In the plasma membrane of eukaryotes
B. In the outer membrane of mitochondria
C. In the inner membrane of mitochondria
D. In the plasma membrane and inner membrane of mitochondria.

68. Which is the end product in the first stage of glycolysis?
A. Fructose – 6 – Phosphate
B. Glucose – 6 – Phosphate
C. Glyceraldehyde – 3 – Phosphate
D. 3 – Phosphosglycerate

69. During muscular exercise anaerobic respiration in humans results in the accumulation of:
A. Pyruvic acid
B. Lactic acid
C. Ethanol
D. Phosphoenol pyruvic acid

70 Starting from pyruvic acid
in the form of:

A. $4NADH + 4H^+ + 1\ FADH_2$
B. $2NADH + 2H^+ + 2\ FADH_2$
C. $5NADH + 5H^+ + 1\ FADH_2$
D. $4NADPH + 4H^+ + 1\ FADH_2$

71. **Succinate being the substrate to be oxidized, the number of ATP's produced during oxidative phosphorylation is:**
A. One
B. Two
C. Three
D. Four

72. **In Krebs cycle the number of ATPs formed per mole of pyruvate oxidized is**
A. 6
B. 15
C. 30
D. 38

73. **Photolysis of water in the light phase takes place in the:**
A. Thylakoid membrane
B. Stroma
C. Interior of thylakoid
D. Photosystems I & II

74. **The actual products of the light reaction are**
A. O_2, ATP & NADPH
B. O_2, ATP & NADH
C. ATP & NADH
D. O_2 & ATP

75. **In C3 pathway the initial CO_2 fixation takes place with:**
A. Phosphoenol pyruvate
B. Ribulose monophosphate
C. Ribulose diphosphate
D. 3-phosphoglycerate

76. **The immediate product of CO_2 fixation in Hatchslack pathway is a:**
A. 3–C Sugar
B. 4–C dicarboxylic acid
C. 5–C sugar
D. 6–C dicarboxylic acid

77. **The ETC pumps H^+ ions**
A. From the mitochondrial matrix into the intermembranal space.
B. From the cytosol into the intermembranal space
C. From the intermembranal space in to the mitochondrial matrix.
D. From the intermembranal space in to the cytosol.

78. **The oxidation of fatty acids takes place in the:**
A. Mitochondrial matrix
B. Intermembranal space
C. Innermembrane of the mitochondria
D. Cytosol

79. Cofactors tightly bound to the enzyme molecule are known as:
A. Holoenzymes
B. Coenzymes
C. Prosthetic group
D. Metal activators

80. For the initiation of transcription the protein factor required is:
A. rho factor
B. Sigma factor
C. Omega factor
D. All of the above factors

81. The enzyme used for removing the ss hair-pin during c-DNA synthesis is:
A. Topoisomerase
B. S1 Nuclease
C. DNA polymerase
D. Alakaline phosphatase

82. Chromosomal aberration in which a set of genes is first removed and then inserted in a reverse order is called:
A. Duplication
B. Deletion
C. Translocation
D. Inversion

83. In presence of nitrous acid adenine is oxidatively deaminated to form:
A. Hypoxanthine
B. Xanthine
C. Thymine
D. Uracil

84. Bacteria become suscpetible to transformation in the presence of:
A. Ca^{++}
B. Mg^{++}
C. K^{+}
D. Na^{+}

85. Blunt end fragments are produced by the restriction enzyme:
A. Eco RI
B. Hind III
C. Sma I
D. Bam HI

86. Activation of amino acids during protein synthesis takes place in:
A. Ribosomes
B. Nucleus
C. Cytoplasm
D. Mitochondria

87. The mobile DNA elements are called as:
A. Fertility factor

B. Mutator elements
C. Transposable elements
D. None of the above

88. Reverse transcriptase makes DNA in:
A. 3′ to 5′ direction
B. 5′ to 3′ direction
C. Both ways
D. None of the above

89. Peptide bond formation takes place on:
A. Amino acyl site
B. Peptidyl site
C. In the cytoplasm
D. None of the above

90. 5-Bromouracil is an analogue of:
A. Cytosine
B. Adenine
C. Thymine
D. Guanine

91. Antibodies are specific:
A. Glycolipids
B. Lipoproteins
C. Glycoproteins
D. None of the above

92. RNA dependent DNA polymerase requires:
A. Both template and primer
B. Only primer
C. Only template
D. None of the above

93. Hypoxanthine pairs with
A. Guanine
B. Xanthine
C. Cytosine
D. Adenine

94. DNA as the "transforming principle" was first established by
A. Avery, Mcleod and Mc Carty
B. Watson and Crick
C. Hershey and Chase
D. Beadle and Tatum

95. cAMP activates a specific enzyme called
A. Phosphorylase kinase
B. Phosphodiestrase
C. Protein kinase
D. None of the above

96. Biosynthesis of glucose from non-carbohydrate sources is called
    A. Glycogenolysis
    B. Glycolysis
    C. Glucone ogenesis
    D. Hexose monophosphate shunt

97. Protenoids are also known as
    A. Viroids
    B. Prions
    C. Microspheres
    D. None of the above

98. Coacervate model of protobiont was given by
    A. Sydney Fox
    B. Oparin
    C. Richard Goldacre
    D. Monod and Jacob

99. A point mutation in which a purine is replaced by a pyrimidine or vice versa is called
    A. Translocation
    B. Transition
    C. Transformation
    D. Transversion

100. T4 DNA ligase specifically requires for its activity
    A. NAD^+
    B. ATP
    C. ADP
    D. NMN

REFERENCES AND SUGGESTED READING

CHAPTER - 1

1. Baker, J.J. and Allen, G.E. (1965). *Matter, Energy and Life*, Addison Wesley Pub. Co., Palo Alto.
2. Baum, S.J. (1987). *Introduction to Organic and Biological Chemistry*, 4th ed., Macmillan, New York.
3. Gardner, E.L. (1965). *History of Biology*, Burgess, Minneapolis.
4. Keeton, W.T. (1980). *Biological Sciences*, Norton, New York.
5. Loewy, A.G. and Siekevitz, P. (1974). *Cell Structure and Function*, 2nd ed., Amerind Pub. Co., New Delhi.
6. Penrose, L.S. (1959). Self-reproducing Machines, *Sci., Amer*. Offprints, W.H. Freemen, San Francisco.

CHAPTER - 2

1. Bourne, G.H. (1962). *Division of Labour in Cells*, Academic Press, New York.
2. Butler, P.J.G., and Klug, A. (1978). The Assembly of a Virus, *Sci. Amer*. 239 (5): 62.
3. Diener, T.O. (1981). Viroids, *Sci. Amer*. 244 (1) : 66.
4. Fawcett, D.W. (1981). *The Cell: Its Organelles and Inclusions*, 2nd ed., Saunders, Philadelphia.
5. Holtzman, E. and Novihoff, A.B. (1984). *Cell and Organelles*, 3rd ed., Saunders, Philadelphia.
6. Rastogi, S.C. (1988). *Cell Biology*, Tata McGraw Hill, New Delhi.

CHAPTER - 3

1. Baum, S. J. (1987). *Introduction to Organic and Biological Chemistry*, 4th ed., MacMillan Pub. Co. New York.
2. Calvin, M. and Pryor, W.A. (1973). *Organic Chemistry of Life*, W.H. Freeman, San Francisco.
3. Campbell, P.N. and Smith, A.D. (1988). *Biochemistry Illustrated*, Churchill, Livingtone, England.
4. Tolbert, N.E., Ed. (1980). *The Biochemistry of Plants*, Vol. I, Academic Press, New York.

CHAPTER - 4

1. Atkinson, D.E. (1990). 'Over View of Metabolism' in Biochemistry. (Zubey, G., Coord. Auth.) 2nd ed., Maxwell Macmillan, Singapore.
2. Fersht, A. (1985). Enzyme Structure and Mechanism, W.H. Freeman, New York.
3. Goldsby, R.A. (1977). *Cells and Energy*, Macmillan, New York.

4. Govindjee, X., Ed., (1982). *Photosynthesis: Energy Conversion by Plant and Bacteria*, Academic Press, New York.
5. Hammes, G.G. (1982). Enzyme Catalysis and Regulation, Academic Press, New York.
6. Parson, W.W. (1990), 'Electron Transport and Production of ATP' in *Biochemistry* (Zubey G., Coord Auth), 2nd ed., Maxwell Macmillan, Singapore.
7. Wood, W.B. (1974). *The Molecular Basis of Metabolism*, McGraw Hill Pub., New York.

CHAPTER - 5

1. Bitter, E.E., Ed. (1980). *Membrane Structure and Function*, Vol. I & II., Wiley Intersci., New York.
2. Cuatrecasas, P. and Greeves, M.F., Ed. (1978). *Receptors and Recognition*, Chapman & Hall, London.
3. Eddman, G.M. (1970). The Structure and Function of Antibodies, *Sci. Amer* 223 (2): 34.
4. Lloyd, C.W. (1982). *Cytoskeleton in Plant Growth and Development*. Academic Press, New York.
5. Rogers, H.J. (1983). *Bacterial Cell Structure*, Van Nostrand Reinhold, England.
6. Roitts, E.M. (1988). *Immunology*, Oxford Univ. Press., England.
7. Shay, J.W., Ed. (1986). *Cell and Molecular Biology of the Cytoskeleton*, Plenum, New York.
8. Turner, A.J., Ed. (1990). *Molecular and Cell Biology of Membrane Proteins*, Harwood, England.
9. Vance, D.E. and Vance, J.E. eds. (1985). *Biochemistry of Lipids and Membranes*. Benjamin, Menlo Park California.

CHAPTER - 6

1. Kornberg, R. and Klug, A. (1981). The Nucleosome, *Sci. Amer*. 244 (2): 52.
2. Lewin, B. (1980). *Gene Expression-2: Eukaryotic Chromosome*, 2nd ed., John Wiley, New York.
3. Mazia, D. (1974). The Cell Cycle, *Sci. Amer*. 230:54.
4. White, M.J.D. (1973). *The Chromosomes*, 6th ed., Chapman & Hall, London.

CHAPTER - 7

1. Burns, G.W. (1976). *The Science of Genetics: An Introduction to Heredity*, 3rd ed., Macmillan Pub. Co. New York.
2. Ingle, M.R. (1985). *Genetic Mechanism*, Basil Blackwell, Oxford, England.
3. Jacob, F. (1975). *The Logic of Life: A History of Heredity*. Pantheon, New York.
4. Orgel, L.E. and Crick, F.H.C. (1980). Selfish DNA: The Ultimate Parasite, *Nature*. 284:604.
5. Stent, G.S. and Calender, R. (1978). *Molecular Genetics: An Introductory Narrative*, 2nd ed., W.H. Freeman, New York.

CHAPTER -8

1. Campbell, J.L. (1986). Eukaryotic DNA Replication, *Ann. Rev. Biochem.* 55: 733.
2. Friedberg, E.C. (1985). *DNA Repair*, W.H. Freeman, New York.
3. Gilbert, W. (1981). DNA Sequencing and Gene Structure, *Science*, 214 : 1305.
4. Holand, R. (1983). Initiation of DNA Replication in Eukaryotic Chromosomes in *DNA Makes RNA/Makes Protein.* (T. Hunt et al. Eds.), Elsevier Biomedical.
5. McClure, W.R. (1985). Mechanism and Control of Transcription Initiation in Prokaryotes, *Ann. Rev. Biochem.* 54:171.
6. Nossal, N. (1983) Procaryotic DNA Replication Systems, *Ann. Rev. Biochem.* 52 : 581.
7. Ogawa, T. and Okazaki, T. (1980). Discontinuous DNA Replication, *Ann. Rev. Biochem.* 49 : 421.

CHAPTER - 9

1. Hunt, T. (1980). The Initiation of Protein Synthesis, *Trends Biochem. Sci.* 5:178.
2. Kozak, M. (1983). Comparison of Initiation of Protein Synthesis in Prokaryotes, Eukaryotes and Organelles, *Microbial Rev.* 47 : 1.
3. Lake, J.A.A. (1981). Ribosome, *Sci. Amer.* 245 : 84.
4. Moldave, K. (Ed.), (1981). *RNA and Protein Synthesis*, Academic Press, New York.

CHAPTER - 10

1. Alberts, B. and Fox, C.F. (1980). *Mechanistic Studies of DNA Replication and Recombination*, Academic Press, New York.
2. Casjens, S. (1985). *Virus Assembly*, Jones & Bartletts, Bostan.
3. Kuchlerlapati, R., Ed. (1986). *Gene Transfer*, Plenum Press , New York.
4. Lewin, B. (1977). *Gene Expression III: Plasmids and Phages*, Wiley Interscience, New York.
5. Stahl, F.W. (1979). *Genetic Recombination*, W.H. Freeman, New York.

CHAPTER - 11

1. Ames. B.W. (1979). Identifying Environmental Chemical Causing Mutations *Science*, 204 : 587.
2. Drake, J.W. (1970). *The Molecular Basis of Mutation*, Holden-Day Nev York.
3. Muhaminad, A. and Brostel, R.C. Von, Ed. (1985). *Basic and Applie Mutagenesis*, Plenum Press, New York.
4. Serris, F.J. de, and Sheridan, W., Ed. (1985). *Genetic Consequences c Nucleotide Pool Imbalance*, Plenum, New York.

CHAPTER - 12

1. Galton, D.J. (1985). *Molecular Genetics of Metabolic Disease*, Edwa Arnold, London.

2. Hawkins, J.D. (1985). *Gene Structure and Expression*, Cambridge Univ. Press, Cambridge.
3. Latchman, D. (1990). *Gene Regulation: A Eukaryotic Perspective*, Unwin Hyman Inc., London.
4. Scandalios, J.G. Ed. (1979). *Physiological Genetics*, Academic Press, New York.
5. Stanbury, J.B., Wyngaarden, J.B., Fredrickson. D.S., Gold Stein, J.L. and Brown M.S. (1983). *The Metabolic Basis of Inherited Diseases*, 5th ed., McGraw-Hill, New York.
6. Vogal, F. and Moltusky, A.G. (1980). *Human Genetics: Problems and Approaches*, Springer Verlag, Berlin.

CHAPTER - 13

1. Abelson, J. and Butz, E. (1980). Recombinant DNA, *Science*, 209:1317.
2. Dillon, J.R., Nasin, A. and Nestmann, E.R. (1985). *Recombinant DNA Methadology*, John Wiley, New York.
3. Gibson, R. (1982). Genetic Manipulation: Principles and Practice, *Biologist*, 29.
4. Glover, D.M. (1984). *Gene Cloning: The Mechanics of DNA Manipulation*, Chapman-Hall, London.

CHAPTER - 14

1. Ayala, F.J., Ed. (1976). *Molecular Evolution*, Sinauer, Sunderland, Massachusetts.
2. Ayala, F.J. and Valentine, J.W. (1979). *Evolving the Theory and Processes of Organic Evolution*, Benjamin Pub. Co., Menlo Park, California.
3. Cairns-Smith, A.G. (1982). *Genetic Takeover*, Cambridge Univ. Press, New York.
4. Crick, F. (1981). *Life Itself: Its Origin and Nature*, Simon and Schuster, New York.
5. Eeign, M. and Schuster, P. (1979). *The Hypercycle*, Springer Verlag, Berlin.
6. Fox, S.W. and Dose, K. (1977). *Molecular Evolution and Origin of Life*, 2nd ed., Mercel Dekker, New York.
7. Hsu, L.H.N.L. (1984). Concepts of Prebiological Evolution, in *Molecular Evolution and Protobiology* (Matsuno, K., Dose, K., Harada, K. and Rohlfing, D.L., Eds.). p. 397, Plenum Press, New York.
8. Kimura, M. (1982). *Molecular Evolution, Protein Polymorphism and Neutral Theory*, Springer- Verlag, Berlin.
9. Miller, S.L. and Orgel, L.E. (1974). *The Origin of Life on the Earth*, Prentice-Hall, Englwood Chiffs, New Jersey.
10. Oparin. A.I. (1964). *The Origin of Life on the Earth*, Academic Press, New York.
11. Rohlfing, D.L. (1984). "The Development of the Proteinoid Model for the Origin of Life", in *Molecular Evolution and Protobiology* (Matsuno, K., Dose, K. Harada, K. and Rohlfing, D.L., Eds.), Plenum Press, New York.

INDEX

Molecular Biology